Genetics, Society, and Decisions

RICHARD V. KOWLES

Saint Mary's College
Winona, Minnesota

Scott, Foresman and Company
Glenview, Illinois
Boston   London

International Standard Book Number: 0-673-18678-4
Library of Congress Catalog Card Number: 84-62585
2 3 4 5 6 7 8 9—91 90 89 88

Preface

The discussions in this textbook focus on many of the important ways in which heredity interrelates with societal activity. Initial treatments of the fundamentals of hereditary transmission, gene function, and genes in populations provide the background for studying these interrelationships. Readers are given not only the basic concepts of genetics, but are also alerted to the course of events in genetics and society that lead to controversial issues and problematic situations. Other objectives are to help the reader realize that biological outcomes are not always definitive, that data are sometimes difficult to interpret, and that in some cases we may not have enough biological information to make sound decisions yet.

The primary concept of this textbook, unlike most of this type, is to consider biology non-majors as its principal audience. The book is intended for students majoring in such disciplines as psychology, sociology, anthropology, philosophy, education, health, and political science, among others. On the other hand, biology majors can certainly gain useful information from reading the book. Within many biology programs, students do not have the time or opportunity to study important social implications because of their heavy involvement with mathematics, chemistry, physics, and core biology courses. This textbook presents background genetic information and contrasting points of view on most controversial social issues. I hope this book will allow students to draw intelligent conclusions on the issues. In only a very few cases, where most scientists are vehemently opposed to any kind of compromise, do I deviate from this format.

Other biology courses are not prerequisite, but are, of course, helpful. The genetics underlying each topic is developed from a very elementary point, to make some of the complicated subjects more understandable to the average college student with little or no biological background. In addition, some seemingly simple subjects emerge as being quite complex.

Many uncomplicated illustrations will help the reader understand complex concepts. Other features include reference lists of the more important studies and sources of data, questions for discussion, problems ranging from elementary to challenging, inclusive summaries, and a complete glossary. There are extensive lists of further readings for each chapter, including those of historical interest as well as current articles and books. Much of the literature in these lists was selected on the basis of appropriateness for non-biology students and to aid students in term paper preparation. Fundamental statistics and probability are explained and demonstrated where they are required to elucidate genetic concepts.

I feel the many graphs, tables, and other forms of data are essential because logical thinking, drawing sound conclusions, and developing a generally scientific approach to problems should not be restricted to science researchers. In this scientific age, all students and responsible citizens need to develop these attributes to the best of their abilities. The students who read this book are among those who will soon participate in making the difficult decisions that relate to the interactions between heredity and society.

All writers probably have support and expertise behind the scenes. In this case, I particularly want to thank Dolores Munson, who is much more than a typist. One can only wish that more scientists were as careful as this irreplaceable person. Others who helped in typing and proofreading segments of the manuscript were Jodine Menden, Thomas Soukup, Mary Lou Willson, and Renée Zameic. Gratitude is also extended to Rose Kowles who was called upon many times to provide reference details. With regard to the publisher's involvement in the project, one could not find better people with whom to work than Robert Lakemacher and his assistants, Cleopatra G. Eddie and Carole Jackson. Their encouragement, and occasional nudging, played a significant role in completion of the project. The expertise of Mary Lou Motl and Molly Kyle in shaping the book into its final form was invaluable.

PREFACE

Throughout the project, Charles E. Merrill Publishing Company proved to be a first-class organization, and I appreciate the efforts of the staff. An author would be remiss not to acknowledge the many reviewers. Gaining very little in return, reviewers spend many tedious hours reading rough manuscript and giving attention to detail in an effort to improve the product. Lastly, I thank the many classes of *Heredity and Society* at Saint Mary's College who were the catalysts and the guinea pigs for this book.

Richard V. Kowles
Winona, Minnesota

Contents

ix

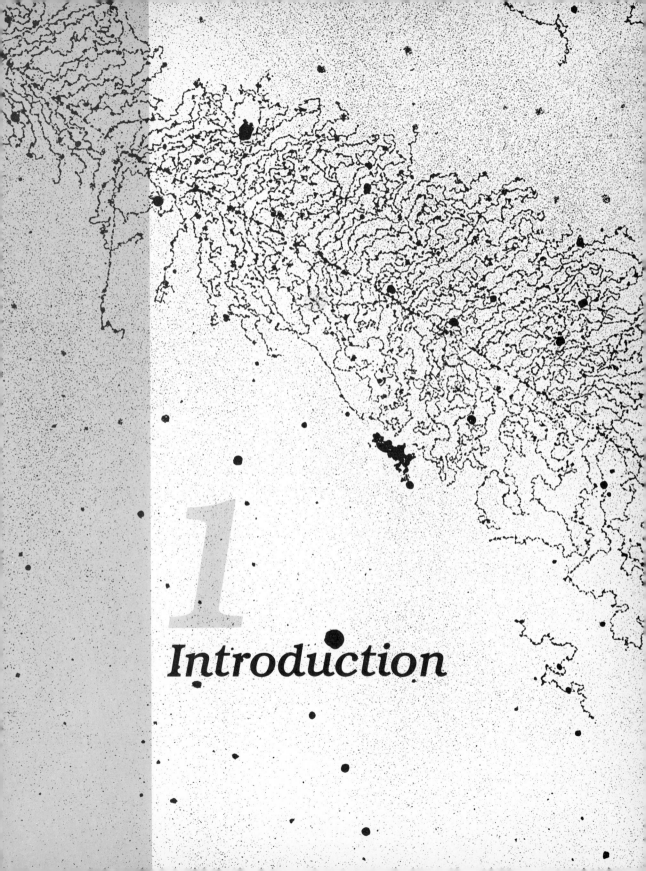

1

Introduction

What factors combine to make us what we are? Why are we male or female, tall or short, fat or thin, blonde or brunette, blue-eyed or brown-eyed? Why do we resemble our parents, aunts, uncles, siblings? Why are some people born with defects of anatomy (such as extra fingers and cleft palates), of biochemistry (such as an inability to metabolize the amino acid phenylalanine), or of mind (such as schizophrenia)?

All these questions—and more—are investigated by researchers in the field of **genetics**, the **biological science** that deals with the **inheritance** of physical and chemical characteristics. Genetics is the study of **heredity** and variation and the effect of the environment on this variation. Because its implications affect everyone, genetics is one of the most fascinating of the sciences. All living creatures, no matter how simple or primitive, are affected by the laws of genetics. Genetics also gives us information about the master plan that bacteria, plants, and animals use to assemble themselves, organizing the lifeless molecules and atoms in the environment into a living whole. It is the study of carefully controlled growth and development. The units of inheritance, the **genes**, hold the information necessary to direct the construction of **enzymes**, which are the protein tools that are active in the cellular reactions required for life itself.

Genes have an important and unavoidable impact on people. They intimately affect health, personality, parenthood, ethics, industry, and other social concerns, sometimes in surprising ways. We will consider many of those effects in detail later in this book. For now, let us look very briefly at the divisions within the science of genetics, its history, and its societal ramifications.

THE TYPES OF DIVISIONS OF GENETICS

Genetics can be subdivided into five main divisions: **cytogenetics, molecular genetics, population genetics, behavioral genetics**, and **developmental genetics**. Cytogeneticists are concerned with the cellular aspects of heredity, focusing especially on the **chromosomes**, which occur in every cell and are the carriers of genetic information. Most organisms have chromosomes that are big enough to be seen under the light microscope. Molecular geneticists are concerned with the chemical and molecular bases of genetics. They study the chemical reactions that the genes themselves undergo. They also investigate all the important chemical reactions initiated and controlled by the genes. Population geneticists deal with inheritance on the population level, looking at the inheritance and distribution of particular traits in large groups of organisms. Their ideas and conclusions are mostly expressed in mathematical terms. Behavioral geneticists are concerned with the genetic bases of different types of behavior. They study the genetic component of various forms of mental disorders, criminal behavior, and certain personality traits. Developmental geneticists study the way in which genes turn on and off on a very precise schedule to control the process of growth and development.

Genetics can also be subdivided according to the organisms being studied. For example, we speak of **animal genetics, plant genetics, microbial genetics, viral genetics**, and **human genetics**, among others. Human genetics is, of course, the study of heredity and variation in human individuals and populations. This field will occupy most of our attention, but we will also study the genetics of other organisms— fruit flies, bread mold, mice, bacteria, and viruses—because of their significance in the history of genetics.

GENETICS AS SCIENCE

Geneticists rely on the **scientific method** in their search for knowledge. The key to this method is objectivity: geneticists, like most scientists, do not decide how things ought to be. Instead, they observe how things really are in the world and then try to understand their observations in the context of existing scientific knowledge. Their fundamental activities in genetic research are controlled experimentation and critical observation.

The scientific method often seems a very rigid, "cut and dried" procedure. As usually presented, it calls for:

1. stating a problem,
2. making observations,
3. forming a hypothesis,
4. performing experiments to obtain results that can be repeated by other scientists, and
5. drawing conclusions.

However, science is actually more of an art than a set of procedures, and there are so many exceptions to the scientific method that some scientists claim that the process as described above is seldom followed. At any rate, as a science, genetics is based on generalizations, deductive reasoning, calculations, analyses, theories, relationships, interconnections, concepts, and principles. Genetics is the process of asking and answering questions about the characteristics and continuity of life. But most of all, genetics, and science in general, is seeking answers through active experimenting.

MILESTONES IN GENETICS

The term ''genetics'' was not coined until the twentieth century, but the idea goes back at least to the pre-Christian Greeks, who understood that inheritance did follow definite patterns. More modern ideas waited upon the development of microscopy, when people learned that reproduction requires a female's **ovum** (egg) and a male's **spermatozoa** (sperm). However, that initial discovery did not lead directly to modern genetics. First came such theories as **preformation,** which held that an entire miniature individual (or **homunculus**) resided in either the egg or the spermatozoon. Figure 1-1 shows several versions of the homunculus. But no one knew how the parental traits were transmitted to the homunculus.

Eventually, biologists realized that the homunculus does not exist and they discovered the actual process of **fertilization,** the union of sex cells that produces a fertilized egg or **zygote.** So too did they observe the later cell divisions and the process of embryonic development, but how the parental traits were actually transmitted to the offspring remained a mystery.

Almost everyone agrees that the real birth of genetics came with Gregor Mendel, an Austrian monk. Mendel devoted years to a study of several hereditary traits, or **characters,** that he observed in

FIGURE 1–1

The homunculus as envisioned by early scientists. (Reprinted with permission of Macmillan Publishing Company from *Genetics* by Monroe Strickberger. Copyright © 1968 by Monroe W. Strickberger.)

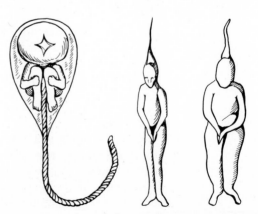

ordinary garden peas. His classic paper explaining his plant hybridization experiments first appeared in journal form in 1866, although no one gave his results serious attention at that time (see Peters, 1959, for an abridged translation of Mendel's paper).

Mendel collected a large amount of data indicating that hereditary characteristics were carried on particles of some kind. He called these particles **factors** and showed that they passed from one generation to the next according to predictable patterns. He noticed that definite hereditary characteristics (such as, short or tall peas) do not blend together into an average form (medium height). The characteristics remain in their original form, generation after generation.

Many students of the history of genetics see 1900 as the true birthdate of genetics, for it was then that three biologists independently rediscovered Mendel's paper. As a result, the basic laws of heredity became known, and their study could then progress at a rapid pace. Still, traditional biologists were slow to accept studies of heredity as a distinct discipline. William Bateson named the field "genetics" in 1905, and it became a respectable science about 1910.

During the period from 1900 to 1910, scientists made significant contributions to genetics. In 1903, W. S. Sutton and T. Boveri independently and simultaneously saw the correlation between the pattern of inheritance of Mendel's factors and the behavior of chromosomes in the cells. In other words, they recognized the chromosomes as the vehicles carrying the hereditary particles. William Bateson and R. C. Punnett (reprinted in Peters, 1959), among others, performed many **hybridization** experiments during this time, with both peas and other organisms. They learned that heredity was more complicated than Mendel had thought. They demonstrated the independence of two gene pairs in cell division, but discovered at the same time that gene expression sometimes depends upon the interdependence of two gene pairs.

By 1911, the fruit fly, *Drosophila*, had become a popular experimental subject, and Columbia University's eight-desk "fly room" had become a center of genetic discovery and the professional home of Thomas Hunt Morgan and other great pioneers of genetics. These researchers chose the fruit fly as an appropriate organism for genetic studies because it is small and easy to culture, it is convenient to handle and observe, determining the sexes of flies is straightforward, crosses can be easily set up, generation time is short, and offspring are numerous. Fruit flies were found to have many different observable hereditary traits, making possible a wide variety of investigations. Furthermore, the giant chromosomes visible in *Drosophila's* salivary glands allowed

5

researchers to observe the detailed morphology of the chromosomes themselves.

In 1927, Herman J. Muller published his famous discovery that genes, as the factors were known by then, could be changed (**mutated**) by the artificial means of irradiation. He used X rays to develop new methods in the search for a better understanding of gene composition and function.

In 1941, George Beadle and Edward Tatum provided a very significant insight into gene function when they linked genes to enzymes. In 1944, Oswald Avery, Colin MacLeod, and Maclyn McCarty published strong evidence that the gene was composed of **deoxyribonucleic acid (DNA)**. In 1953, James Watson and Francis Crick used their own models and measurements made by other researchers to solve the basic chemical structure of DNA. The structure they identified became a milestone in genetics because it suggested an obvious and elegant process for DNA to make identical replicas of itself for future generations. Indeed, their short paper revolutionized genetics, if not all of biology. These latter achievements, along with a number of other important contributions, form the basis of the still-exploding field of molecular genetics.

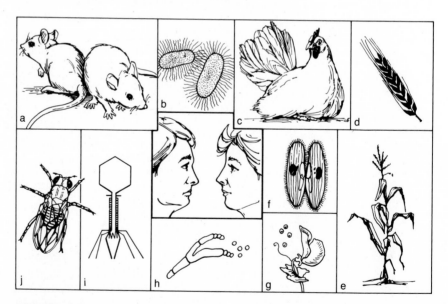

FIGURE 1–2

Some of the organisms crucial to the development of genetics: (a) mice, (b) bacteria, (c) chickens, (d) barley, (e) corn, (f) paramecia, (g) peas, (h) pink bread mold, (i) viruses, and (j) fruit flies.

Clearly, the history of genetics is a long series of fascinating discoveries. The science was not even conceived until 1866, or born until 1900, but in the forty years since the identification of the hereditary material in 1944, the study has grown into a giant. Figure 1-2 shows many of the organisms that have been significant in genetic research. Today, many people are greatly excited and others are gravely concerned about the potential applications of the discoveries that remain to be made in the fields of genetics and molecular biology.

THE STATUS OF HUMAN GENETICS TODAY

Early geneticists worked with plants, fruit flies, and a variety of small organisms. Very little research into human genetics took place, other than a few simple observations and unsophisticated experiments. Much of the work in human genetics concerned **eugenics**. Briefly, eugenics refers to the application of genetics to improving the hereditary qualities of future human generations. Its aim is laudable, but in practice, its social ramifications have encouraged mistaken assumptions, prejudice, racism, and unscientific attitudes. Many fine and idealistic geneticists were reluctant to become involved with human genetics because of the eugenics connection.

However, the greatest progress in human genetics has been made where it specifically relates to medicine. Between 1902 and 1909, the physician Archibald Garrod was the first to identify diseases such as **alkaptonuria, albinism**, and **cystinuria** as hereditary traits in humans. He called these conditions **inborn errors of metabolism**, indicating that the genetic make-up of an individual was responsible for a specific biochemical defect. However, Garrod's diligence and perception had little effect at the time. Human genetics did not begin to flourish until after World War II. By the 1950s, human genetics was part of the curriculum in only a few medical schools. It was not until 1956 that we learned the correct number of chromosomes in the human cell: 46. In 1959, the discovery that **Down syndrome (mongolism)** was due to an extra chromosome had a tremendous impact upon human genetics. Since then, the advances in clinical genetics have been rapid, especially in the application of genetics to diagnosis. The importance of human genetics in medicine has grown steadily. Partly because of the development of experimental techniques, human genetics is becoming more precise. The field has capitalized on many genetic discoveries made using other organisms. But human genetics has now become sophisticated enough to contribute to the overall knowledge of genetics. Human geneticists have

7

made great strides in the biochemistry of metabolism, the molecular biology of human cells, the tissue culturing of human cells, chromosome analyses, enzyme therapy, immunology, cell hybridization, fetal research, behavior, and genetic engineering. Curricula at all levels are increasing their emphasis on human genetics.

— Stop

SOCIETAL RAMIFICATIONS

Heredity affects everyone—personally and on the levels of the family, the community, the nation, and the world. Certainly this is true for most genetic diseases, which directly affect the patient and the patient's family. But beyond them, taxpayers throughout the nation may pay for research on, and treatment of, genetic diseases. In fact, with the decline of infectious diseases, one can see that many of our most serious remaining health problems—for example, some kinds of heart disease and cancer—have an important genetic component. National and international agencies concerned with family planning, the rights of the handicapped, biomedical ethics, and other serious issues also play a role in questions involving genetics.

Genetics and society interact in numerous ways. Medical science confronts genetic disorders, cancers, fetal diagnosis, genetic counseling, mutation, and chromosome aberrations. Society is concerned about genetic damage resulting from radiation, drugs, and other chemicals, and about reproductive issues such as abortion, inbreeding, artificial insemination, *in vitro* (test-tube) fertilization, the possibility of parents selecting the sex of offspring, and even cloning. Many behavioral traits have been assigned genetic components. Schizophrenia, mental depression, mental retardation, intelligence, and even alcoholism and criminality are the subjects of genetic research.

In agriculture, straightforward genetic techniques consist of selective breeding and hybridization. Many of the newer techniques being applied to agriculturally important plants and animals constitute genetic engineering, a subject of very wide interest. This is especially true of recombinant DNA work, which puts human genes, for instance, into bacteria. This worries many people, and genetics has found itself embroiled in controversies with the public, government, industry, and the law. More than ever before, genetics has become a political issue.

Genetics has also met religion and politics in the recently reawakened creationism vs. evolution debate, and in the new concept and discipline of **sociobiology**, which attempts to explain much of behavior and societal structure in terms of heredity. Genetics meets politics again in such worldwide tragedies as Hitler's infamous execution of six million Jews, gypsies, and East Europeans in the misguided interests of eugenics. Or one can point to the Lysenko incident in the Soviet Union, in which too great an ideological reliance on environmental shaping caused a nearly complete collapse of Soviet agriculture and excessive dependence on North American wheat supplies. The mere mention of wheat brings to mind the wheat breeder, Norman Borlaug, who won the Nobel Prize for peace because of the close relationship of his improved wheat varieties to world food supply and world peace. These and other interactions of genetics and society are expanded upon later in this text.

Controversy surrounds almost every topic listed above. The moral, ethical, and legal aspects of these societal problems are not judged in the same way by all individuals. Many questions are asked, but few are answered. Sometimes, we simply do not have enough information to make sound decisions. Very often, the achievements of technology are far ahead of society's capacity to accommodate them, and we do not know the extent of the social impact that will result from the information already available. Almost everyone agrees that the problems are too important and their consequences too great to be left to scientists alone. Society as a whole must take an interest in them. Many critical decisions regarding the interaction of genetics and society will have to be made in the near future, and the direction of the decisions could greatly affect humanity. Such concern is certainly not an overstatement—these decisions are apt to change the course of history.

Before we can delve into the social aspects of heredity and genetics in any depth, we must learn what genes are, how genes are transmitted, and how they function. An elementary command of these fundamentals of genetics will make the study of its sociological aspects possible.

SUMMARY

Genetics is the study of heredity and variation. The mechanisms of heredity result in offspring who tend to resemble their parents. Genetics can be divided into subfields according to the mode of study and the organisms studied. Numerous lower organisms, each possessing certain research advantages, have been used to elucidate the concepts of genetics in general, and to some extent, those of human genetics.

Gregor Mendel's 1866 publication stands alone as the beginning of our understanding of the rules of heredity. The rediscovery of that classic paper in 1900 set the stage for modern genetics. Crucial among

Mendel's ideas was the conclusion that heredity was due to particles, or factors, now called genes. Soon after 1900 came the association of the hereditary particles with chromosomes, and additional genetic concepts soon emerged from tireless work with fruit flies. Research had shown by 1941 how genes might function, by 1944 that genes were composed of DNA, and by 1953 that this "genetic stuff" of life had a particular molecular structure. Since the early 1950s, genetics has progressed at an incredible pace.

Today, human genetics has become an active and strong subdiscipline of genetics. Although eugenics initially dominated the field, human genetics has reached maturity and is very important in medicine. Genetics also interacts with society in numerous other ways, many of which engender controversy. The interactions, ramifications, and controversies surrounding genetics and society are the theme of this book.

DISCUSSION QUESTIONS

1. Heredity has sometimes been described as a tendency of offspring to resemble their parents. Why is the word "tendency" so important in this definition?

2. What characteristics make "classic" research "classic"?

3. Some scientists have been said to be ahead of their time. What does "ahead of one's time" mean?

4. The scientific method is the usual framework for research. On the other hand, many discoveries have come about as a result of chance. Explain how chance events would tend to detract from the scientific method.

5. If Mendel's 1866 publication had not been lost and/or neglected in an obscure journal, where would genetics be today?

6. What is wrong with the expression, "blood relative"?

7. Much of our knowledge of human genetics has actually come from studies of microbes, plants, and lower forms of animals. Why do you suppose that human beings have been so difficult to study in terms of genetics?

8. What does the fact that our knowledge of human genetics has come from the study of other organisms tell you about the application of the laws of heredity? Explain.

9. Can you think of any current or recent events that might be examples of eugenics?

REFERENCES

Avery, O. T., C. M. MacLeod, and M. McCarty. 1944. Studies on the chemical nature of the substance inducing transformation of pneumococcal types. *Journal of Experimental Medicine* 79:137–158.

Bateson, B., ed. 1928. *William Bateson: Essays and addresses*. Cambridge: Cambridge University Press.

Bateson, W., E. Saunders, and R. C. Punnett. 1904–1908. Experimental studies in the physiology of heredity. *Reports to the Evolution Committee of the Royal Society* 2:1–154; 3:1–53; 4:1–60.

Beadle, G. W., and E. L. Tatum. 1941. Genetic control of biochemical reactions in *Neurospora*. *Proceedings of the National Academy of Sciences* 27:499–506.

Boveri, T. 1904. *Ergebnisse uber die Konstitution der chromatischen Substanz des Zellkerns*. Jena: Gustav Fisher Verlag.

Garrod, A. E. 1902. The incidence of alkaptonuria: A study in chemical individuality. *Lancet* ii:1616–1620.

Garrod, A. E. 1908. Inborn errors of metabolism. *Lancet* 2:1–7, 73–79, 142–148, 214–220.

Garrod, A. E. 1909. *Inborn errors of metabolism*. London: Oxford University Press.

Lejeune, J., M. Gautier, and R. Turpin. 1959. Etude des chromosomes somatiques de neuf enfants mogoliens. *Comptes Rendus des Seances de l'Academie des Sciences* 248:1721–1722.

Muller, H. J. 1927. Artificial transmutation of the gene. *Science* 66:84–87.

Peters, J. A., ed. 1959. *Classic papers in genetics*. Englewood Cliffs, NJ: Prentice-Hall, Inc.

Strickberger, M. W. 1968. *Genetics*. New York: Macmillan Publishing Co. Inc.

Sutton, W. S. 1903. The chromosomes in heredity. *Biological Bulletin* 4:231–251.

Tjio, J. H., and A. Levan. 1956. The chromosome number of man. *Hereditas* 42:1–6.

Watson, J. D., and F. H. C. Crick. 1953. Molecular structure of nucleic acids. *Nature* 171:737–738.

2
The Genetic Material

Most living things are made up of one or more **cells**. The best known exceptions, the **viruses**, are obligate parasites that function as living organisms only when they are inside the cells of their hosts.

All cells have certain characteristics in common with each other. They are minute masses of living material enclosed within a **cell membrane** made of phospholipid and protein. Plant and bacterial cells add a **cell wall** outside the cell membrane. Animal cells often have a cell coat made of a material related to mucus.

The carriers of heredity, the genes, are located inside every cell. The cells of higher organisms concentrate most of this genetic material within the **nucleus**, surrounded by the nuclear envelope. Cells also contain a variety of **organelles**, loosely analogous to the organs of the human body. Figure 2-1 shows a cell with its organelles. Organelles

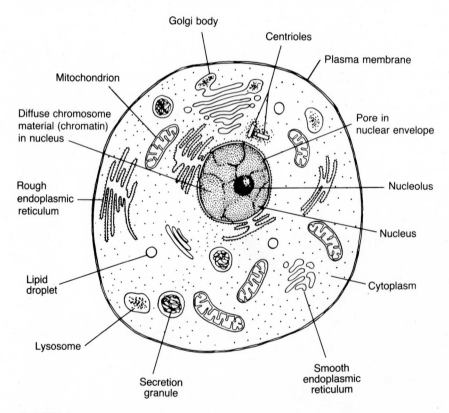

FIGURE 2–1
A generalized animal cell.

14

have specific functions. For example, some help make proteins, others release energy, break down foreign materials, package secretions, and perform other functions essential to the life of the cell.

In a basic sense, each cell is composed of lifeless molecules that obey all the physical and chemical laws that apply to inanimate matter. Yet these molecules—the parts of a cell—are so intricately organized, each one with its own specific function, that the whole displays some remarkable features. Living things can extract and transform energy from their environment. They can incorporate atoms and molecules from the environment into their own structure. And they can produce other organisms like themselves.

The most remarkable feature of living things is the process of self-reproduction. This is the key to life that lies at the heart of genetics, and it is the mechanism that generates the hierarchy of being shown in Figure 2-2. At the top of this hierarchy is the **biosphere**, consisting of all the **biomes** and **ecosystems** on Earth. Ecosystems, in turn, are comprised of **populations** of various species, and a population consists of **organisms** that are a representative sampling of a species. Organisms are made of cells, and cells of **molecules**. At the base of the hierarchy we see the **atoms** that go into molecules and the sub-atomic particles that make up atoms. The groups of atoms bonded into molecules are held together by forces of attraction. Chemists and physicists study these forces and their behavior—both biology and genetics are a kind of superchemistry.

The automatic process by which organisms build molecules, organelles, and cells is **biosynthesis**. The organism relies on instructions stored in the genes, and it uses cellular machinery to turn raw materials (**precursors**) into different, more complex substances that are essential to life. Many of these substances are special large molecules called **macromolecules**. They include proteins and the nucleic acids of the genes themselves. In this chapter, we will learn more about the genetic material.

IDENTIFYING THE GENETIC MATERIAL

We know today that the genetic material is **DNA**, or **deoxyribonucleic acid**, in all organisms except for a few viruses that use the related material, **RNA (ribonucleic acid)**. However, the identity of this material was a mystery before the 1940s.

Until then, scientists had spent many years wondering just what the substance of inheritance might be. They knew that it must be

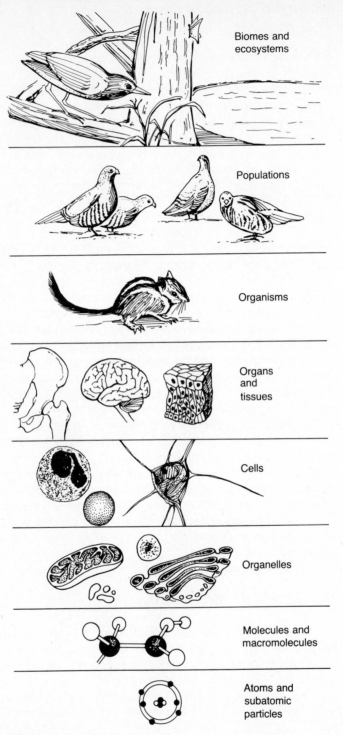

Biomes and ecosystems

Populations

Organisms

Organs and tissues

Cells

Organelles

Molecules and macromolecules

Atoms and subatomic particles

FIGURE 2–2
The hierarchy of being.

able to replicate as accurately as the cells that bear it, be stable, carry a phenomenal amount of biological information, and transmit that information accurately and usefully. DNA was only one of the possibilities they investigated.

DNA consists of only five elements, carbon, hydrogen, oxygen, nitrogen, and phosphorus. These elements are arranged into sugar molecules (**deoxyribose**), phosphate molecules, and four **nitrogenous bases** (**adenine**, **thymine**, **cytosine**, and **guanine**), all bonded together in a specific pattern to form DNA.

Friedrich Miescher, a Swiss chemist, discovered DNA in 1869. He isolated the impure substance from the pus he obtained from used bandages. Since he located it in the nuclei of the white blood cells in the pus, he called it **nuclein**. He also found nuclein in salmon sperm, and was able to conclude that it consisted largely of nitrogen and phosphorus. Later, other chemists were able to purify nuclein, and they changed its name, first to **nucleic acid**, and then to deoxyribonucleic acid (DNA). Even then, the substance attracted very little biological attention. Many years passed before conclusive evidence showed Miescher's crude nuclein to be the genetic material of life.

The first relevant evidence came from an experiment performed by Fred Griffith in 1928. Griffith was studying the bacterium *Streptococcus pneumoniae*, which exists in two forms. In one form, the cell is covered with a slime layer or **capsule**. The cell of the other form has no capsule. The first type is called "smooth," the second, "rough." Mice injected with smooth bacteria die of pneumonia, because these bacteria are **virulent**. Mice injected with the rough bacteria survive, because these bacteria are **avirulent**. Evidently, the white blood cells of a mouse's immune system can cope with the rough bacteria but not with the smooth ones.

Griffith made his discovery when he injected heat-killed smooth or rough bacteria into his mice. As expected, dead bacteria did not harm the mice. However, when he injected heat-killed smooth bacteria *plus* live rough bacteria, some of the mice died. Two substances, neither of which killed the mice separately, killed the mice when they were combined. Figure 2-3 summarizes what happened in this experiment.

When researchers found that the dead mice contained **live** smooth bacteria as well as the rough bacteria injected, they interpreted Griffith's experiments to mean that some substance from the heat-killed bacteria actually entered some of the live bacteria and transformed them into the virulent type. Since no one knew what this bacterial substance was, it was called simply the **transforming principle**. Whatever it was,

FIGURE 2–3
Griffith discovered that some factor in dead smooth bacteria could transform live rough bacteria into live smooth bacteria. He called this the transforming principle.

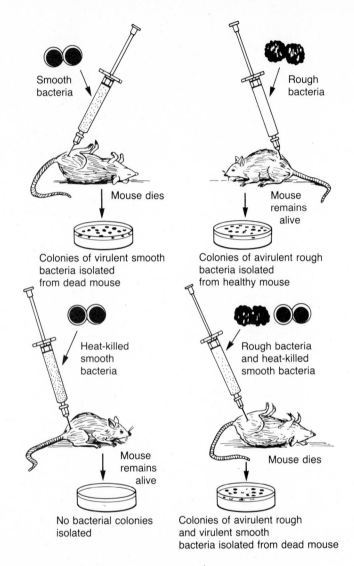

Smooth bacteria

Mouse dies

Colonies of virulent smooth bacteria isolated from dead mouse

Rough bacteria

Mouse remains alive

Colonies of avirulent rough bacteria isolated from healthy mouse

Heat-killed smooth bacteria

Mouse remains alive

No bacterial colonies isolated

Rough bacteria and heat-killed smooth bacteria

Mouse dies

Colonies of avirulent rough and virulent smooth bacteria isolated from dead mouse

it had the capacity to change an organism's heredity and cause it to reproduce itself as a different form.

In the early 1940s, Oswald T. Avery, Colin MacLeod, and Maclyn McCarty extended Griffith's work on transformation. Their overall plan appears in Figure 2-4. By extracting and purifying individual substances found in the bacteria and testing them for the power to transform rough bacteria into smooth in a test tube (not in mice), they pro-

18

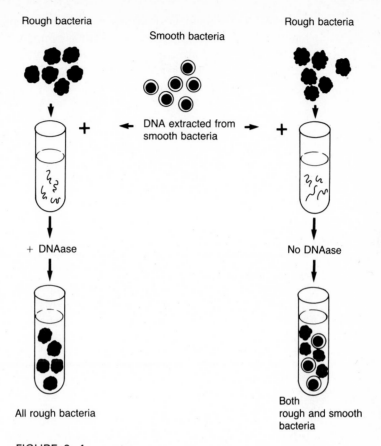

Rough bacteria

Smooth bacteria

Rough bacteria

+ ← DNA extracted from smooth bacteria → +

+ DNAase

No DNAase

All rough bacteria

Both rough and smooth bacteria

FIGURE 2–4
The experimental plan of Avery, MacLeod, and McCarty (1944) showing the transformation of rough bacteria to smooth bacteria with isolated DNA from smooth bacteria.

vided very strong evidence that the transforming principle was DNA. No other substance tested could transform the bacteria. Furthermore, when the DNA was exposed to the enzyme that destroys DNA, **DNAase**, the treated DNA lost its ability to transform the bacteria.

Clearly, DNA was shown to be the genetic material. In transformation, it could change the heredity and the characteristics of a bacterium. In a sense, natural transformation was the first crude example of genetic engineering, and it was intriguing enough to prompt thoughts of what could be accomplished with a more controlled, precise version of the phenomenon. As soon as Avery and his colleagues published their

19

work in 1944, scientists began to think about changing the heredity of higher forms of life, especially agriculturally important species.

The transformation results were not the only signs that DNA was the genetic material, but most people thought of the others as mostly circumstantial evidence. For example, DNA was found in all organisms except a few RNA viruses. In a given organism, all of the cells contained the same amount of DNA. The only exception was observed in reproductive cells, or gametes (sex cells), which contained half the DNA found in **somatic** or body cells. Table 2-1 shows the comparison of amounts of DNA contained in somatic cells of several organisms.

TABLE 2–1
DNA comparisons among cells within an organism

Organism	Relative DNA Amounts		
	Spermatozoa	Erythrocytes (Red Blood Cells)	Ratio of Somatic Cell/Sex Cell
Domestic fowl	1.26	2.34	1.86
Toad	3.70	7.33	1.98
Carp	1.64	3.49	2.13

Source: Mirsky, A. E., and Ris, H. 1949. *Nature,* 163: 666–667.

In spite of these facts, some scientists insisted that protein was a better candidate to be the genetic material. DNA was too simple a molecule, for it had only four varying components, the nitrogenous bases. How could DNA possibly be responsible for the tremendous genetic diversity apparent among and within species? On the other hand, proteins came in a myriad of types, and they were composed of twenty different amino acids. However, enough excellent experiments were eventually performed to establish that DNA is the hereditary material. Probably the most conclusive of these experiments was done by Alfred D. Hershey and Martha Chase in 1952. They showed that the genetic material had to be DNA, and not protein, by following both substances through the life cycle of a virus that infects bacteria. By labeling the virus's protein with radioactive sulfur and its DNA with radioactive phosphorus, they were able to show that only DNA entered the bacterial cell to direct the manufacture of new viruses. The main steps of the Hershey-Chase experiments are diagrammed in Figure 2-5. This experiment convinced even the most skeptical scientists.

FIGURE 2–5
Hershey-Chase experiments indicating that DNA is the genetic material. (From Goodenough, U. 1978. *Genetics*. Holt, Rinehart, and Winston, New York.)

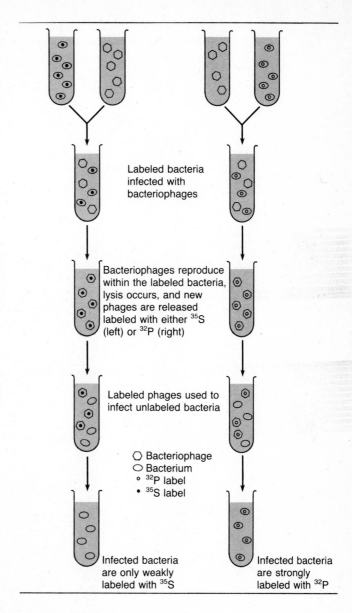

Labeled bacteria infected with bacteriophages

Bacteriophages reproduce within the labeled bacteria, lysis occurs, and new phages are released labeled with either ^{35}S (left) or ^{32}P (right)

Labeled phages used to infect unlabeled bacteria

○ Bacteriophage
○ Bacterium
∘ ^{32}P label
• ^{35}S label

Infected bacteria are only weakly labeled with ^{35}S

Infected bacteria are strongly labeled with ^{32}P

SOLVING THE STRUCTURE OF DNA

By the early 1950s, biologists were calling DNA the master molecule of life. Most agreed that unraveling its structure was the next essential

21

step toward understanding how the molecule replicates and functions as the genetic material. When the necessary discovery was made in 1953, it revolutionized all of biology. It was, thus, a landmark in the history of science. It is also an intriguing human interest story.

Many scientists were actively engaged in the quest for DNA's secrets. Two of the main characters were James Watson, an American studying in the Cavendish Laboratory of Cambridge University in England, and Francis Crick, an Englishman. They used current information on DNA, wooden blocks, and wire to build models until they found the one, shown in Figure 2-6, that fitted all the data. Their success, reported in a 1953 issue of *Nature*, transformed biology into a molecular science almost overnight.

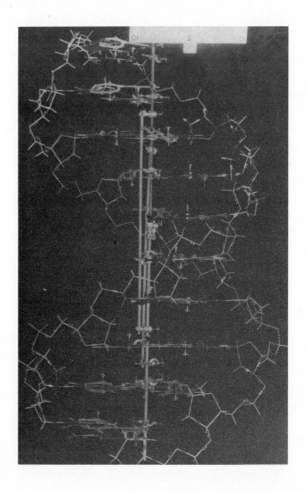

FIGURE 2–6
The original model of the DNA double helix constructed by Watson and Crick. (From Watson, J. D. 1968. *The Double Helix*. Atheneum, New York.)

Watson and Crick knew that DNA consisted of the sugar deoxy-ribose, phosphate groups, and the four nitrogenous bases, adenine, cytosine, guanine, and thymine. The chemical structures of these components are illustrated in Figure 2-7. Adenine and thymine both belong to the group of chemicals called **purines**; thymine and cytosine are **pyrimidines**.

Watson and Crick also knew that the DNA molecule was very long, very narrow, and built of repeating units. This information came from **X-ray diffraction** photographs of DNA crystals obtained mostly by Maurice Wilkins and Rosalind Franklin. These photographs, an example of which appears in Figure 2-8, recorded the angles at which X rays were scattered from various regular features of crystallized DNA molecules.

In addition, they knew of Erwin Chargaff's observations in 1950 that DNA contained equal amounts of adenine and thymine, and equal amounts of guanine and cytosine. As a result, the amount of adenine

FIGURE 2–7
The chemical components of DNA.

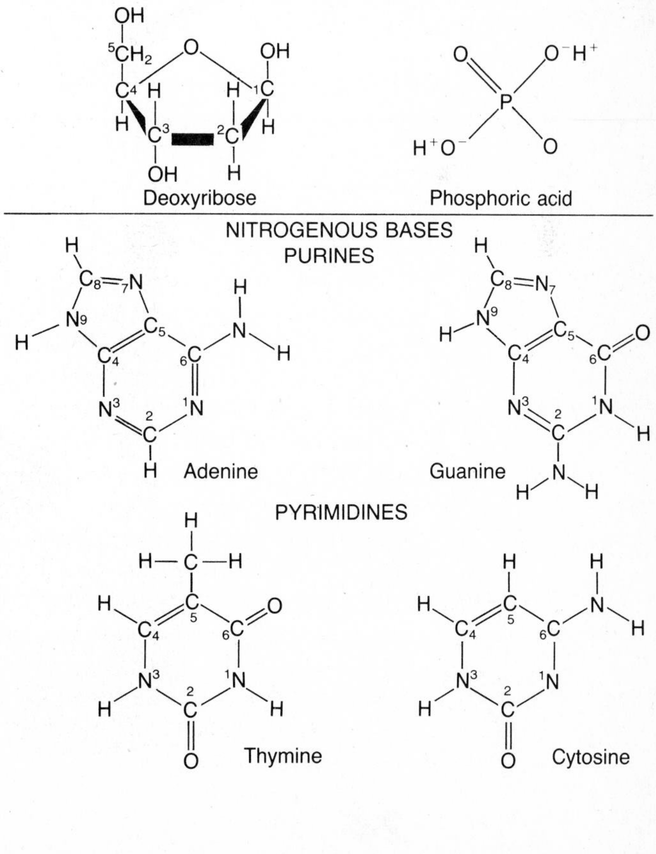

Deoxyribose

Phosphoric acid

NITROGENOUS BASES
PURINES

Adenine

Guanine

PYRIMIDINES

Thymine

Cytosine

23

FIGURE 2-8

An X-ray diffraction photograph of DNA, viewed from the end. (Reprinted by permission from *Nature*, Vol. 171, No. 4356, pp. 738–740. Copyright © 1953 Macmillan Journals Limited. By courtesy of the Biophysics Department, King's College London.)

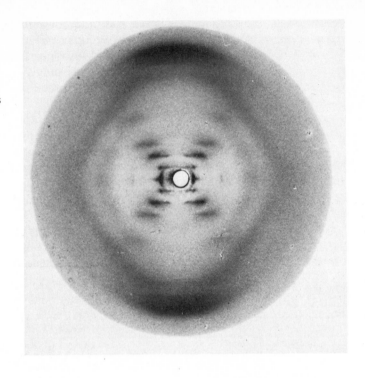

plus guanine always equaled the amount of thymine plus cytosine. However, adenine plus thymine showed no such simple relationship to guanine plus cytosine; DNAs from different organisms showed very different ratios. These relationships, known as **Chargaff's Rules**, were crucial to the eventual insight into DNA's structure.

Watson and Crick made many measurements and built many of their makeshift molecular models before they decided upon what they thought was the correct structure. Although the existing data seemed to dictate certain features of the molecules, they continually had difficulty achieving good fits between some of the components. Like all matter, atoms and molecules have shape and take up space, and Watson and Crick had to rearrange the parts of their model until they found a configuration that permitted all the right **chemical bonds** to hold the parts close together. Eventually, they considered a model that looked much like a twisted ladder. When this structure did not quite work, they had their monumental insight—they turned one of the two strands (one of the ladder sides) upside down. With this innovation, the model fit well with the existing data. The fundamental structure of life's most important molecule was solved.

24

THE NATURE OF THE GENETIC MATERIAL

The DNA molecule is an extremely long **double helix**, or **duplex**. The repeating units of this helical structure, the **deoxyribonucleotides**, each consist of a deoxyribose group, a phosphate group, and one of the four nitrogenous bases, all chemically bonded together. The four common deoxyribonucleotides of DNA are shown in Figure 2-9. Sometimes called simply **nucleotides**, these subunits bond together to form a linear structure known as a **polynucleotide strand**, shown in Figure

FIGURE 2–9
The four common deoxyribonucleotides found in DNA, each consisting of a deoxyribose molecule, phosphoric acid, and one of the nitrogenous bases.

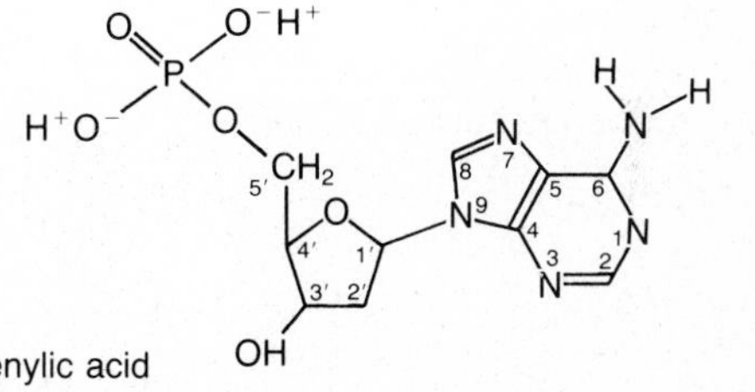

Deoxyadenylic acid

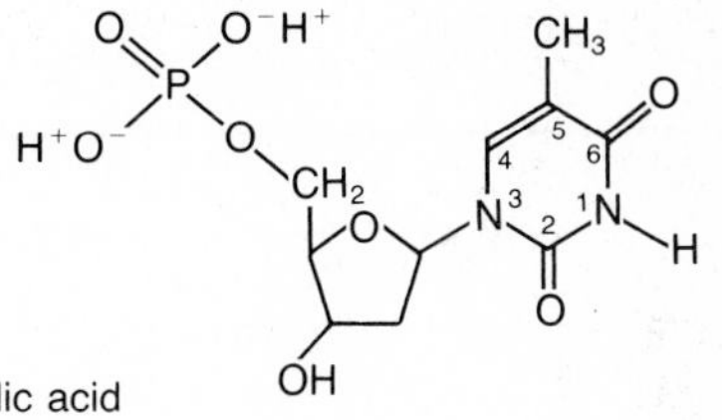

Deoxythymidylic acid

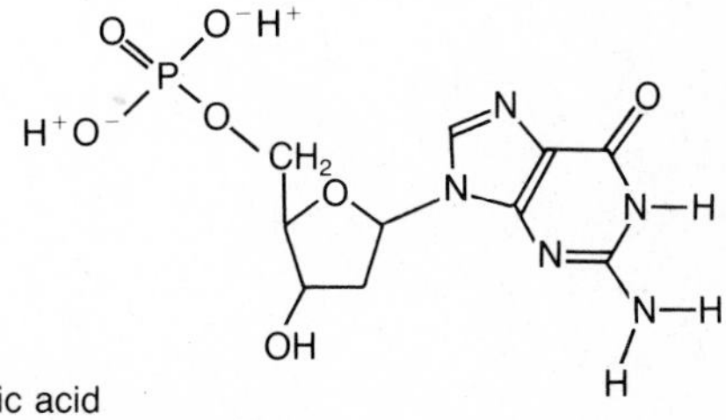

Deoxyguanylic acid

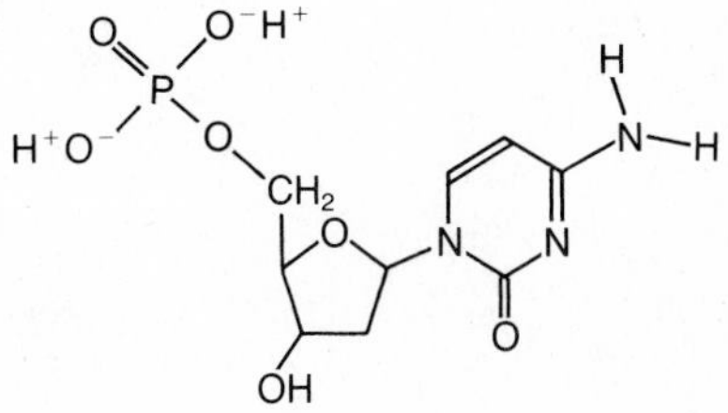

Deoxycytidylic acid

25

2-10. Two of these polynucleotide strands chemically bond with each other along their lengths to form the ladderlike structure in Figure 2-11. Normally, each nitrogenous base bonds with only one other. Adenine pairs with thymine by means of two chemical bonds and cytosine uses three bonds to pair with guanine. This is the only way that permits a good chemical fit between nucleotides, and for this reason the pairs are described as being **complementary**. Complementarity is the reason underlying Chargaff's Rules.

The two linked polynucleotide strands run in opposite directions. In Figure 2-11, the pentagon structures of the sugars in the right-hand

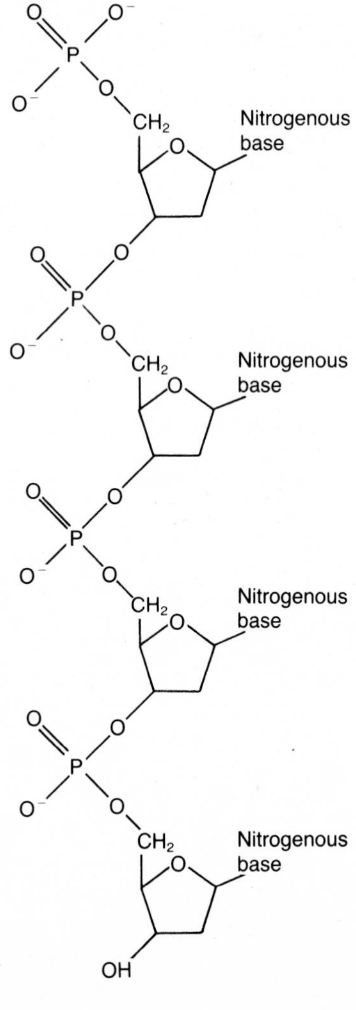

FIGURE 2–10

A polynucleotide strand consisting of only four deoxyribonucleotide sub-units.

strand point downward. Those in the left-hand strand point upward. This is an **antiparallel** alignment, as opposed to a **parallel** one. This feature was one of Watson and Crick's final realizations that made the entire DNA structure chemically feasible. The antiparallel structure allows the necessary chemical bonding to occur between bases. All that remains to finalize the actual fundamental structure of the DNA molecule is to twist it to bring about the double helix effect shown in Figure 2-12.

Researchers have developed much biochemical and genetic data concerning DNA since 1953. We now know that DNA molecules are indeed very long relative to most molecules. They extend for many thousands of nucleotide base pairs, with the base pairs spaced 3.4 **ang-**

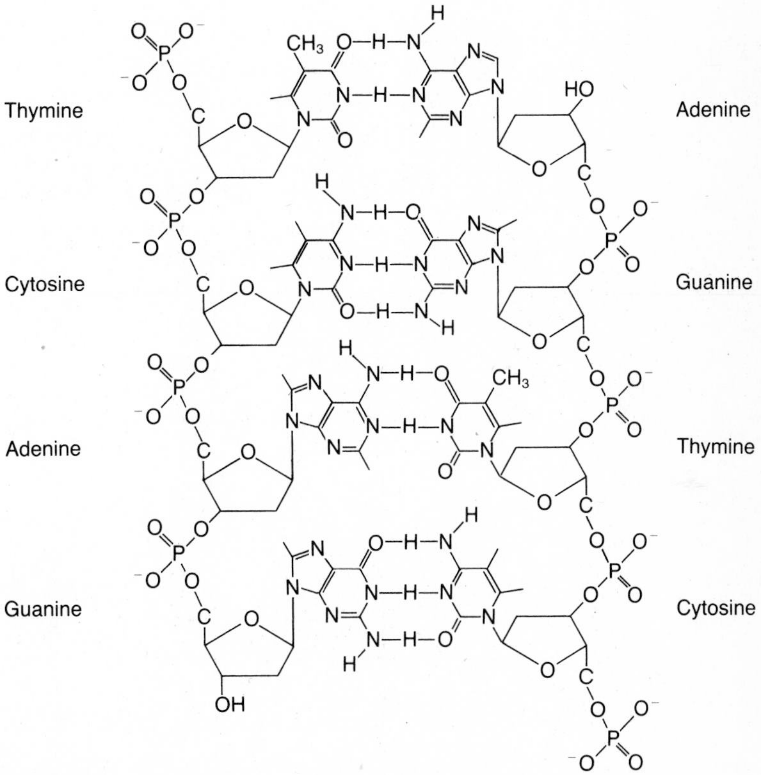

FIGURE 2–11
A double-stranded segment of DNA showing four deoxynucleotide pairs. Note the complementary pairing and the antiparallel orientation of the two polynucleotide strands. The chemical designations refer to the nitrogenous bases involved in each of the deoxyribonucleotides.

27

FIGURE 2–12
The DNA double helix according to Watson and Crick's model.

stroms apart (an angstrom is 10^{-10} meter, or one ten-billionth of a meter). The double helix makes one complete turn every ten nucleotide base pairs. Therefore, each complete turn is 34 angstroms long.

Only four different nitrogenous bases are normally incorporated into the molecule, and they must bond only with their complements (adenine-thymine; guanine-cytosine). One might wonder how such simplicity can yield so much variation within and among species. The answer is that although a DNA molecule is simple, it is also very long, and a tremendous number of different *sequences* of nucleotides can exist.

Look at one strand of a DNA helix (the other is, of course, determined by the rule of complementarity). There are four different possibilities for just one nucleotide:

However, there are 16 different possible sequences of two nucleotides:

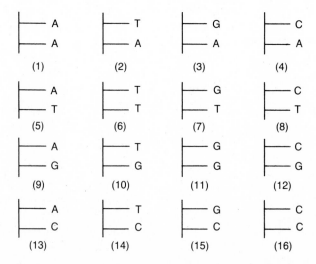

28

The mathematical relationship can be expressed as 4^n, where n is the total number of nucleotides in the sequence. There are thus 4^5 or 1024 different possible sequences of 5 nucleotides, and 4^{20} or 1,099,511,500,000 (over a trillion) sequences of 20 nucleotides. If a DNA molecule has 100,000—or more—nitrogenous base pairs, the number of possible sequences becomes astronomical. Therein lies the vast genetic diversity possible from the DNA molecule. And that diversity determines the genetic differences between humans and petunias, squirrels, bacteria, chimps, and other human beings. The variety in these sequences within DNA molecules is the biological basis of life.

RIBONUCLEIC ACID

DNA is not the only nucleic acid in cells. Where DNA is concentrated in the nucleus, ribonucleic acid (RNA) is dispersed throughout the cell. RNA differs from DNA in that its backbone alternates **ribose**, not deoxyribose, with the phosphate groups; ribose contains one more oxygen atom than deoxyribose. RNA also differs in that it uses the nitrogenous base **uracil** in place of DNA's thymine; these differences are illustrated in Figure 2-13. Complementarity between two RNA strands will therefore show the same cytosine to guanine relationship, but adenine will complex with uracil, not thymine; when DNA and RNA strands complement, DNA's adenine will complex with RNA's uracil and DNA's thymine with RNA's adenine. In addition, RNA is usually single stranded, while DNA is usually double stranded. This is not the clearest of differences, however, for both single-stranded DNA and double-stranded RNA do exist. The differences between ribose and deoxyribose, and uracil and thymine, are more significant.

Cells contain several types of RNA, each playing an important part in carrying out gene functions as shown schematically in Figure 2-14. **Messenger RNA**, or **mRNA**, is the chemical message for which the DNA codes. DNA also codes for two other types of RNA. **Transfer**

FIGURE 2–13

The molecular structures of ribose and uracil found in RNA. Note the similarity of ribose to the deoxyribose shown in Figure 2-7 (a difference of one oxygen atom). Also note the similarity of uracil to thymine.

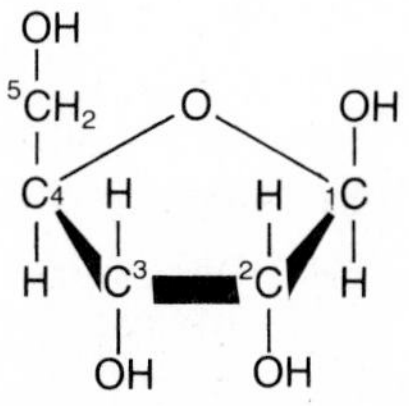

Ribose

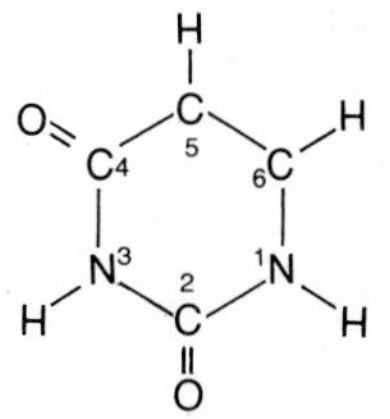

Uracil

FIGURE 2–14
The RNAs are the tools by which the messages stored in DNA are translated into action in the form of proteins.

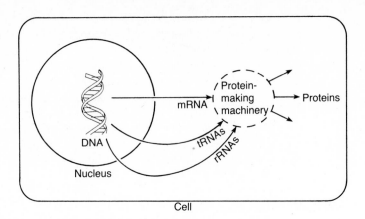

RNA, or tRNA, is intimately involved in translating the message carried by the mRNA. **Ribosomal RNA, or rRNA,** is one of the components of small cellular organelles called **ribosomes** that also help translate the RNA messages carried from the DNA. The end result of the translation process is protein. These events are discussed more thoroughly when we describe how genes work.

REPLICATION OF THE GENETIC MATERIAL

The specific structure of the DNA that Watson and Crick worked out suggested to them a mechanism by which the molecule could replicate. Obviously, some sort of replication is necessary, because all the cells of a body are descended from a single fertilized egg, they all contain the same amount of DNA, and the hereditary information of the genes must pass from generation to generation. The replication process must be accurate since mistakes could interfere with an individual's health and successful reproduction.

Many experiments have been done to elucidate the process of replication. The process begins with the unwinding and separation of a segment of the double helix. The chemical bonds that hold the nitrogenous bases together at the central axis of the molecule are relatively weak and can be ruptured under the right conditions.

Once the polynucleotide strands have separated, new strands can be assembled from single nucleotides, as shown in Figure 2-15. Normally, the cell contains an adequate supply of additional nucleotide units that have been synthesized in advance. The nucleotides move into the appropriate positions according to complementarity. Adenine pairs

30

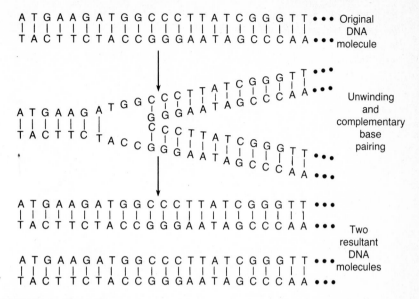

FIGURE 2–15
DNA replication, showing the complementary base pairing format.

only with thymine, thymine with adenine, cytosine with guanine, and guanine with cytosine. Two new polynucleotide strands appear, and as they grow the original double helix forks into two double helices, each one having an original strand and a new strand. The result is the construction of two DNA molecules from the initial one. These two resultant DNA molecules are identical to each other and their parent molecule.

Let us emphasize the outcome of this type of replication. As we can see from Figure 2-16, each of the two resultant molecules consists of one polynucleotide strand that existed prior to the replication event and one that has been newly synthesized. This mode of replication is called **semiconservative** because each daughter molecule conserves half of the original.

Molecular fit is the underlying basis of accurate replication, but one or more **enzymes** are necessary to make each of the steps take place. Enzymes are organic catalysts (composed of proteins) that make specific chemical reactions proceed rapidly. They form a crucial part of the biochemical machinery of DNA replication. Like all other essential machinery of the cell, they are synthesized according to information carried in the DNA. The DNA is the significant part of the cellular components called **chromosomes**. Later in this book, we will see that DNA replication and chromosome replication are very closely correlated.

FIGURE 2-16
Semiconservative mode of DNA replication.

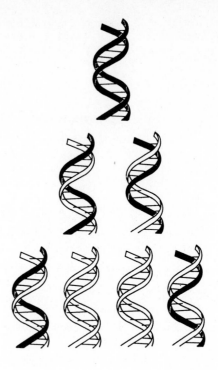

SUMMARY

Living things are made up of the same inanimate matter as nonliving things. Unlike the latter, living things can extract and transform energy and materials from the environment and can reproduce. The complexity of organization steadily increases from subatomic particles to atoms, molecules, macromolecules, organelles, cells, tissues, and organisms. Individual organisms, in turn, make up populations and higher ecological groupings.

The genetic material that forms the basis of life is almost always DNA. DNA was first isolated more than a century ago and designated as the genetic material in the mid-twentieth century. Although many scientists thought protein was the genetic material because of its vast diversity, transformation and several other phenomena eventually confirmed that DNA was the master molecule of life.

Biochemical analyses showed that DNA is composed of deoxyribose, phosphate groups, and four different nitrogenous bases: adenine, thymine, guanine, and cytosine. Research techniques such as X-ray diffraction showed that the molecule was long, narrow, and had repeating units. Chargaff's observations (Rules) revealed that DNA contains equal amounts of adenine and thymine and equal amounts of guanine and cytosine. Using all this information and more, Watson and Crick described the structure of the DNA molecule. With that discovery, molecular biology and genetics were catapulted to the forefront of biology.

The long DNA molecule is a double helix consisting of two polynucleotide strands twisted around each other. The strands alternate deoxyribose sugar units with phosphate groups; paired nitrogenous bases form cross-rungs between the two strands in a molecule. A nucleotide is made up of one deoxyribose, a phosphate group, and one of the bases. Ten nucleotide pairs are found in each 34-angstrom long turn of the helix. Tremendous variation characterizes DNA molecules because of the many possible different nucleotide sequences.

Ribonucleic acid (RNA) differs from DNA in that it contains uracil in place of thymine and the sugar ribose in place of deoxyribose. Cells contain three different types of RNA: messenger (mRNA), transfer (tRNA), and ribosomal (rRNA). All three are necessary to convert DNA information into the cell's chemical products.

Replication of the genetic substance occurs with the aid of a battery of enzymes that unwind the DNA, separate the polynucleotide strands, and bond new nucleotides into place according to the rules of base complementarity. The result is two DNA molecules, each having a nucleotide sequence identical to the original molecule.

DISCUSSION *QUESTIONS*

1. Genetic engineering is a major topic later in this book. It is usually regarded as a recent development, but one could point out that it was first performed in 1928. Explain.

2. When Avery, MacLeod, and McCarty were showing that DNA is the transforming principle, they mixed smooth bacteria DNA with DNAase before exposing rough bacteria to it. Why was this such an important part of their experiment?

3. In the same experiment, why were *both* rough and smooth bacteria observed as a result of mixing the smooth bacteria DNA with live rough bacteria?

4. Could you make DNA simply by mixing phosphoric acid (a phosphate source), deoxyribose, adenine, guanine, thymine, and cytosine? Explain.

5. Pre-1950 researchers expected the DNA molecule to contain A–A, T–T, C–C, and G–G base pairs. What data told them they were wrong? Explain the data.

6. Imagine yourself as Watson or Crick. Assume that you are very close to solving the structure of DNA. You have all the accumulated data, and you have built a model such as the one below. Nonetheless, the two polynucleotide backbones do not fit properly when brought together. Explain why the simple inversion of the entire backbone as a unit on the right side of the molecule will not result in a perfect fit. What else must you do to succeed?

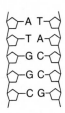

7. Some scientists severely criticized the model-building approach used by Watson and Crick to determine the DNA structure. What shortcomings of the technique might elicit such reactions?

8. The famous DNA molecule displays both an underlying simplicity and an astronomical complexity. Discuss both the simplicity and the complexity of DNA in as much detail as you can.

9. One of the concluding statements in Watson and Crick's landmark paper on DNA structure was: "It has not escaped our notice that the specific pairing we have postulated immediately suggests a possible copying mechanism for the genetic material." To what mechanism were Watson and Crick referring?

10. Many say that the discovery of the structure of DNA was the first step in the biological revolution. Why should this event have had such a profound effect on the biological sciences?

PROBLEMS

1. Give the nucleotide sequence complementary to:
   C C C G G T A A T G C

2. What percent of any organism's DNA must consist of
   a. purines?
   b. pyrimidines?

34

3. Analyses of four samples of double-stranded DNA yielded the following information:
   Sample 1: 20% cytosine
   Sample 2: 14% guanine
   Sample 3: 30% thymine
   Sample 4: 28% adenine
   a. What are the percentages of the other nitrogenous bases in each sample?
   b. Could any of these samples have come from the same organism? Which ones?

4. How many guanines would you expect to find in a piece of double-stranded DNA 40 nucleotide pairs long?

5. How many deoxyribose groups would you find in a piece of double-stranded DNA 20 nucleotide pairs long?

6. Given that the $\dfrac{A + G}{T + C}$ ratio in one polynucleotide strand of a DNA molecule is 0.8,
   a. What is the $\dfrac{A + G}{T + C}$ ratio in the complementary polynucleotide strand?
   b. What is the $\dfrac{A + G}{T + C}$ ratio of the entire DNA molecule?

7. Most DNA is double stranded, but single-stranded DNA is also possible. Which of the samples analyzed below are single stranded and which are double stranded? Can you be absolutely sure of these answers, even if the data listed below are extremely accurate? Explain.

DNA Samples	Adenine	Guanine	Thymine	Cytosine
A	15%	35%	15%	35%
B	21	21	29	29
C	12	38	38	12
D	18	27	30	25

8. How many different nucleotide sequences are theoretically possible in a DNA segment six nucleotides long?

9. The more closely related two organisms are, the more similar are the sequences of nucleotides in their DNA. Based on what you know about evolutionary relationships, place the organisms in order of their similarity to your own DNA, from the most similar to the least similar.
   a. petunia
   b. chimp
   c. squirrel
   d. another human

35

10. Consider that one DNA molecule has become 16 DNA molecules by replicating four successive times. How many of the "daughter" molecules incorporated a polynucleotide strand from the original "parent" molecule after:
  a. two replications?
  b. three replications?
  c. four replications?

REFERENCES

Avery, O. T., C. M. MacLeod, and M. McCarty. 1944. Studies on the chemical nature of the substance inducing transformation of pneumococcal types. *Journal of Experimental Medicine* 79:137–158.

Chargaff, E. 1950. Chemical specificity of nucleic acids and mechanism of their enzymatic degradation. *Experientia* 6:201–209.

Griffith, F. 1928. The significance of pneumococcal types. *Journal of Hygiene* 27:113–159.

Hershey, A. D., and M. Chase. 1952. Independent functions of viral protein and nucleic acid in growth of bacteriophage. *The Journal of General Physiology* 36:39–56.

Judson, H. F. 1979. *The eighth day of creation*. New York: Simon & Schuster.

Miescher, F. 1871. On the chemical composition of pus cells. *Hoppe-Seyler's medizinische-chemische Untersuchungen* 4:441–460. In *Great experiments in biology*, ed. M. L. Gabriel and S. Fogel, 1955, 233–239. Englewood Cliffs, NJ: Prentice-Hall, Inc.

Mirsky, A. E., and H. Ris. 1949. Variable and constant components of chromosomes. *Nature* 163:666–667.

Olby, R. 1974. *The path to the double helix*. Seattle: University of Washington Press.

Portugal, F. H., and J. S. Cohen. 1977. *A century of DNA*. Cambridge: The MIT Press.

Watson, J. D. 1968. *The double helix*. New York: Atheneum Publishers.

Watson, J. D., and F. H. C. Crick. 1953. Genetical implications of the structure of deoxyribonucleic acid. *Nature* 171:964–967.

Watson, J. D., and F. H. C. Crick. 1953. Molecular structure of nucleic acids. *Nature* 171:737–738.

Wilkins, M. H. F., A. R. Stokes, and H. R. Wilson. 1953. Molecular structure of deoxypentose nucleic acids. *Nature* 171:738–740.

3

How Hereditary Transmission Works

*I*n 1866 Gregor Mendel published the results of the various crosses he had made with garden peas. No one realized it until 1900, but clearly this Austrian monk, pictured in Figure 3-1, had discovered the basis for the transmission of hereditary traits from one generation to another. With amazing insight, he had formulated the fundamental laws of heredity.

Mendel used several different strains of *Pisum sativum* (garden pea) that bred **true to type**; that is, the progeny were always like the parents. His peas were stable because peas normally have but one parent; they do not receive traits from two parents. Each pea flower is closed, as shown in Figure 3-2, such that bees and other insects cannot enter to cross-fertilize it. As a result, each pea flower fertilizes itself; the pollen produced in its anthers must reach its own pistil. Flowers that are fertilized by pollen from other flowers are subject to an exchange of hereditary material and usually do not breed true. Their progeny combine the traits of two parents and can differ from either one.

FIGURE 3–1
Gregor Mendel formulated the fundamental principles of genetics with experiments using the garden pea (*Pisum sativum*). (Photograph courtesy of The Bettman Archives.)

38

FIGURE 3–2
The closed structure of the pea flower normally allows only self-fertilization, but cross-fertilization can be done manually.

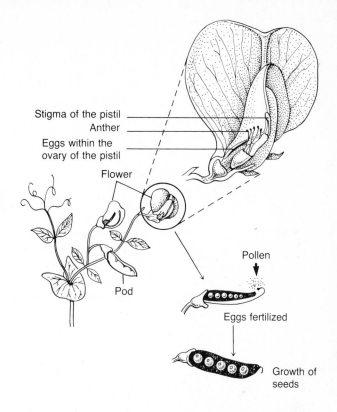

Stigma of the pistil
Anther
Eggs within the ovary of the pistil
Flower
Pod
Pollen
Eggs fertilized
Growth of seeds

Mendel could obtain more peas of any of his strains simply by allowing self-fertilization to proceed as usual. However, he could also cross-breed or **hybridize** types to obtain **hybrid** plants. An essential step was to **emasculate** one flower by removing its anthers before they could form pollen. He would then transfer pollen from another flower to the pistil of the emasculated flower.

Mendel's hybridization experiments led him to believe that the inheritance of traits involved discrete, particulate factors of some kind. This was a crucial concept, for it suggested a physical basis for heredity. Some tangible unit was actually being transmitted.

Mendel also concluded that his **unit factors** came in pairs, and that the members of the pairs segregated from each other during the formation of **gametes** (sex cells—eggs and sperm nuclei of pollen grains). This concept became his **law of segregation**.

Mendel also concluded that the factors formed pairs again when male and female gametes fused in the process of **fertilization**. Finally, he pointed out that each particular pair of factors segregated independently of other pairs of factors. This concept became his **law of independent assortment**.

Mendel's laws form the basis of **classical genetics**. They apply not only to peas, but also to mice, fruit flies, and human beings. They are even indispensable to today's rapidly moving study of molecular genetics. It is essential that we understand them and the related chromosome mechanics of cell division before we become involved in any deeper study of genetics. Once you have read and studied this chapter, you should understand these important concepts.

DOMINANCE AND RECESSIVENESS

Mendel's tremendous accomplishment was partly due to his care in collecting quantitative data. He counted the numbers of the various types of progeny resulting from his crosses and self-pollinations. The particular ratios that occurred revealed patterns to him. Other researchers had made similar crosses long before Mendel, but they did not count progeny or calculate the ratios of one type of progeny to the others. As a result, they failed to discover the mechanisms of heredity.

We can follow Mendel's work most easily if we start with his conclusion that hereditary factors occur in pairs. That is, a pure-breeding tall pea plant must have two factors for tallness, *TT*. A pure-breeding short pea plant must have two for shortness, *tt*. When each plant makes gametes, the pairs are separated, into *T* and *T*, and *t* and *t* respectively. In a cross between the two types, the sex cells combine by fertilization; one *T* comes from the tall parent and one *t* from the short parent. Thus, the offspring are all *Tt*

$$\begin{array}{ccccc} \text{parent 1} & & \text{parent 2} & & \text{offspring} \\ TT & \times & tt & \rightarrow & Tt \end{array}$$

Mendel observed that the hybrid offspring of such a cross are all tall, just like their tall parent. He then called tallness, the version of the height trait that is expressed in a tall × short hybrid, **dominant** to shortness. He called shortness **recessive**. Mendel found similar dominant-recessive relationships for other traits. For instance, *round* seed shape is dominant to *wrinkled* seed shape, and *yellow* seed color is dominant to *green* seed color. It is conventional to use capital letters for dominant versions of a trait and lowercase letters for recessive versions.

Modern geneticists still use the terms dominant and recessive, but they no longer call the hereditary particles "factors." The modern

40

concept states that a trait such as height in peas is controlled by a **gene**. Alternative forms of a trait (tallness or shortness) are specified by alternative versions of the gene, called **alleles**. An organism that always breeds true to type for a trait is called **homozygous** (*TT* or *tt*). A hybrid for a trait is **heterozygous** (*Tt*).

Because of dominance, a homozygote (*TT*) and a heterozygote (*Tt*) may look quite similar. That is, even though they have different sets of alleles (**genotypes**), they have the same appearance, or **phenotype**. When the situation is this clearcut and the heterozygote does not show a phenotype that is intermediate between the two homozygotes, the trait is said to show **complete dominance**. In humans, we can see an example of complete dominance in the inheritance of one of a number of genes that affect pigment in the skin and hair. Let *A* be the allele for normal skin and hair pigmentation and *a* be the allele for the complete lack of pigmentation, or **albinism**. Both *AA* and *Aa* genotypes produce normal phenotypes. An *aa* individual is an **albino**. This pattern of Mendelian inheritance is illustrated in Figure 3-3.

Incomplete dominance is indicated when the appearance of the heterozygote's phenotype falls between the phenotypes of the two homozygotes, as when a hybrid of a red-flowered and a white-flowered plant produces pink flowers. **Codominance** is the case when both alleles of a heterozygote express themselves individually, as when a person with type A blood, having two *A* alleles, mates with one with type B blood, having two *B* alleles. Their child has one *A* and one *B* allele and, therefore, type AB blood.

Incomplete dominance and codominance are the case for many traits, rather than complete dominance. However, it was complete dominance that Mendel was fortunate enough to observe. It made the results of his experiments more clearcut and the hereditary relationships far easier to discover. Incomplete dominance and codominance entered our knowledge of genetics only after 1900, when they helped in unraveling how genes work.

SEGREGATION

When Mendel crossed pure-breeding homozygous tall and short pea plants, he found all tall plants among the offspring, the **F_1 (first filial)** generation. He observed similar dominant-recessive relationships when he crossed yellow-seeded with green-seeded and round-seeded with wrinkled-seeded plants.

FIGURE 3–3
Simple Mendelian inheritance of albinism in humans. For convenience, the probabilities are illustrated for families of four children.

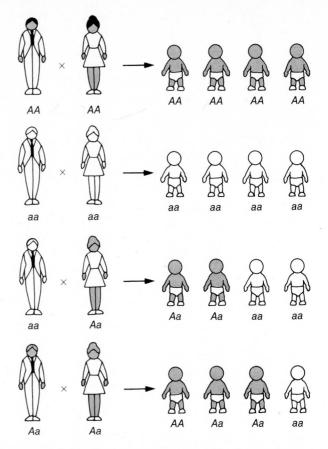

When he crossed his F_1 heterozygotes with themselves, he made another interesting discovery: the recessive character appeared, though only in one-fourth of the F_2 **(second filial)** progeny. Clearly, the recessive alleles were not lost in the F_1 generation. Although they did not show in the F_1 phenotype, they were present in the F_1 genotype. When the F_1 plants made their gametes, the recessive alleles of a particular gene could **segregate** from the dominant alleles to reappear in the next generation of progeny.

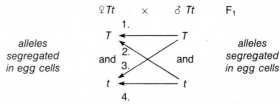

fertilization rebuilds pairs of alleles
F_2 progeny results:

1. *TT* ⎫
2. *Tt* ⎬ 3/4 tall
3. *tT* ⎭
4. *tt* ⎬ 1/4 short

Tall (*T*) is dominant to short (*t*). Hence, the F_2 genotypes are 1 *TT*, 2 *Tt*, and 1 *tt* and the F_2 phenotypes are 3 tall and 1 short, the classic 3:1 ratio of the **monohybrid** (involving only one trait) cross.

Mendel observed 787 tall and 277 short F_2 plants in his cross. This is very close to the theoretically expected 3:1 ratio. Today, such a ratio indicates that a trait is inherited as a **Mendelian trait**. That is, the trait is governed by one gene with one pair of alleles, in this case, one dominant and one recessive. Human albinism is just such a trait, as we can see by diagramming its pattern of inheritance (*A* is the allele for normal pigmentation; *a* is the allele for albinism):

parents: ♀ *AA* × ♂ *aa*
F_1: all *Aa*
      ♀ *Aa* × ♂ *Aa*
F_2: 1 *AA* : 2 *Aa* : 1 *aa*
or 3 normal : 1 albino

The key is the *segregation* of the two different alleles in the F_1 individuals. This is also the key to the results of other possible crosses. The mating of a heterozygous parent with a homozygous normal parent (non-albino) will produce no albino children:

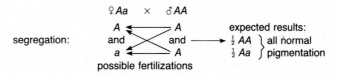

On the other hand, a cross between a heterozygous parent and a homozygous (albino) parent will yield half normally pigmented children and half albino children:

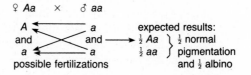

The cross between two heterozygotes will come up again and again in our discussions. It can also prove very useful in real-life situations, for we usually do not know whether a person is homozygous for a dominant allele or heterozygous while expressing only the dominant trait. However, if two parents of dominant phenotype have a child of recessive phenotype, we know immediately that both parents are heterozygotes. Under those circumstances, we can identify the genotypes of both parents. Our knowledge is less precise when the character we examine does not follow strict Mendelian patterns. For example, certain traits are controlled by the interaction of many genes, not just one pair.

INDEPENDENT ASSORTMENT

Considering one character at a time, Mendel found dominance, recessiveness, and segregation in his cross-breeding experiments. His next discovery came when he considered *two* traits at a time. He found that two different pairs of his factors assort independently of each other. That is, each pair of alleles follows the patterns Mendel observed for a single pair of alleles, and one pair does not influence the other.

This stage of Mendel's work required that he use **dihybrids**, organisms that were heterozygous for both characters under study. In one case, he crossed pure-bred tall, yellow-seeded (*TTYY*) pea plants with pure-bred short, green-seeded (*ttyy*) plants. The F_1 progeny were all tall, yellow-seeded dihybrids (*TtYy*).

What would the F_2 progeny be like? Mendel hypothesized that each trait individually would show the 3:1 ratio of phenotypes and that the two together would show a 9:3:3:1 ratio. How did he predict these numbers? A **Punnett square diagram** is one way geneticists work out such predictions. To use this system, draw a checkerboard and label each side with all the allelic combinations possible in the gametes of one parent after independent assortment. Then fill in the squares of the checkerboard with the genotypes of the offspring resulting from each combination of gametes. In the Punnett square on page 45, the result is 9 tall, yellow-seeded, 3 tall, green-seeded, 3 short, yellow-seeded, and 1 short, green-seeded. This ratio is very close to what Mendel found.

The same process can be illustrated using two traits that assort independently in humans. Consider the blood types. Blood type A is codominant with B, as we mentioned earlier, but it is dominant to O (A, B, and O are all specified by alleles of a single gene). The Rh (Rh positive) blood type is dominant to the rh (Rh negative) blood type. It is easy

	♀			
	TY	Ty	tY	ty
TY	TTYY	TTYy	TtYY	TtYy
♂ Ty	TTYy	TTyy	TtYy	Ttyy
tY	TtYY	TtYy	ttYY	ttYy
ty	TtYy	Ttyy	ttYy	ttyy

to find people with the heterozygous genotypes *AORhrh*; they have type A, Rh positive (usually written as A+) blood. Their children, however, will have A+, A−, O+, and O− blood, in the ratio 9:3:3:1. Observations of actual cases confirm this ratio.

The 9:3:3:1 ratio is a very standard one. Whenever it appears among the progeny of two parents who are both heterozygous for two different characteristics, it means that the two pairs of alleles probably assort independently of each other. That is, the alleles are not connected to each other in any way. They come together in the gametes purely according to chance. However, although Mendel thought independent assortment was a general rule, numerous exceptions were observed after 1900. As we will see, these exceptions are due to the close physical attachment of genes to neighboring genes on the same chromosome. Today, we take the 9:3:3:1 ratio as an indication that two gene pairs assort independently.

Trihybrid crosses and more complex situations can be handled in the same way as dihybrids by appropriately extending the problem. For example, given independent assortment, the heterozygote *AaBbCc* can produce eight different gametic types, each with a probability of one in eight (1/8). Therefore, the Punnett square would require eight columns and eight rows, with boxes for 64 different genotypes. Eight different phenotypes would appear among the F_2 progeny in a 27:9:9:9:3:3:3:1 ratio. To convince yourself, perform the Punnett square analysis yourself.

VEHICLES OF HEREDITY—
THE CHROMOSOMES

The twentieth century began with the knowledge that heredity was due to a material transmitted from generation to generation according to definite basic laws. In addition, biologists had known of the existence of **chromosomes** since 1865. Chromosomes are easily stained structures visible in the **nucleus** of the **cell**. The cell is the basic unit of all living things; multicellular organisms such as human beings are composed of trillions of them. Figure 2-1 shows the nucleus, the plasma membrane, and many other cell structures.

In the early 1900s, Sutton and Boveri independently realized that Mendel's factors behaved in the same way as the chromosomes in the cell. That is, in both cases:

1. They are found in pairs;
2. One member of each pair has a maternal origin, the other has a paternal origin;
3. Members of the pairs segregate from each other during the reproductive process; and
4. They form pairs again following fertilization.

FIGURE 3–4
A human male's chromosomes at mitotic metaphase. Note the sister chromatids that are visible following replication; most show a constriction marking the position of the centromere. (Courtesy of C. Korte.)

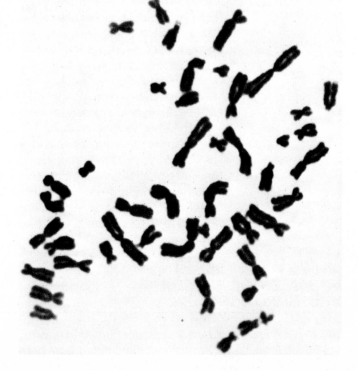

Sutton's and Boveri's conclusion that Mendel's factors were carried by the chromosomes became known as the **chromosome theory** of inheritance. Numerous later experiments showed this theory to be correct. Chromosomes are indeed the vehicles of heredity.

The principal substance of the chromosome is deoxyribonucleic acid, or DNA. The chromosome's structure also includes various proteins and a few ions. Most investigators of chromosomes believe that only one DNA molecule comprises each chromosome. The DNA molecule is about 20 angstroms wide—only 1/500,000 of a millimeter (and a millimeter is only 1/25 of an inch). However, the chromosome itself is relatively large, and each DNA molecule is very long. Each human cell contains 1736 millimeters (5.7 feet) of DNA divided among 46 chromosomes (23 pairs) of various sizes. The average length of DNA per chromosome is 37.74 millimeters, or 1.5 inches. Individual DNA molecules in the human range from perhaps 0.5 to 3 inches.

The chromosomes are long and threadlike in cells that are not

FIGURE 3–5

Although the exact mechanisms of chromosome condensation are not fully known, the process obviously requires several levels of coiling and folding.

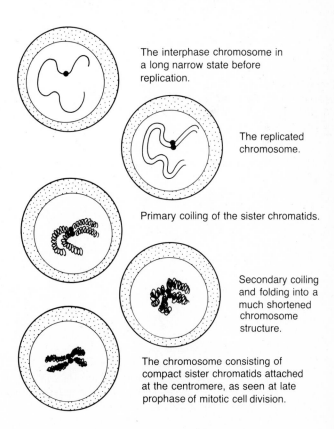

The interphase chromosome in a long narrow state before replication.

The replicated chromosome.

Primary coiling of the sister chromatids.

Secondary coiling and folding into a much shortened chromosome structure.

The chromosome consisting of compact sister chromatids attached at the centromere, as seen at late prophase of mitotic cell division.

actively dividing. In this condition, they are called **chromatin**. When the cell is dividing, the long threads of chromatin coil tightly into very short, compact structures, that we can see with a good microscope. These compact chromosomes are shown in Figure 3-4. The steps in the process of chromosome coiling are shown in Figure 3-5.

In humans, the chromosome pairs are given numbers from 1 to 22 according to their relative sizes, with number 1 being the longest and number 22 being the shortest. The two different kinds of sex chromosomes comprising the remaining pair are called X and Y, without a

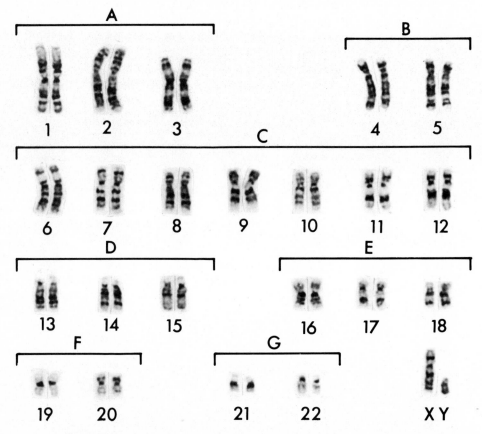

FIGURE 3–6
The 23 pairs of chromosomes of a human male, numbered and arranged into a karyotype. (From James Thompson and Margaret W. Thompson, 1980. *Genetics in Medicine*. W. B. Saunders Company, Philadelphia. Used with the permission of the publisher.)

FIGURE 3-7

The chromosomal basis of sex determination in humans.

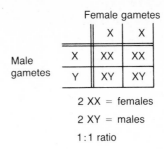

Female gametes

Male gametes		X	X
X		XX	XX
Y		XY	XY

2 XX = females

2 XY = males

1:1 ratio

number designation. The paired chromosomes, organized in a numbered karyotype, appear in Figure 3-6.

It was suggested very early in the history of genetics that there was a chromosomal basis for the differences between the two sexes in animals. Many animal species show a fundamental difference in the chromosome constitutions of males and females. Such organisms, including humans, have two general types of chromosomes: **sex chromosomes** and **autosomes**. The sex chromosomes (X and Y) play a direct role in sex determination in addition to transmitting genes for other traits. The autosomes include all the chromosomes of the cell, except the two sex chromosomes. In humans, males have an XY combination, while females have a pair of X chromosomes (XX). Segregation of the XY pair during gamete formation yields two different gametes: those with the X chromosome and those with the Y chromosome. Females can produce only gametes with an X chromosome. The chromosomal basis for sex determination in this system, called XX/XY, is diagrammed in Figure 3-7.

CELL DIVISION

Cell division is the process by which one cell gives rise to two, two to four, four to eight, and so on. In one form, **mitosis**, it is the process by which a fertilized egg gives rise to an entire multicellular organism. This is the basic phenomenon behind normal growth, development, and wound healing. In a second form, **meiosis**, it is the process by which an organism's **germ cells** give rise to gametes.

The two forms of cell division share certain features. Since all cells carry a full supply of the hereditary information of the genes, cell division must always involve the making of duplicate copies of the parent cell's DNA. One copy ultimately resides in each of two **daughter cells**. The duplicate copies are made in the stage of the cell's life cycle

49

FIGURE 3–8
The cell cycle.

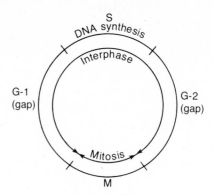

called **interphase**; to be more specific, they are made in the **S (synthesis)** phase of interphase, which is shown in Figure 3-8. The copying mechanism is essentially **semiconservative** DNA replication. The periods of interphase before and after the replication of DNA are devoted to cellular growth and metabolism.

Because both mitosis and meiosis require an efficient parceling out of all the cell's constituents, not only the DNA copies, both have two stages. **Karyokinesis** is the division and separation of the chromosomes. **Cytokinesis** is the division of the **cytoplasm**, that portion of the cell's contents outside the nucleus. Here, we will dwell on karyokinesis, for chromosome behavior in meiosis is where Mendel's laws are worked out mechanically on the **cytological** (cellular) level. We will look first at mitosis, for it is a simpler process, and easier to understand.

Mitosis—Equational Cell Division

Mitosis is often called equational cell division because its essence is the construction of two exact copies of each chromosome and the equal distribution of the resulting two identical sets of chromosomes, one to each of two daughter cells. Semiconservative DNA replication ensures that the chromosome copies are *exact* (barring **mutations**, changes that occasionally occur in DNA). The specific mode of chromosome distribution during mitosis ensures that, if no errors occur, the chromosome numbers and gene contents are identical in the two daughter cells. Mitosis involves a series of drastic changes in cell structure and in chromosome structure to accomplish these objectives.

The human **somatic** (body) cell contains 23 pairs of chromosomes, and the two members of each pair are called **homologues**. Homologous chromosomes are extremely similar to each other in size,

50

structure, and gene content. Nevertheless, the two homologues of a pair often carry different alleles for a single gene. The two alleles occupy the same relative location, or **locus**, on each homologue.

When the cell is ready to begin a mitotic cell division, the chromosomes replicate during interphase. Each chromosome now exists as a pair of exact copies, **sister chromatids**, joined to each other in the area of the **centromere** or **kinetochore**. During interphase, the chromatids are so long and spread out within the nucleus that none of these structural relationships can be readily seen with ordinary microscopy.

During the first stage of mitotic division, **prophase**, the chromosomes (each consisting of two sister chromatids) shorten and condense by coiling and folding. The process of condensation, which is shown in Figure 3-9, continues until the chromosomes become very compact rod-shaped bodies.

Several other events are proceeding at the same time. At the beginning of prophase, the cell's **centriole**, a small organelle composed of fibrous protein, is duplicated. Protein fibers radiate from the duplicates and seem to push them apart. They move toward opposite ends of the cell, reaching their final positions at the end of prophase. Near the end of prophase, the nuclear envelope disintegrates. The group of fibers radiating from the centrioles take on the tapered look that gives them the name **spindle apparatus**. Some of the spindle fibers reach from centriole to centriole. Others reach to the chromosomes' centromeres and end there.

With the formation of the spindle, the chromosomes move to the central plane of the cell, midway between the two centrioles. This marks the stage of mitosis called **metaphase**, quite clear in both Figures 3-10 and 3-11.

The next stage is **anaphase**. The centromeres split, and the sister chromatids of each chromosome move apart, seemingly towed by shortening spindle fibers. The separating chromatids can now be called chromosomes again.

When the migrating chromosome groups are as far apart as they will get, they congregate into compact groups. This marks the start of the **telophase** stage of mitosis. The individual chromosomes gradually become longer, thinner, and more diffuse. They uncoil to a large extent, the spindle degenerates, and a nuclear envelope reappears around each of the two daughter nuclei. Finally, the cytoplasm divides in animal cells by constricting between the two nuclei or in plant cells by forming a cell plate between the nuclei. This final process is called cytokinesis. Its two products are daughter cells, entering a new interphase. Each

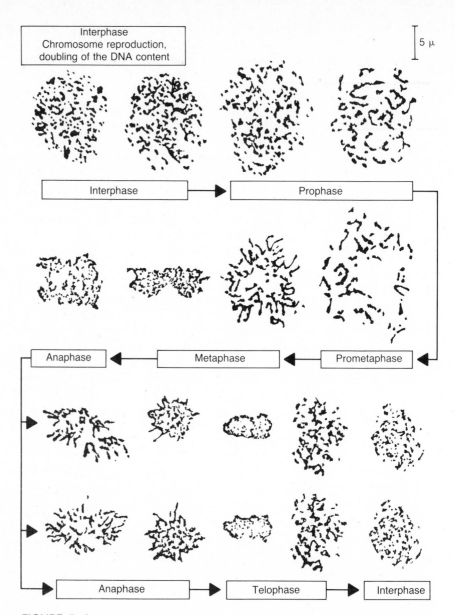

FIGURE 3–9
The sequence of events in the mitotic cycle of human white blood cells cultured under laboratory conditions. (Adapted from Eeva Therman, 1980. *Human Chromosomes*. Springer-Verlag, New York. Reproduced by permission of the publisher.)

52

FIGURE 3–10
Mitosis illustrated in a cell with only two pairs of chromosomes. For convenience, A and a mark one pair of homologous chromosomes, and B and b mark the other.

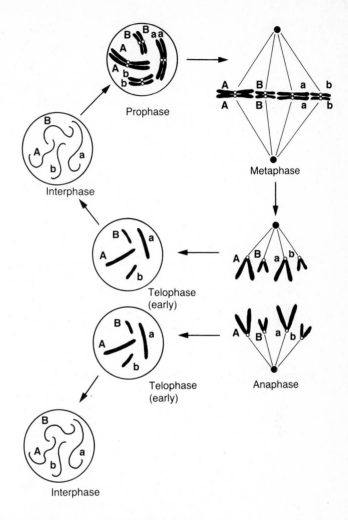

Prophase

Metaphase

Interphase

Telophase
(early)

Telophase
(early)

Anaphase

Interphase

one has the same type and number of chromosomes, and hence a genetic constitution identical to the other body cells and to the now nonexistent parental cell.

Meiosis—Reductional Cell Division

Meiosis is often called reductional cell division because it results in daughter cells with *half* the number of chromosomes of the parental cell. These daughter cells are the gametes or sex cells of sexual reproduction. When the gametes from separate parents fuse in the process of fertilization, the normal chromosome number is restored. The ferti-

FIGURE 3–11

Chromosomes at prophase I in the meiosis of a human male. Note that the 23 bivalents can be easily discerned, and that the X and Y chromosomes associate end-to-end, not side-by-side like the others. (From James Thompson and Margaret W. Thompson, 1980. *Genetics in Medicine*. W. B. Saunders Company, Philadelphia. Used with the permission of the publisher.)

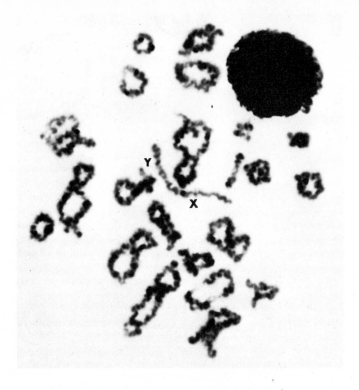

lized egg or **zygote** then divides repeatedly by mitosis to become a new individual.

The normal somatic cell has a set number (called *n*) of pairs of chromosomes. The cell thus has 2*n* individual chromosomes, and it is called **diploid**. A gamete has one chromosome from each diploid pair. Therefore, it has *n* chromosomes and it is called **haploid**. Meiosis is called reductional division, because it is the process that reduces the number of chromosomes in the cell from the diploid to the haploid number. This reduction is one of the key differences between meiosis and mitosis, and its chromosomal mechanism underlies many important genetic concepts.

There are two main benefits of meiosis. The first is that it keeps the number of chromosomes constant from generation to generation; even small changes in chromosome number can have disastrous effects on development. The second is that it produces genetic variation. Meiosis separates the members of a pair of chromosomes into single units. Fertilization combines these single chromosomes from separate parents into new pairs, making many new combinations of alleles pos-

54

sible. This phenomenon is very important, and we will return to it later in this chapter.

Meiosis is a long and complex cytological process. It includes two successive chromosome segregations and cytokineses (cell divisions), **meiosis I** and **meiosis II**. The result is that *four* daughter cells are produced, not two, as in mitosis.

The interphase before meiosis I is the time when the chromosomes replicate. Meiosis I itself begins with **prophase I**. During this stage, the homologous chromosomes line up side by side in an intimate pairing called **synapsis**. This pairing of homologous chromosomes is a critical difference between meiosis and mitosis. The paired chromosomes are called **bivalents**. Because each chromosome is actually a pair of sister chromatids at this time, each bivalent is a **tetrad** of four chromatid strands. This arrangement can be seen clearly in Figures 3-12 and 3-13. It is at this point that **crossing over**, an important mechanism of **genetic recombination**, occurs between homologous chromosomes. While they are closely paired, the non-sister chromatids of homologous chromosomes exchange segments one or more times, as shown in Figure 3-13. Usually, the *closer* two genes are to each other on the chromatid, the *less likely* it is that an exchange (or break and reunion) will occur between them. Genes that rarely recombine as a result of crossing over are said to be **linked**. Along with segregation and independent assortment, crossing over is a mechanism of recombination that contributes to genetic variation in gametes.

In prophase I, homologues are often visible in a bivalent, but the individual chromatid strands are not, for they are too intimately meshed.

Metaphase I is marked by the alignment of the bivalents on the cell's equatorial plane and by the formation of the spindle apparatus. The centromeres of the two homologues in each bivalent face opposite poles of the cell. These centromeres do not divide during **anaphase I**; rather, the paired homologues separate and move to the opposite poles. This means that the homologous chromosomes, each consisting of two joined sister chromatids, move away from each other. In each homologous pair, the chromosome of maternal origin goes to one pole and the chromosome of paternal origin goes to the other. The combination of maternal and paternal chromosomes ending up at each pole is random, further evidence of independent assortment. **Telophase I** and cytokinesis then complete meiosis I.

Meiosis II begins just as meiosis I finishes, but with no intervening DNA replication. In **prophase II**, the chromosomes, still consisting of joined sister chromatids, again condense. In **metaphase II**, they

55

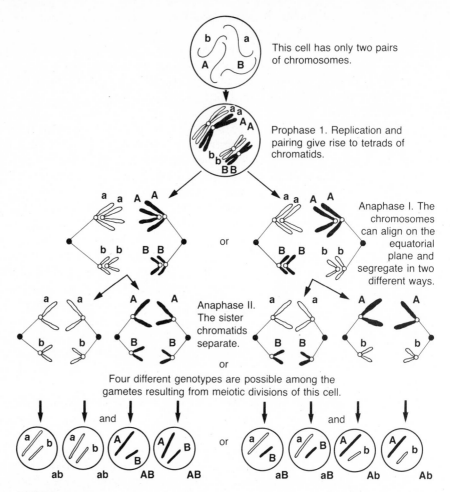

This cell has only two pairs of chromosomes.

Prophase 1. Replication and pairing give rise to tetrads of chromatids.

Anaphase I. The chromosomes can align on the equatorial plane and segregate in two different ways.

Anaphase II. The sister chromatids separate.

Four different genotypes are possible among the gametes resulting from meiotic divisions of this cell.

FIGURE 3–12
Meiosis illustrated in a cell with only two pairs of chromosomes. To aid you in following the movements of the chromosomes, **A** and **a** mark one pair of homologues, and **B** and **b** mark the other.

align on the equatorial planes of both daughter cells, and a spindle apparatus forms in each cell. **Anaphase II** marks the division of the centromeres and their migration to opposite ends of the cells. **Telophase II**, the second cytokinesis, and a return to interphase in all four final daughter cells rapidly follow. Meiosis II is now complete.

56

FIGURE 3-13
Crossing over. (a) Replication and pairing produces a tetrad; (b) Exchange of segments between two non-sister chromatids; (c) The result of the cross-over event is recombination.

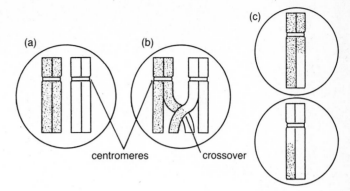

centromeres   crossover

The crucial end result of this special type of cell division is that the four resulting nuclei are all haploid; that is, each nucleus contains half the number of chromosomes initially present in the parental cell. In fact, the two meiotic divisions have distributed one of the four chromatids of each tetrad in prophase I to one of the four daughter cells. Figure 3-12 shows these critical chromosome relationships. Their relation to Mendel's ideas about inheritance being transmitted by discrete particulate factors is very significant.

Meiosis has not been easy to study in humans. Such investigations require the removal of small amounts of tissue for analysis, and only a few researchers have successfully observed the process in humans. Figure 3-14 depicts the major events of meiosis in a male grasshopper, whose tissue is relatively easier to study and whose cells contain fewer chromosomes.

THE ROLE OF PROBABILITY IN GENETICS

To predict the frequencies of offspring having different phenotypes, we must have both a working knowledge of Mendel's laws and the correct information about dominance and recessiveness. In addition, we must understand the fundamentals of probability. One of Mendel's laws required that the two genes of a pair have an equal chance of entering any gamete. Another required that the male gametes have an equal chance to fertilize the female gametes independently of the specific alleles they contained. These requirements and the corresponding probability theory allow us to predict the expected genotypes and phenotypes of offspring.

57

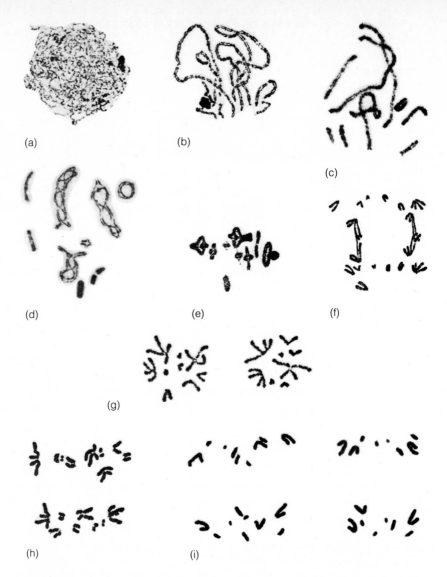

(a)

(b)

(c)

(d)

(e)

(f)

(g)

(h)

(i)

FIGURE 3–14

Meiosis in a male grasshopper. The 2n chromosome number is normally 17 (not an even number) because the male has only one X chromosome (the female has two). (a) Very early prophase I; (b–d) later stages in prophase I (Note the cross-like figures in (d); they result from crossing over.); (e) metaphase I, showing eight bivalents and one unpaired X chromosome; (f) anaphase I, when the unpaired X chromosome must migrate to one pole of the cell or the other; (g) prophase II in both daughter cells; the X chromosome is in only one; (h) metaphase II, polar view; (i) anaphase II, which will result in four cell products, two with eight chromosomes and two with nine (including the X). (Reproduced with permission from B. John and K. R. Lewis, *The Meiotic Mechanism*, 2nd ed., 1983. Published and copyright by Carolina Biological Supply Company, Burlington, North Carolina, USA.)

We have already had several chances to work with probabilities, but a few important points deserve review at this time. They are best explained with simple examples. In many organisms, the probability of having a male offspring is one in two, as is the probability of having a female offspring. Therefore, the probability that the first offspring is a male is a one in two chance. The probability that the second—or tenth—offspring is a male is also one chance in two. The probability does not change with successive events because in this case the events are independent of each other. One event does not influence another.

What is the probability that two successive offspring will both be males? If two or more events are independent, the probability that they will occur together is the *product* of their separate probabilities. That is, the probability of two males in succession is $\frac{1}{2} \times \frac{1}{2} = \frac{1}{4}$.

To demonstrate further, we can assign genders to two offspring in four different ways:

Outcome	First offspring	Second offspring
1	male	female
2	female	male
3	female	female
4	male	male

Only one of the four outcomes is the one in question: two males in succession.

What is the probability of obtaining three successive *aabb* offspring from parents who are both *AaBb*? This is a dihybrid cross to obtain F_2 offspring, one in 16 of whom will be *aabb*. The probability of three such offspring in a row is $\frac{1}{16} \times \frac{1}{16} \times \frac{1}{16} = \frac{1}{4096}$.

Many of the discussions in this treatment of genetics and human society will use the basic concepts of probability. We will repeatedly apply the few concepts above, as well as others. We will explain each new rule of probability carefully as we come to it. For now, let us be aware of the tremendous part played by probability in heredity.

SUMMARY

Gregor Mendel established the laws of heredity in 1866 through his work with garden peas. He pointed out that physical hereditary factors exist in pairs, and these pairs of factors segregate during gamete formation, assort independently of each other, and pair again as a result of reproduction. Segregation, independent assortment, and probability are the concepts that form the backbone of classical genetics. These concepts, which Mendel developed by observing the ratios of offspring of controlled crosses, are further substantiated by the behavior of chromosomes during meiosis. Mendel's laws, together with probability theory, allow us to predict the progeny types when the genotypes of the parents are known, whether we are working with monohybrids, dihybrids, trihybrids, or even more complicated crosses.

We now call Mendel's factors genes, and their alternative forms alleles. The list of alleles in an organism is its genotype; the alleles' visible expression is the organism's phenotype. When both alleles of a pair are the same, the organism is homozygous for that gene. When they are different, it is heterozygous. An allele that determines phenotype in a heterozygote is a dominant allele. One that has no obvious effect on phenotype except when homozygous is a recessive allele. Some alleles show incomplete dominance or codominance.

The genes were identified with the chromosomes early in the twentieth century. Chromosomes are chiefly DNA and protein; the DNA is the genetic component. Cell division of body cells, called mitosis, keeps chromosome number and gene constitution the same from the parental cell to the resulting daughter cells. Cell division of germ line cells to produce gametes, called meiosis, reduces the number of chromosomes in half. Both types of cell division are necessary to keep the chromosomes constant from one generation to another, and meiosis further affords a good opportunity for new combinations of alleles to be formed when gametes fuse in fertilization.

DISCUSSION QUESTIONS

1. One early scientist severed the tails of numerous mice and crossed these tailless mice with each other. The offspring all had tails, which the scientist again severed. After 22 generations treated in this way, all the offspring were still born with tails. Give an explanation and point out the significance that these results might once have had.

2. If Mendel had not performed his experiments, would the course of genetics have been different? Why or why not?

3. Much of Mendel's work with garden peas involved dihybrids. Which of his basic laws of heredity could not have been demonstrated with monohybrids? Explain.

4. The law of independent assortment emerged as Mendel worked with seven different characteristics of peas, testing each of the possible pairs of characteristics in dihybrid crosses without ever noting an exception to independent assortment. Much later, cytologists found that peas have seven pairs of chromosomes. At this point, many people thought Mendel had been very fortunate in his results. Why do you suppose they reached this conclusion?

5. The study of genetics and the theory of probability are inseparable. Point out why the two subjects go together so well.

6. What does it mean to call a particular pattern of inheritance "Mendelian"?

7. Name several major differences between mitosis and meiosis. Explain why both types of cell division are essential to the life processes of most organisms.

8. A human geneticist wishes to learn whether a particular birth defect is a genetic recessive trait governed by simple Mendelian rules. To do this, she locates a very large number of families in which both parents are normal and at least one child has the defect. She finds that 36 percent of the children in these families have the defect. Even if the trait is caused by one allele of a gene that obeys the Mendelian rules, why would the researcher not expect 25 percent of the children to have the defect? That is, why should she not expect a 3:1 ratio?

9. Assume that in observing the natives of a large island, you find that very nearly three-fourths of the population show one form of a trait. The rest of the natives show a second form of the trait. Is the trait Mendelian? Does it follow a straightforward dominant/recessive mode of inheritance? Explain.

PROBLEMS

1. How many different gametic genotypes are possible from the diploid genotypes:
   a. *AA BB CC Dd EE FF GG HH II*
   b. *Mm Nn*
   c. *PP QQ rr ss TT*
2. A zygote with the four chromosome pairs *AA'*, *BB'*, *CC'*, and *DD'* will produce what chromosome combinations
   a. in the organism's body cells during growth and development?
   b. in the mature organism's gametes?

61

3. Two allelic pairs, *Gg* and *Hh*, each affect a different characteristic and assort independently of each other. If *G* and *H* are dominant to *g* and *h*, what is the probability of obtaining
   a. a *Gh* gamete from a *Gg Hh* individual?
   b. a *gh* gamete from a *gg Hh* individual?
   c. a *gg hh* zygote from a *Gg Hh* × *Gg Hh* cross?
   d. a *Gg HH* zygote from a *GG HH* × *gg Hh* cross?

4. In the cross *Aa Bb DD* × *AA Bb dd*, all allelic pairs segregate independently and show complete dominance.
   a. What proportion of the offspring will show the B phenotype?
   b. What proportion of the offspring will be *aa bb dd*?
   c. What proportion of the offspring will be heterozygous at all three loci?
   d. How many genetically different gametes are theoretically possible from each of the parents?

5. A woman has a son with a recessive genetic disorder. Her four brothers also have the disorder. Her husband blames her for their son's defect. Is the defect due to heredity from her side of the family only?

6. We often hear that someone has inherited characteristics from one parent or the other.
   a. In reality, how much of your gene complement comes from your father?
   b. Your mother?
   c. Is it *theoretically* possible to have all the same alleles as your brother or sister?
   d. Is it *theoretically* possible to have none of the same alleles as your brother or sister?
   e. On the average, how many alleles would you have in common with your brother or sister?

7. In humans, woolly hair (*W*) is dominant to non-woolly hair (*w*), and normal pigmentation (*A*) is dominant to albinism (*a*); the two allelic pairs assort independently. A woman with a *Wwaa* genotype marries a man with a *wwAa* genotype.
   a. What are the man's and woman's phenotypes?
   b. What are the allelic combinations in their gametes?
   c. What genotypes, and in what proportions, should appear among their children?

8. After observing 3 dominant : 1 recessive phenotypic ratios in the F_2 offspring of his monohybrid crosses, Mendel sometimes let his F_2 plants with the dominant character produce F_3 offspring by self-fertilization. He found that about two-thirds of the self-pollinations yielded both dominant phenotype and recessive phenotype offspring. One-third yielded only offspring with the dominant phenotype. Diagram the appropriate crosses of genotypes to show why Mendel obtained those results. Use, for example, the *tall* vs. the *short* characters.

62

9. If the recessive disorder albinism occurs only once in 20,000 human births, and if the first child of two normally pigmented parents is an albino, what is the probability that their second child will also be an albino?

10. Polydactyly (six-fingered hands) is a genetic condition attributable to an allele (*P*) dominant to the allele (*p*) for five-fingered hands. If a six-fingered woman and a five-fingered man have a normal child, what is the woman's genotype? If they have nine children, all with six fingers, what must the genotype of the woman be?

11. Two young brothers have a genetic disease that is, 40 percent of the time, fatal before age 20.
    a. What is the probability that neither brother will survive to age 20?
    b. What is the probability that both brothers will survive to age 20?
    c. What is the probability that at least one of the brothers will survive to age 20?

12. What is the probability that all of seven children in a family will be males? Of 1000 families with seven children, how many would you expect to have all female children? Of 3800 families with seven children, how many would you expect to have all male children?

13. If the normal diploid cell has two amounts of DNA (called 2C), how many C amounts of DNA are there in cells in:
    a. meiotic metaphase I?
    b. meiotic prophase II?
    c. mitotic telophase?
    d. meiotic telophase II?
    e. mitotic prophase?
    f. meiotic prophase I?
    g. mitotic interphase?

REFERENCES

Boveri, T. 1904. *Ergebnisse uber die Konstitution der chromatischen Substanz des Zellkerns*. Jena: Gustav Fisher Verlag.

John, B., and K. R. Lewis. 1976. The meiotic mechanism. In Carolina Biology Reader 65. Burlington, NC: Carolina Biological Supply Co.

Sutton, W. S. 1902. On the morphology of the chromosome group in *Brachystola magna*. *Biological Bulletin* 4:24–29.

Sutton, W. S. 1903. The chromosomes in heredity. *Biological Bulletin* 4:231–251.

Therman, E. 1980. *Human chromosomes: Structure, behavior, effects*. New York: Springer-Verlag.

Thompson, J. S., and M. W. Thompson. 1980. *Genetics in medicine*. Philadelphia: W. B. Saunders Co.

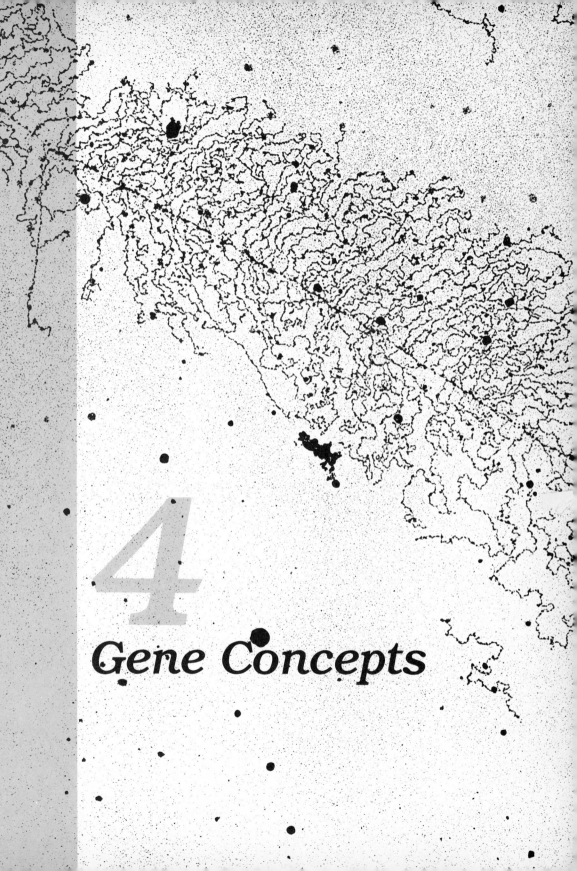

4

Gene Concepts

G regor Mendel's insight allowed him to think in a new way about heredity. As a result of this fresh approach, he discovered the reality of the gene. Once this fundamental idea was suggested, other researchers could learn that the chromosomes carried the genes. Still others could elucidate the biochemical components of the gene.

Although all these discoveries were essential to an understanding of heredity, they did not provide a complete explanation. There still remained the very basic question of *how* the genes exerted their effects on phenotype. Some answers were found in the biochemistry of the cell. Simply stated, the **central dogma** of biology specifies that genetic information usually flows from DNA to RNA to protein, as illustrated in Figure 4-1.

We have already discussed the basics of DNA replication, which is the way genes are copied for transmission in cellular and organismal reproduction. Now we must consider two further stages in the flow of genetic information: **transcription** and **translation**. In transcription, enzymes use DNA as a template or mold to make a message composed of RNA, but specified from the structure of the DNA molecule. In translation, the RNA message directs the synthesis of protein which, in turn, directly affects the phenotype.

At one time, geneticists believed that information flowed only in one direction, namely, from DNA to RNA to protein. Today, we know of a few cases in which the flow of information can go from RNA to DNA. This process, called **reverse transcription**, has been discovered in certain tumor viruses. It requires that we rephrase the central dogma, but only to say that in all organisms the information always flows from nucleic acids (DNA and RNA) to nucleic acids and from nucleic acids to protein—never from protein to nucleic acids or from protein to protein. In this chapter, we will concentrate on this information flow and the concepts underlying the role of proteins in shaping the phenotype.

FIGURE 4–1
The central dogma. Reverse transcription, that is, the flow of information from RNA to DNA, has been shown to exist in a few systems.

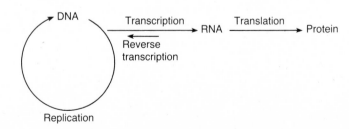

INBORN ERRORS OF METABOLISM

One of the first clues leading to an understanding of gene function appeared almost as soon as Mendel's work was rediscovered. Like Mendel's work, this research was not fully appreciated in its own time, but proved to be fundamentally important to genetics, medicine, and biochemistry.

Archibald E. Garrod, a physician and scientist, first made the connection between genes and enzymes in 1902. Some of his patients had **alkaptonuria**, an arthritic condition characterized by the excretion of a dark pigment in the urine. The disease is sometimes called the "black urine" disease. Garrod noticed that many of these patients were related in one way or another, and he assembled data such as that in Table 4-1. He found that in nine families, 19 members had alkaptonuria and 29 did not. This observation reminded him of the Mendelian 3:1 ratio expected for an autosomal recessive trait (an exact 3:1 ratio would be 12 alkaptonurics and 36 normals). The difference between the observed and the expected ratio comes about partly because the data are biased. Garrod could neither notice nor count families in which the parents were capable of having alkaptonuric children but did not; all his families had at least one alkaptonuric member.

Garrod assumed that his data were biased and thought that if he could have counted the families in which both parents were heterozygotes with no affected children, he would have seen a true 3:1 ratio.

TABLE 4-1

The proportions of alkaptonuric and normal persons in nine families studied by Garrod

Family studied	Members with alkaptonuria	Members without alkaptonuria	Total family members
1	4	10	14
2	3	1	4
3	3	4	7
4	1	1	2
5	2	0	2
6	1	0	1
7	1	9	10
8	2	3	5
9	2	1	3
	19	29	48

Source: Garrod, A. E. 1902. The incidence of alkaptonuria. *The Lancet*, ii: 1616–1620.

He therefore concluded that alkaptonuria was a metabolic disorder—an "inborn error of metabolism"—inherited as a recessive Mendelian trait. Thus, he gave the first substantive account of Mendelian inheritance in humans.

More importantly, Garrod suggested that the condition was **congenital** (present at birth) and that it was due to a deficiency of an enzyme. He was quite correct, although the precise deficiency was not determined until 50 years later. Garrod also noted correctly that other human diseases might have similar causes. We will discuss many of these diseases in detail later in this text.

THE ONE GENE-ONE POLYPEPTIDE CONCEPT

By 1941, George Beadle and Edward Tatum had further investigated the role of the gene. Their elegant experiments with the pink bread mold, *Neurospora crassa*, shown in Figure 4-2, fortified the idea that genes exert their effects through enzymes.

Neurospora was an excellent organism for the purpose. Normal

FIGURE 4–2
The vegetative form of *Neurospora crassa* consists of a network of branched filaments containing haploid nuclei. (Photograph courtesy of Stewart Brody, University of California/San Diego.)

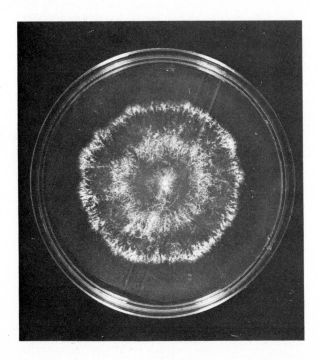

(wild-type) *Neurospora* stocks can grow on a liquid or gel (agar) chemical medium containing a carbon source such as sugar, a growth factor called **biotin**, and a number of specific inorganic salts. This material is called a **minimal medium**; from its simple constituents, wild-type *Neurospora* can synthesize all the substances (such as amino acids) they need to survive.

Beadle and Tatum treated wild-type *Neurospora* with radiation or **mutagenic** chemicals to cause changes in their genetic material. Then they isolated **mutant** strains of *Neurospora* by first growing the treated *Neurospora* on a **complete medium** containing many of the substances wild-type *Neurospora* synthesize to survive. They screened the treated *Neurospora* by means of biochemical testing that showed that some mutants could not survive on the minimal medium, but could survive if some specific amino acid or vitamin were added to the minimal medium. The overall plan of these experiments is illustrated in Figure 4-3.

Beadle and Tatum inferred that each of their *Neurospora* mutants had lost the ability to synthesize a specific, necessary nutrient or growth factor. When they hybridized the mutant strains with each other and with the wild-type strain, and studied the biochemistry of the growth factors, they were able to conclude that most of the growth factor requirements they had observed were due to mutations of single genes, and resulted from changes in the activity of single enzymes. This conclusion became known as the **one gene-one enzyme hypothesis**.

The one gene-one enzyme hypothesis endowed the gene with a specific function: to provide the information for the synthesis of specific enzymes. However, as knowledge of protein structure improved, we have had to modify this description of gene function. All enzymes are proteins. (Sometimes a small nonprotein substance may be associated with the enzyme, but the bulk of the molecule is always protein.) The basic structure of a functional protein consists of one or more **polypeptide** chains, each of which is a sequence of **amino acids**. The 20 different amino acids that are known to exist in the proteins of living organisms are listed, along with their symbols, in Table 4-2. Amino acids that are linked together interact with each other to give each polypeptide chain a specific three-dimensional shape, and that shape determines the chain's function. This dependence of function upon shape is exemplified by the phenomenon of **denaturation**. Heated proteins lose their shapes (denature) and their functions. On cooling, some proteins regain both shape and function.

The amino acids are often regarded as the ''building blocks'' of life. There are many different polypeptides of various lengths, and each

69

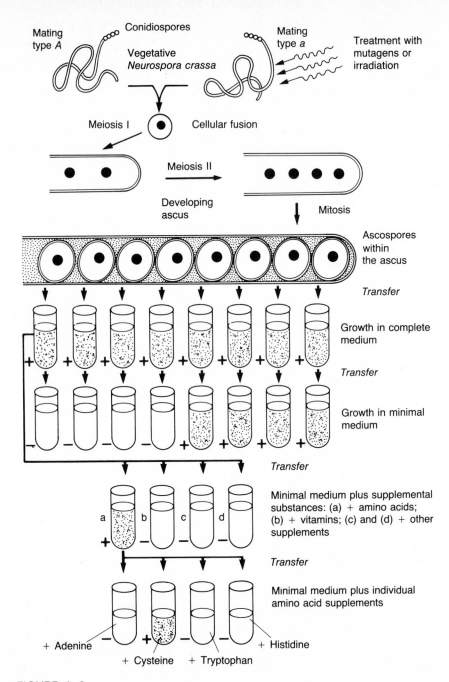

FIGURE 4–3

The scheme used by Beadle and Tatum in their *Neurospora* experiments to elucidate the role of genes in the organism. In this example, the mutation prevented the organism from synthesizing cysteine.

TABLE 4–2

The 20 amino acids found in the proteins of living organisms

Amino acid	Symbol
glycine	Gly
alanine	Ala
valine	Val
leucine	Leu
isoleucine	Ile
serine	Ser
threonine	Thr
aspartic acid	Asp
glutamic acid	Glu
lysine	Lys
arginine	Arg
asparagine	Asn
glutamine	Gln
cysteine	Cys
methionine	Met
phenylalanine	Phe
tyrosine	Tyr
tryptophan	Trp
histidine	His
proline	Pro

one is composed of a different sequence of the 20 amino acids. A gene contains the directions for forming a particular polypeptide, and an enzyme can be a single polypeptide, a group of several identical polypeptides, or a group of several different polypeptides, as shown in Figure 4-4. The important point is that genes code for polypeptides, not necessarily enzymes. The function of the gene is more correctly summed up by **one gene-one polypeptide**. The basic concept, that each gene is responsible for some biochemical step in the synthesis of a substance, remains unchanged.

TRANSCRIPTION—THE SYNTHESIS OF RNA

How does the information encoded in the DNA become expressed as protein? The transmission of this information to the protein-synthesizing machinery of the cell takes place with the aid of RNA molecules. Certain RNA molecules are chemical messages, and they are produced by the process called transcription.

In transcription, the two strands of the DNA double helix sepa-

rate at some point. The enzyme **RNA polymerase** then assembles single **ribonucleotides** (containing the sugar ribose) into an RNA molecule
complementary to one strand of the DNA (the **sense strand**; the other
is the **antisense strand**). The enzyme moves from the 5′ end to the 3′

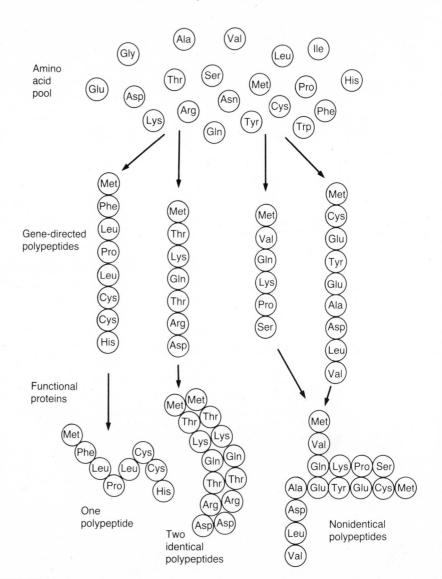

FIGURE 4–4
Some of the relationships among amino acids, polypeptides, and proteins.

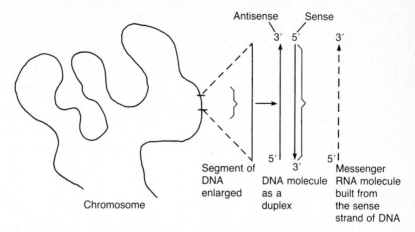

FIGURE 4–5
The relationship of DNA and RNA in transcription.

end of the DNA strand shown schematically in Figure 4-5. The positions of the five carbon atoms in the sugar (deoxyribose) are designated by numbers. Since the two strands of DNA run in opposite directions (anti-parallel), one ends with the 5′ carbon atom and the other with the 3′ carbon. As RNA polymerase moves, it lines up ribonucleotides complementary to the bases of DNA and links them together. For each G of DNA, it uses a C; for each C, a G; for each T, an A; and for each A, a U (**uracil**). In RNA, uracil takes the place of the thymine found in DNA. You can see these relationships clearly in Figure 4-6.

The RNA polymerase can direct the synthesis of RNA from either of the two DNA strands, but only on one side in any one region. It works with incredible speed and accuracy, adding ribonucleotides to the RNA at a rate of 40 to 50 per second (at 37°C). In addition, it can make many copies of the RNA message in rapid succession, for as soon as one RNA polymerase molecule moves away from the DNA's **initiation site** where transcription begins, another RNA polymerase molecule can begin to transcribe the DNA, and then another, and another.

In the context of transcription, it is interesting to note that genes are sometimes split, or interrupted by intervening DNA sequences called **introns**. These introns are interspersed among the segments of sense DNA, called **exons**, making the gene discontinuous. During transcription, the entire length of this DNA is copied and then the segments of RNA corresponding to the introns are subsequently spliced out. After the RNA segments transcribed from the exons of the

73

FIGURE 4-6

Transcription. The DNA helix unwinds, opens, and RNA is transcribed from one side of the molecule.

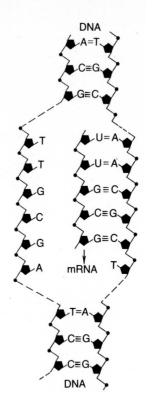

DNA

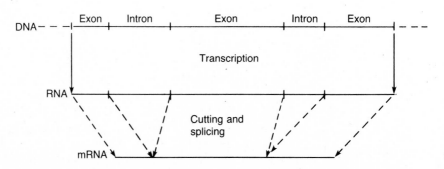

DNA — — –|

Exon | Intron | Exon | Intron | Exon

Transcription

RNA

Cutting and splicing

mRNA

FIGURE 4-7

Many genes in higher organisms contain introns that are also transcribed into RNA. The resultant RNA is subsequently cut and spliced into a message that eliminates the regions corresponding to the introns.

DNA are rejoined, the message is ready to be translated; this is shown diagrammatically in Figure 4-7. Very little is known about these intervening DNA sequences; nonetheless, genes are apparently not the discrete units that they were once thought to be.

TRANSLATION—FROM RNA TO PROTEIN

The RNA produced by transcription can be any of three different types. **Messenger RNA (mRNA)** carries the information that actually specifies a polypeptide. **Transfer RNA (tRNA)** and **ribosomal RNA (rRNA)** are molecules the cell uses to translate the mRNA message into a polypeptide.

Translation begins when an mRNA molecule transcribed from a gene leaves the nucleus. Once in the cell's cytoplasm, it forms an association with an organelle called a **ribosome**. Ribosomes are very small particles composed of protein and rRNA. We can call them the *workbench* upon which polypeptide synthesis occurs, but their role is far more intricate than this term suggests.

The end result of translation is a string of chemically linked amino acids in a sequence specified by DNA by means of its RNA message. Transferring the amino acids in the cytoplasm to their appropriate positions in the polypeptide calls for the third type of RNA, that is, tRNA. There is a pool of specific tRNA molecules in the cytoplasm, one kind for each amino acid. Each tRNA molecule has a shape similar to a cloverleaf; see Figure 4-8. One end of the tRNA attaches to its specific amino acid with the aid of the enzyme **aminoacyl-tRNA synthetase**. This **activating enzyme** uses **adenosine triphosphate (ATP)**, a common source of energy for cellular reactions, to activate the amino acid and catalyze its attachment to the tRNA:

(1)

| Amino acid | ATP | Aminoacyl-tRNA synthetase | Enzyme-bound aminoacyl adenylate | 2 Phosphate groups |

(2)

| Enzyme-bound aminoacyl adenylate | tRNA | Aminoacyl-tRNA complex | Adenosine monophosphate | Aminoacyl-tRNA synthetase |

75

FIGURE 4–8

A generalized tRNA molecule. In many of the tRNA molecules, the configuration resembles a cloverleaf. Although the molecule is single stranded, note the double-strandedness that normally exists as a result of intramolecular complementary base pairing.

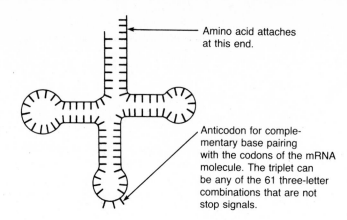

Amino acid attaches at this end.

Anticodon for complementary base pairing with the codons of the mRNA molecule. The triplet can be any of the 61 three-letter combinations that are not stop signals.

The tRNA molecule develops its characteristic three-dimensional shape because, although RNA molecules are only single stranded, portions of the single strand fold back on themselves and undergo complementary base pairing with other portions. At the opposite end of the molecule from the amino acid attachment site is a triplet of nucleotides whose bases are unpaired and exposed. By means of this triplet, called an **anticodon**, the tRNA "recognizes" where to place its amino acid in the polypeptide sequence of amino acids. It is complementary to a triplet of bases in the mRNA, called a **codon**. Each codon, in turn, is complementary to a triplet of bases in the DNA, so that the sequence of bases in the DNA, taken three at a time, specifies the sequence of amino acids in a polypeptide.

The 64 (4^3) different possible codons that make up the **genetic code** of living organisms, often called the dictionary of life, are spelled out in Table 4-3. This genetic code is **universal**, for almost all organisms use precisely the same set of code words (there are very few exceptions). Since there are 64 possible code words designating only 20 amino acids, scientists reasoned that there are several different codons for each amino acid. For instance, GGU, GGC, GGA, and GGG all stand for glycine. Three codons, UGA, UAG, and UAA, do not stand for any amino acid. Translation stops when the process reaches one of these codons in the mRNA; they are called **nonsense codons** and serve as stop signals. Each of the remaining codons stands for one and only one of the 20 amino acids, but keep in mind that each amino acid may be represented by several three-letter codons. The sequence of codons in an mRNA spells out the sequence of amino acids in a polypeptide.

How does the mRNA sequence become the amino acid sequence in the process of translation? The ribosome has two adjacent

TABLE 4–3
Genetic code

		Second letter				
		U	C	A	G	
First letter	U	UUU ⎱ Phe UUC ⎰ UUA ⎱ Leu UUG ⎰	UCU ⎱ UCC ⎱ Ser UCA ⎰ UCG ⎰	UAU ⎱ Tyr UAC ⎰ UAA Stop UAG Stop	UGU ⎱ Cys UGC ⎰ UGA Stop UGG Trp	U C A G
	C	CUU ⎱ CUC ⎱ Leu CUA ⎰ CUG ⎰	CCU ⎱ CCC ⎱ Pro CCA ⎰ CCG ⎰	CAU ⎱ His CAC ⎰ CAA ⎱ Gln CAG ⎰	CGU ⎱ CGC ⎱ Arg CGA ⎰ CGG ⎰	U C A G
	A	AUU ⎱ AUC ⎱ Ile AUA ⎰ AUG Met	ACU ⎱ ACC ⎱ Thr ACA ⎰ ACG ⎰	AAU ⎱ Asn AAC ⎰ AAA ⎱ Lys AAG ⎰	AGU ⎱ Ser AGC ⎰ AGA ⎱ Arg AGG ⎰	U C A G
	G	GUU ⎱ GUC ⎱ Val GUA ⎰ GUG ⎰	GCU ⎱ GCC ⎱ Ala GCA ⎰ GCG ⎰	GAU ⎱ Asp GAC ⎰ GAA ⎱ Glu GAG ⎰	GGU ⎱ GGC ⎱ Gly GGA ⎰ GGG ⎰	U C A G

Third letter

sites, the **A-site** and the **P-site**. Once the ribosome is associated with an mRNA molecule, the A and P sites align precisely with two adjacent codons. Each codon is complementary to the anticodon of just one type of tRNA (with its attached amino acid). The ribosome allows for these tRNAs to move into position. It then bonds one amino acid to the other with a **peptide bond**, releases one tRNA, and moves one codon along

77

the mRNA. A new tRNA and amino acid then move into position. The processes of transcription and translation are summarized in Figure 4-9.

To understand how the process works step-by-step, imagine a ribosome with its A-site "reading" a GCA mRNA codon. A tRNA with a CGU anticodon and an attached alanine amino acid moves into position. The ribosome moves one codon along the mRNA, so that the alanine-tRNA and the GCA codon are at the P-site. The A-site is now at a new codon, say GUU. A CAA tRNA with its valine amino acid moves into position. The alanine bonds to the valine, and the P-site releases its tRNA. The CAA tRNA at the A-site is now attached to a pair of linked amino acids. The ribosome moves one more codon along the mRNA, and another amino acid, such as phenylalanine, joins the growing polypeptide chain, as illustrated in Figure 4-10.

Once the ribosome releases a tRNA, that tRNA can collect another amino acid and be used again. Once the ribosome has moved away from the beginning of the mRNA strand, another ribosome can

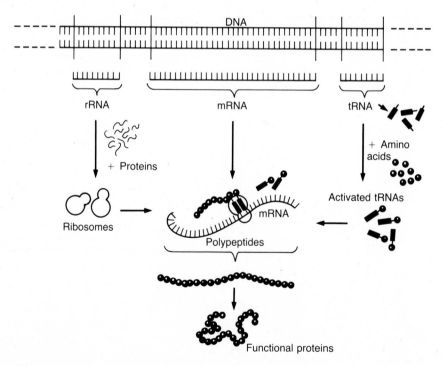

FIGURE 4-9
An overall view of protein synthesis as a result of transcription and translation.

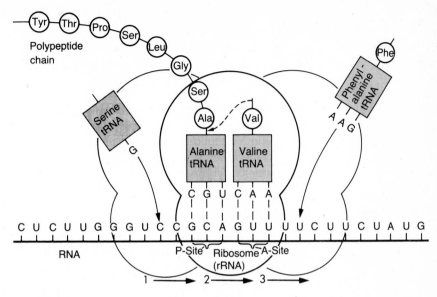

FIGURE 4-10

Polypeptide synthesis taking place on the surface of a ribosome. (Adapted from The genetic code III by F. H. C. Crick. Copyright © 1966 by *Scientific American*, Inc. All rights reserved.)

associate with the strand and begin assembling a second, identical polypeptide. Many ribosomes can be reading a single mRNA strand simultaneously, and the complex of many ribosomes and one mRNA strand is called a **polyribosome**.

Polypeptide assembly stops when the ribosome encounters a stop signal in the mRNA. It then releases the polypeptide to become a fully functional protein by itself or in combination with others. Thousands of polypeptides and proteins are assembled and released by the cell as products of transcription and translation. The process is simple in its outlines, but it requires marvelous timing to bring together all the components—DNA, mRNA, tRNA, ribosomes, enzymes, the energy transfer by ATP, and other molecular factors.

METABOLIC BLOCKS

Now that we understand the basics of how the information in the genes is translated to protein, we are left with questions about how proteins affect an individual's visible traits, or phenotype, and how mutations in

79

genes can alter those traits. The answers to these questions have come from studies of **intermediary metabolism**.

Each trait of an organism is due to some constituent of the organism's body—a pigment, a hormone, an enzyme, or possibly a gene-regulating substance. Many such traits require that **precursors** (obtained in the diet) will be built up by a series of discrete chemical steps. Each of these steps is mediated by a specific enzyme, and each enzyme is specified by one or more genes. Figure 4-11 shows how three genes, three enzymes, and three metabolic steps might convert a precursor, A, into a body constituent, D. Substances B and C are "intermediate" substances. In actual cases, there may be many more than three steps, and each enzyme may consist of several different polypeptides, each with its own gene.

Intermediary metabolism refers to the intermediate chemical steps between precursor and final product, and it is what must be probed when some genetic change interferes with the synthesis of that product. Consider a mutation in gene c. It prevents the production of a functional enzyme c. Therefore, substance C cannot be converted to substance D. As a result of the gene c mutation, the phenotype is altered to *D-less*. There is a **metabolic block** in the sequence of steps of intermediary metabolism. With this specific block, we might also expect a buildup of unmetabolized substance C.

The *D-less* phenotype can also result from mutations in genes *a* and *b*. In the first case, *D-less* is also *C-less* and *B-less*, and there may be a buildup of precursor. In the second case, *D-less* is also *C-less*, and the buildup is only of B. In each case, the phenotypes are *D-less*. The consequence of being *D-less* may be hardly noticeable. On the other hand, it may be as extreme as the difference between life and death.

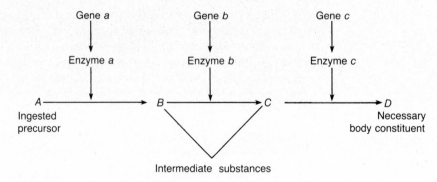

FIGURE 4–11
A hypothetical example of intermediary metabolism.

Consider a real-life example. Phenylalanine is an amino acid obtained in the diet. The body converts it to the amino acid tyrosine, and then to the skin pigment melanin. Blocks in the metabolic pathways that convert phenylalanine to other products can cause phenylketonuria, a buildup of toxic material that causes brain damage in infants, albinism, tyrosinosis, or alkaptonuria. The blocks in the metabolism of phenylalanine are shown diagrammatically in Figure 4-12. Many other such examples have been documented in humans, and we will study some of them later.

The various mutations that stop a particular enzyme from working offer us an opportunity to study **allelism**, which in turn allows us to map out the steps of intermediary metabolism. Consider two *D-less* individuals who mate and produce offspring as outlined in Figure 4-13. If the resulting offspring are all phenotypically *D-less* like their parents, we can assume that the two parents have the same metabolic block. Their respective mutations seem to be allelic; that is, they seem to be mutations of the same gene. However, if the offspring are all normal instead of *D-less*, we can assume that the parental mutations are not allelic. The normal phenotypes of the offspring indicate that the metabolic blocks occur at different steps in the pathway. Based on this information, we can understand how two albino parents can occasionally have

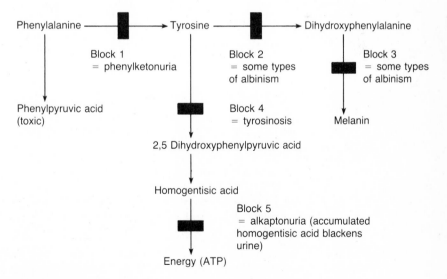

FIGURE 4–12
Blocks in the intermediary metabolism of phenylalanine and the serious disorders associated with them.

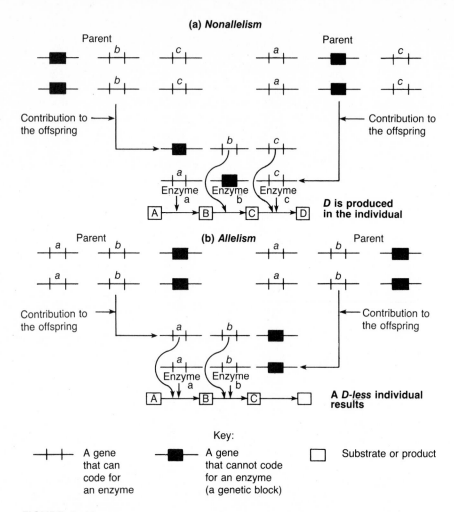

FIGURE 4–13

Two parents can both be *D-less* as a result of their homozygosity for a recessive allele that blocks an intermediate step in the production of substance D. (a) Their blocks are *nonallelic* when they relate to different steps in the intermediary metabolism; their children will be heterozygous for both steps and able to produce D. (b) Their blocks are *allelic* when they relate to the same step; their children will also be *D-less*.

normally pigmented children, or how two parents with dwarfism can have children of normal stature. Such results can occur even though albinism and dwarfism are recessive disorders, and both parents must be homozygous for the trait in order to express it.

GENES AND MUTATION

Figure 4-14 summarizes what we have discussed about the concepts of genes and their fundamental workings. Starting with the final gene product and moving backwards, we see that a phenotypic trait or body constituent is due to some enzyme-mediated chemical reaction. The en-

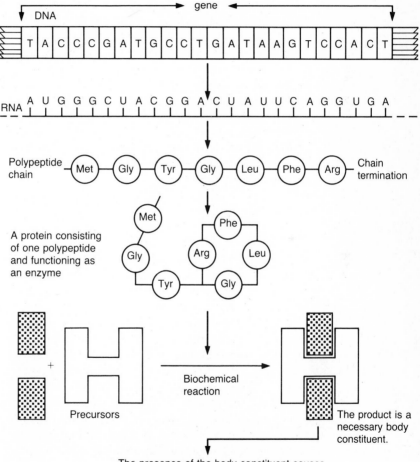

FIGURE 4–14
A diagrammatic summary of gene function illustrating the relationships among genes, polypeptides, proteins, and intermediary metabolism.

83

zyme consists of one or more polypeptide chains whose functioning depends on their three-dimensional shapes, which are fixed by the sequences of amino acids in the chains. The specific sequence of amino acids in a chain is determined by the information carried in a molecule of messenger RNA, which in turn is dictated by the sequence of nucleotides in a certain region of the master molecule, DNA. Such a region of DNA is a gene.

A change or **mutation** in the nucleotide sequence of a gene can happen in three ways. First, one or more nucleotides may be replaced by others (**base substitutions**); second, a nucleotide may be lost (**base deletion**); and third, a nucleotide may be added to the sequence (**base insertion**). The genetic code consists of words that are three nucleotide bases long, and any change in the sequence of bases in the DNA will result in a complementary change in the nucleotide sequence of mRNA. Such changes may alter the amino acid sequence of a polypeptide chain, possibly resulting in a difference in shape and function. The altered polypeptide chain may be unable to catalyze a biochemical reaction that would normally occur, and hence the body may be deficient in some essential substance. This deficiency may result in a noticeable difference in phenotype.

SUMMARY

Genes and their functions have been studied thoroughly. The central dogma states that genetic information flows from nucleic acids to nucleic acids or from nucleic acids to proteins, but never from proteins to nucleic acids or from proteins to proteins.

One of the first investigators to link genes and body chemistry was A. E. Garrod, who believed that many congenital disorders were due to a missing enzyme. He thought these were heritable by means of a Mendelian pattern. Garrod called these human disorders inborn errors of metabolism. Beadle and Tatum worked with the pink bread mold, *Neurospora crassa*, to extend Garrod's work significantly. Their results prompted the one gene-one enzyme hypothesis, which was later modified to the one gene-one polypeptide hypothesis, stating that a particular gene is responsible for the production of a particular polypeptide strand. Proteins, and hence enzymes, consist of one or more polypeptide chains that may be identical or different. A polypeptide is a sequence of amino acids, of which 20 different kinds exist.

Transcription is the process in which one strand (the sense strand) of the DNA acts as a template for an mRNA molecule via complementary base pairing. The mRNA molecules associate with ribosomes in the cell's cytoplasm. There, with the aid of tRNA molecules, they are translated. The tRNA molecules each attach to a specific amino acid at one end. At the other end, they bear a specific triplet of exposed nitrogenous bases (an anticodon) complementary to a triplet of bases (a codon) in the mRNA. The tRNA-mRNA interaction on the ribosome ensures the correct order of amino acids in a polypeptide. The 64 codons of the mRNA comprise the universal genetic code.

Necessary body constituents are usually synthesized by a series of discrete chemical steps that change precursor substances into the substances needed by the body. Each step requires a specific enzyme. Mutations—changes in the DNA such that transcription and translation yield different products—can produce nonfunctional enzymes, resulting in metabolic blocks and consequent variations in phenotype.

DISCUSSION QUESTIONS

1. In what way does this chapter support the claim by some scientists that biology is actually a branch of chemistry?

2. Why did the one gene-one enzyme hypothesis have to be revised to the one gene-one polypeptide hypothesis?

3. Exactly what is meant by base pair complementarity? Consider the entire process of protein synthesis, and point out all the places where complementarity of bases plays a role.

4. Describe briefly the relationships among amino acids, polypeptides, proteins, and enzymes.

5. What substances would be required for an *in vitro* (performed in glass) system that incorporated amino acids into polypeptides? Briefly discuss the role of each substance in the polypeptide synthesis.

6. Elaborate and discuss the idea that the essence of life is having correctly shaped molecules.

7. Develop a hypothetical or real situation in which a very small change at one point in the DNA has an ultimately lethal effect.

8. You should now be comfortable with the gene concept. Explain what a gene is, with as much precision as you can.

9. Considering our discussion of intermediary metabolism, enzymes, and gene function, explain why so many hereditary disorders are recessive. How might the same molecular mechanisms explain hereditary disorders that are due to dominant genes?

10. Beadle and Tatum developed strains of *Neurospora* with various gene-controlled dysfunctions. These strains could not grow without specific growth factors in their medium. How is this situation similar to the human inability to synthesize vitamin C (ascorbic acid) even though most animals can do so?

PROBLEMS

1. G. B. S. Roberts (1960. *Annals of Human Genetics*, 24: 127–135) obtained the data below on 65 families in which both parents were normal, and one or more children had cystic fibrosis. Study the data and decide whether it is possible for this disorder to be a Mendelian trait.

Children per family	Number of families	Total number of children with cystic fibrosis
2	16	25
3	23	38
4	15	28
5	4	4
6	5	12
7	2	6

2. Consider an enzyme molecule consisting of one polypeptide 1000 amino acids long.
   a. If the amino acids are represented equally in the molecule, how many tryptophan amino acids would you expect to find in it?
   b. What is the smallest number of tryptophan amino acids possible in the molecule?

3. How many different combinations of bases would be possible in the genetic code if the bases were read two at a time, rather than three at a time? Four at a time?

4. A particular DNA region has as part of its sense strand the base sequence: GGGTACCCGATGCTACCCTTC.
   a. What is the corresponding base sequence in the antisense strand?
   b. What is the base sequence in the corresponding mRNA?
   c. What is the base sequence in the series of anticodons needed to translate the message?
   d. What sequence of amino acids will appear in the resulting polypeptide strand?

5. A particular enzyme consists of two polypeptide strands. There exist both A and B strands, equally effective, and able to combine with equal prob-

86

ability into an enzyme molecule. Each molecule consists of only two poly-peptide strands. If the A and B alleles yield equal numbers of polypeptide strands, what kinds of enzyme molecules, and in what relative proportions, will be formed in an AB heterozygote?

6. In an *in vitro* system for polypeptide synthesis, what polypeptides would be formed with each of the mRNAs below? (Translation begins with the first triplet on the left and progresses to the right.)
   a. UUUUUUUUUUUUUUU
   b. GAGUGGAGUGCUGGG
   c. UAAUAGUAAUGAUAA

7. The codon UCA specifies the amino acid serine.
   a. How many single ribonucleotide substitutions in this codon could change the amino acid?
   b. How many such substitutions fail to change the amino acid?
   c. How many such substitutions result in chain termination?

8. The metabolic pathway $Q \rightarrow R \rightarrow S \rightarrow T \rightarrow U$ makes the necessary substance U from the precursor Q.
   a. *At least* how many gene pairs are needed to accomplish this pathway?
   b. Could the number of gene pairs necessary be more than your answer to (a)?
   c. Explain your answer to (b).

9. One type of dwarfism results from homozygosity for a recessive gene. Matings between dwarfs of this type sometimes produce children of normal stature. Use symbols and a diagram to show how this can happen.

10. Carefully think out the genetics involved in the following *hypothetical* situation: man A marries woman B, but although both have the recessive genetic defect *d*, all three of their children are normal. Husband C and wife E also both have the recessive defect *d*, and so do their two children. Both marriages end in divorce, and ultimately A marries E and B marries C. Each of these second marriages produces two more children, and all the children are normal. What is the smallest number of different genes involved in causing the *d* defect, as shown by these data? Explain your conclusion.

REFERENCES

Beadle, G. W., and E. L. Tatum. 1941. Genetic control of biochemical reactions in *Neurospora. Proceedings of the National Academy of Sciences* 27:499–506.

REFERENCES

Crick, F. H. C. 1966. The genetic code III. *Scientific American* 215:55–62.
Garrod, A. E. 1902. The incidence of alkaptonuria: A study in chemical individuality. *Lancet* ii:1616–1620.
Garrod, A. E. 1908. Inborn errors of metabolism. *Lancet* 2:1–7, 73–79, 142–148, 214–220.
Garrod, A. E. 1909. *Inborn errors of metabolism.* London: Oxford University Press.

5

Genes In Populations

*I*n the previous chapters, we considered single genes, gene pairs, and individuals. Now we will turn our attention to genetic composition of entire **populations**. A population can be any group of organisms that we wish to describe. But in genetic terms, we think of a population as being made up of organisms of the same species that can interbreed with each other. The study of a population's breeding structure, its genetic composition, and the proportions (or **frequencies**) of various alleles among its members is **population genetics**. We must understand the principles of population genetics in order to consider intelligently how genetics and society interact. For example, population genetics is a factor in questions that concern counseling for persons with rare genetic diseases, and the selection of new strains of agricultural plants and animals. We will use what we learn about population genetics in this chapter many times later in this book.

GENE POOLS AND GENE FREQUENCIES

We already know that an individual organism has a genotype, or a particular set of alleles. An organism usually does not contain *all* the possible alleles for a gene at any of its loci. Generally speaking, the diploid individual has a maximum of two alleles for every gene locus in its genotype—and in homozygous individuals, these two alleles are the same. Yet the population to which the individual belongs *can* contain all possible alleles, or at least many more than the individual because the population contains other individuals, each with its own allelic combination.

We can speak of a population as a **gene pool**, composed of all the alleles in all the population's members at any given time. Similarly, we can consider the population as a **gametic pool** of all the different gametes the individuals in the population can generate. If mating occurs at random within the population, then any male gamete in the pool has an equal probability of combining with any female gamete. In the human population, mating is random for some traits and nonrandom for others. For example, human mating would probably be random with respect to blood group genes, but nonrandom for traits such as height, IQ, and skin color. This kind of nonrandom mating is called **assortative mating** and can often be observed in human populations.

The probability that any two alleles existing in separate gametes in the gene pool will come together to form a pair of alleles in a particular zygote is the product of their allelic frequencies. Scientists use these fundamental probabilities extensively in population genetics. Let us consider how they do this in a very simple case of one gene with two alleles, the dominant *E* and the recessive *e*.

Population geneticists use the symbol *p* to stand for the frequency of a dominant allele in a population and *q* for the frequency of a recessive allele. In our example, *p* is the frequency of *E*, and *q* is the frequency of *e*. Since there are no other alleles in this example, *E* and *e* comprise 100 percent of the population's alleles and:

$$p + q = 1$$

Since *p* is also the frequency of *E*-bearing gametes, and *q* is the frequency of *e*-bearing gametes, we can use a Punnett square to calculate the frequency of *EE*, *Ee*, and *ee* zygotes in a randomly mating population:

	E(p)	*e(q)*
E(p)	*EE* (p^2)	*Ee* (pq)
e(q)	*Ee* (pq)	*ee* (q^2)

Since the sum of zygotic frequencies in each generation must be 1:

$$p^2 + 2pq + q^2 = 1$$

This is simply $(p + q)^2$, the sum of the individual frequencies of the two possible gametes, squared. From this equation, we see that the **genotype frequencies** resulting from the gene frequencies *p* and *q* are very simply calculated. The proportion of *EE* genotypes in the population is p^2; the proportion of *Ee* genotypes is $2pq$; and the proportion of *ee* genotypes is q^2. Such a distribution of genotypes leads to the *p* and *q* frequencies for *E* and *e* gametes shown in Figure 5-1.

Let us calculate the allele frequencies for a simple population containing the following numbers of individuals with three possible genotypes:

EE: 632
Ee: 330
ee: 38

FIGURE 5–1

In a randomly mating population, gene frequencies and zygote frequencies are dependent upon each other. They remain stable from one generation to the next.

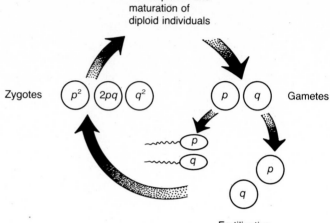

We are assuming that we can distinguish *Ee* individuals from *EE* and *ee* individuals based on their phenotype. That is, we are postulating incomplete or codominance.

Our hypothetical population has a total of 1000 individuals and 2000 alleles. We can calculate the allele frequencies by first counting the alleles that comprise each genotype:

632 diploid *EE* individuals contain 1264 *E* alleles;
330 diploid *Ee* individuals contain 330 *E* and 330 *e* alleles;
38 diploid *ee* individuals contain 76 *e* alleles.

We must then add all the *E* and *e* alleles:

1264 + 330 = 1594 *E* alleles
76 + 330 = 406 *e* alleles

The allele frequencies are then the ratios of the total for each individual allele divided by the total number of all alleles:

1594/2000 = 0.797 for *E*, and
406/2000 = 0.203 for *e*

In most real-life cases, we cannot distinguish heterozygotes from homozygous dominant individuals. To calculate allele frequencies, we must then work backward from the two equations:

$$p + q = 1 \text{ and } p^2 + 2pq + q^2 = 1$$

Suppose we observe that homozygous recessives make up 36 percent of a population. Then:

$$q^2 = 0.36 \text{ and } q = \sqrt{0.36} = 0.60$$

Since $p + q = 1$:
$$p = 1 - q = 1 - 0.60 = 0.40$$

Therefore, the frequency of homozygous dominants is $p^2 = 0.16$, and the frequency of heterozygotes is $2pq = 0.48$. Of course, $p^2 + 2pq + q^2 = 0.16 + 0.48 + 0.36 = 1$. Remember, however, that $q^2 = 0.36$ is the only figure we actually observed. We calculated p^2 and $2pq$ from theory and our initial one observation.

THE CONSERVATION OF GENE POOLS

We have just seen how allele and genotype frequencies follow a very simple algebraic pattern. This pattern is known as the **Hardy-Weinberg principle** (or law). It was first described independently by G. H. Hardy, an English mathematician, and W. Weinberg, a German physician, in 1908. Since then, it has become the cornerstone of population genetics.

Hardy and Weinberg explained mathematically why experimenters were not finding 3:1 ratios for Mendelian traits in populations. The experimenters had accepted Mendel's laws with great enthusiasm after the rediscovery of the 1866 paper in 1900, and they had studied many crosses whose F_2 progeny showed 3:1 ratios. Many of them believed that randomly breeding populations must always reach a state where there were three individuals showing the dominant trait for every individual showing the recessive trait. When they failed to find such patterns, they began to question the validity of Mendel's laws.

Hardy and Weinberg showed that geneticists should not expect the ratio of phenotypes for a Mendelian trait to change gradually from, say, 10:1 to 3:1 in a population. With nothing more than straightforward algebra, they each demonstrated that allele and genotype frequencies, and therefore phenotype frequencies, *will not change* from one generation to another, *unless* some additional force is acting on the population.

The Hardy-Weinberg principle proved immensely important. At the time, some biologists were thinking that hereditary disorders

93

caused by dominant alleles must increase from rare to very common indeed. Such disorders would include **brachydactyly** (short, thick fingers and/or toes), **polydactyly** (supernumerary fingers and/or toes), **Huntington's disease** (severe neurological deterioration), and **achondroplasia** (one kind of dwarfism). Hardy and Weinberg showed that in the absence of disruptive forces acting on the population, the frequencies of dominant disorders will not increase from their present low levels. Rare alleles will tend to remain rare, and the frequency of common alleles will also tend to remain the same.

The situation is easier to understand if we concentrate on entire populations, and not progeny within particular families. Only when heterozygotes mate with heterozygotes do we expect an overall 3:1 ratio. Populations contain homozygotes as well, and in populations six different types of matings are possible (the number greatly increases if we consider more than one pair of alleles):

homozygous dominant × homozygous dominant
homozygous dominant × homozygous recessive
homozygous dominant × heterozygous
heterozygous × heterozygous
homozygous recessive × heterozygous
homozygous recessive × homozygous recessive

Only one of these matings will result in a 3:1 ratio. The results of the others are shown in Table 5-1, which also demonstrates how genotype frequencies are conserved from the parental generation to the progeny. These frequencies stay the same as long as each type of mating produces the same average number of progeny per mating.

When the genotype frequencies approximate p^2, $2pq$, and q^2, and do not change, the population is in **Hardy-Weinberg equilibrium**. They change only if some disruptive factor acts to change them. Populations whose genotype frequencies do not approximate p^2, $2pq$, and q^2 are not in Hardy-Weinberg equilibrium. In such cases, we cannot calculate genotype frequencies from a knowledge of allele frequencies. In addition, the genotype frequencies will usually change from generation to generation. They will fail to change only if some force is holding them at their nonequilibrium levels. As an extreme example, population 7 in Table 5-2 may be composed only of heterozygotes because no homozygotes survive beyond birth. This would be an example of a selective force.

TABLE 5-1

The proportions of progeny genotypes resulting from random mating in a population consisting of 0.49 *AA*, 0.42 *Aa*, and 0.09 *aa* genotype frequencies

Parental generation = 0.49 *AA*, 0.42 *Aa*, and 0.09 *aa*					
Mating Types		Proportions of Each Mating Type	Proportions of the Progeny Genotypes from Each Mating Type		
Female	Male		*AA*	*Aa*	*aa*
AA	*AA*	0.49 x 0.49 = 0.240	0.240	0	0
AA	*Aa*	0.49 x 0.42 = 0.206	0.103	0.103	0
AA	*aa*	0.49 x 0.09 = 0.044	0	0.044	0
Aa	*AA*	0.42 x 0.49 = 0.206	0.103	0.103	0
Aa	*Aa*	0.42 x 0.42 = 0.176	0.044	0.088	0.044
Aa	*aa*	0.42 x 0.09 = 0.038	0	0.019	0.019
aa	*AA*	0.09 x 0.49 = 0.044	0	0.044	0
aa	*Aa*	0.09 x 0.42 = 0.038	0	0.019	0.019
aa	*aa*	0.09 x 0.09 = 0:008	0	0	0.008
Totals		1.000	0.490	0.420	0.090

Progeny generation = 0.49 *AA*, 0.42 *Aa*, and 0.09 *aa*

TABLE 5-2

Genotype frequencies for several different hypothetical populations

Population	Gene Frequencies		Genotype Frequencies		
	A	*a*	*AA*	*Aa*	*aa*
1	0.5	0.5	0.50	0	0.50
2	0.5	0.5	0.40	0.20	0.40
3	0.5	0.5	0.30	0.40	0.30
4	0.5	0.5	0.25	0.50	0.25
5	0.5	0.5	0.20	0.60	0.20
6	0.5	0.5	0.10	0.80	0.10
7	0.5	0.5	0	1.00	0

Note: Only population 4 is in Hardy-Weinberg equilibrium. The genotype frequencies are mathematically p^2, $2pq$, and q^2, based upon p and q both equal to 0.5.

95

CHANGES IN THE GENE POOL

Several kinds of factors can cause populations, over a period of time, to deviate from Hardy-Weinberg equilibrium. These include **migration**, **mutation**, **genetic drift**, **natural selection**, and combinations of them.

Migration, the movement of individuals from one geographic area to another, can change allele frequencies for either or both of the populations involved. This is clear when the movement of individuals introduces an allele to or removes one from a population, but more subtle changes have also been seen in plant and animal populations. Migration also affects human populations, and examples include voluntary emigration and immigration, the past transport of slaves, military invasions, and evictions of so-called undesirables.

Mutations, changes in the nucleotide sequences of genes, are fairly rare events, in terms of their frequency in the gene pool, but they can result in large changes in allele and genotype frequencies. Mutations are also the sole source of new variability in populations. Although mutations necessarily produce rare alleles at first, the new alleles that are successful in evolutionary terms can become established and common.

Genetic drift refers to the changes in allele frequencies of populations from one generation to another that result from chance fluctuations. It is most apparent in small populations, where the only bearer of an allele may just happen not to reproduce. Its simplest version appears in the **founder principle**, which applies when a small portion of a large population is isolated by circumstance (as when a few individuals are stranded on an island) and *founds* a new population. Because the alleles borne by these founders are not likely to be represented in the same frequencies present in the original population, the new population will have different allele and genotype frequencies. This kind of event is illustrated in Figure 5-2.

Natural selection occurs when some genotypes reproduce themselves more successfully than others. It is one of the bases of evolution, which we will discuss in more detail in Chapter 18. In plants and animals—including humans—more successful reproduction may mean having greater numbers of offspring, greater success in finding mates, or a greater probability of surviving to reproductive age.

Nonrandom mating, because it deviates from the basic assumption underlying Hardy-Weinberg equilibrium, can lead to changes in genotype frequencies in a population. One kind of nonrandom mating appears when mating genotypes are more closely related than average for the population; this is **inbreeding**. Another appears when mating genotypes are less closely related than average; this is **outbreeding**. In

FIGURE 5–2
The founder principle. The gene frequencies in the new population are very different from those in the original population.

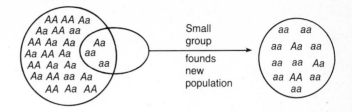

these cases, allele frequencies do not change, but genotype frequencies do.

All the factors that cause changes in alleles and genotype frequencies probably act, to varying degrees, in all populations. However, they can also counteract each other, so that the populations typically show a relatively stable genetic composition, an equilibrium. Migration, genetic drift, mutation, and natural selection are the processes that determine the genetic constitution of a population. The population's genetic constitution, in turn, determines whether the population can meet the challenges of its environment and survive.

THE CONCEPT OF RACE

Biologists often separate species into subpopulations or groups based on characteristic frequencies of certain genes. These groups are called **races**. Although the term race is not easy to define, the consensus is that a race is a group whose members, by reason of culture, geography, or other factors, mate only with each other. Such groups tend to have allele *frequencies* that differ from those of other such groups. They have *the same* genes and alleles of the genes, but in different proportions. The concept of race in human populations is loaded with controversy and emotion, largely because of its political and sociological connotations. But race is a biological phenomenon, and it can be discussed in terms of genetics.

Biologists and geneticists have used several different systems to divide the human species into races. The most complex lists a hundred or more races. The simplest lists only three, the **Caucasoid**, **Mongoloid**, and **Negroid** races. In any scheme, however, all the human races belong to the same species, *Homo sapiens*. Technically, there is but one human gene pool. This means that a mating between any man and any woman can produce fertile offspring. This reproductive capability is what defines a species.

97

TABLE 5–3
Gene frequencies of blood types among three racial groups

Alleles of P Blood Type	Gene Frequencies		
	Caucasoid (N = 1166)	Negroid (N = 900)	Mongoloid (N = 293)
P_1	0.5161	0.8911	0.1677
P_2	0.4839	0.1089	0.8323

Alleles Xg Blood Type	Caucasoid of Northern Europe (N = 5388)	Negroid of New York and Jamaica (N = 219)	Mongoloid of China (N = 549)
Xg^a	0.66	0.55	0.54
Xg	0.34	0.45	0.46

Source: Race, R. R. and Sanger, R. 1968. *Blood Groups in Man*. Philadelphia: F. A. Davis Co.

Races differ in their distribution of blood types, as shown in Table 5-3 and Figure 5-3, fingerprint types, and patterns of urinary excretion. They differ as well in the frequency of alleles that let one taste the chemical **phenylthiocarbamide (PTC)**, that determine whether one's ear wax is dry and crumbly or moist, and that specify various hereditary disorders. For example, **Tay-Sachs disease** is common in

FIGURE 5–3
Map of *B* blood type allele distribution in Europe and western Asia. The B antigen is believed to have had its origin in southern China. Note the distribution of B type blood in Europe and part of Asia. (From A. M. Winchester and T. R. Mertens. 1983. *Human Genetics*, 4th ed. Charles E. Merrill Publishing Company, Columbus, Ohio.)

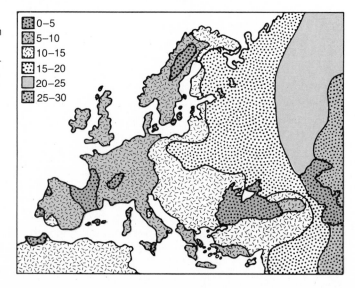

0–5
5–10
10–15
15–20
20–25
25–30

European Jews; in the United States, 62 Jews in every 100,000 suffer from it, while very few other United States citizens have it. The incidence of **cystic fibrosis** is 26 per 100,000 in the United States, and only 1 per 100,000 in Japan; that of **sickle cell anemia** is 200 per 100,000 among blacks in the United States and nearly zero among whites.

In analyzing races, we generally emphasize obvious features such as skin color, hair type, and facial structure. However, such clear-cut phenotypic differences reveal only part of the genetic relationship. Additional comparisons can be made when we use special techniques to examine the body's proteins, whose structural differences reveal both obvious and subtle genetic differences.

One such technique is **electrophoresis**. The process, which is illustrated in Figure 5-4, is based on the fact that different protein molecules move through a column of gel at different rates when subjected to an electrical field. The different structures of the proteins mean that

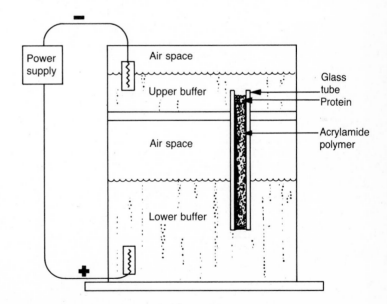

FIGURE 5–4

An electrophoretic apparatus designed such that electricity can flow from one buffer solution to the other through the acrylamide polymer gel. Protein molecules migrate through the gel at rates determined by their charge relationships. (*Genetics, Evolution, and Man*, by W. F. Bodmer and L. L. Cavalli-Sforza. Copyright © 1976 by W. H. Freeman and Company. All rights reserved.)

99

the electrical charges they carry are different, and as a result their responses to the electrical field vary. Single proteins move through the gel at the same rate, and mixtures of proteins separate into a series of bands in the gel column. Researchers use various stains to reveal these bands, as shown in Figure 5-5. Differences in band location indicate differences in structure and, therefore, in genes.

Using these techniques, researchers can show that the genetic differences between individuals from the same population are as great as or greater than the differences between races. In fact, scientists estimate statistically that individuals of different races typically differ in only one to three percent of their codons (mRNA base triplets), at least for the proteins studied. Clearly, there is a great deal of genetic diversity *within* races.

Human beings are very mobile today. Therefore, **gene flow** from one population to another may be increasing. What may be surprising is the extent of *past* gene flow. In 1953, Bentley Glass and C. C. Li reported that the proportion of Caucasian alleles in the American black population was 20 to 30 percent, and possibly higher. Completely random mating among individuals in the United States would almost eliminate distinct races. But as you can see from Figure 5-6, such a change would require many generations.

Phenotype	Genotype	Hemoglobin electrophoretic pattern Origin +	Hemoglobin types present
Normal	*AA*		HbA HbA
Sickle cell trait	*AS*		HbA HbS
Sickle cell anemia	*SS*		HbS HbS

FIGURE 5–5
Electrophoretic patterns observed for normal hemoglobin (HbA HbA), sickle cell hemoglobin (HbS HbS), and the heterozygous type (HbA HbS). (From *Heredity, Evolution, and Society* by I. M. Lerner. Copyright © 1968 by W. H. Freeman and Company. Photograph by A. C. Allison.)

FIGURE 5–6

The area under the dotted line indicates the distribution of skin colors among whites in the United States. The area under the first solid line indicates the distribution of skin colors in the American black population. The shaded area gives the expected distribution of skin colors in the United States after many generations of random mating. (From *Principles of Human Genetics*, 3rd ed., by Curt Stern. Copyright © 1973 by W. H. Freeman and Company. All rights reserved.)

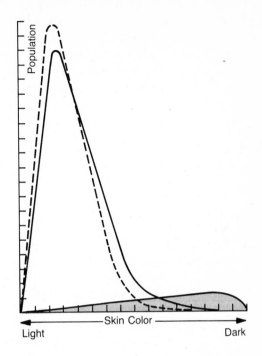

APPLICATIONS OF HUMAN POPULATION GENETICS

Some people believe that there is no scientific or biological value in the concept of human race. They also say that because of abuses of the concept, human races should not even be studied. However, race is a legitimate aspect of population genetics, and the broader subject has proved very useful in the study of human beings.

Using population genetics, scientists analyze genetic traits on the population level, unravel the structures of populations to some extent, make sense of **demographic** data (demography is the science that deals with the distributions and densities of populations), and study natural selection, mutation, migration, and population equilibria. Population genetics has proved especially useful in agriculture and medicine; in the latter it has helped us understand the consequences of inbreeding and the role of **carriers** of recessive alleles in the population. Population genetics also contributes many ideas to the study of evolution and **ecology** (the interrelationship of living things with each other and their environment).

101

SUMMARY

Thinking about genes in populations requires broader, more mathematical ideas than thinking about genes in individuals and families. A population is a group of interbreeding organisms of the same species that makes up a gene or gamete pool. Each allele for a particular locus occurs in a specific proportion or frequency in the gene pool. Genotype frequencies reflect these gene (or allele) frequencies.

Two very important relationships for population genetics are:

1) $p + q = 1$, and
2) $p^2 + 2pq + q^2 = 1$

where p is the frequency of the dominant allele of a pair, q is the frequency of the recessive allele, and p^2, $2pq$, and q^2 are the frequencies of the dominant homozygote, heterozygote, and recessive homozygote, respectively. This Hardy-Weinberg principle uses basic algebra to explain why gene frequencies do not change from one generation to another unless disturbed by forces such as migration, genetic drift, mutation, and natural selection. In addition, G. E. Hardy and W. Weinberg showed in their classic 1908 demonstrations that undisturbed populations should be found in equilibrium. That is, regardless of the frequencies p and q, the three possible genotypes should have the frequencies p^2, $2pq$, and q^2. This relationship is known as the Hardy-Weinberg equilibrium.

Genetically, races are breeding populations or subpopulations that differ in their gene frequencies; however, they do not differ in the genes or alleles in the population. Some investigations have shown a greater amount of protein diversity within races than among races. Also, gene flow from one race to another tends to blur the phenotypic distinctions between them. Considerations of race and genetics have sometimes led to social and political controversy.

The basic objective of the discipline called population genetics is to study the genetic structures of populations, mainly in mathematical terms. Demography, evolution, ecology, and, more recently, clinical genetics, are only a few of the areas depending heavily on genetic information about populations.

DISCUSSION QUESTIONS

1. Some people think race is a biological phenomenon, but others argue strongly that race should be thought of only as a cultural phenomenon. Consider these two views.

2. What are some of the reasons why much work in human genetics has involved the concepts of population genetics and the Hardy-Weinberg principle?

3. Ever since the discovery of genetic dominance and recessiveness, people have been concerned that deleterious dominant alleles would increase to staggering proportions in the population. Carefully explain why such fears are unfounded.

4. Assume that as a genetic investigator you have applied the Hardy-Weinberg principle and found that a population is not in equilibrium for a particular gene locus. What would be your next step in attempting to learn more about the structure of the population?

5. Where in the United States might we be able to demonstrate the concept of genetic drift in the form of the founder principle?

6. Scientists use electrophoresis to determine whether proteins are the same or different in two individuals, races, or species. Whatever the results, they are always related to the genotypes of the organisms involved. Explain in detail why it is scientifically sound to relate an organism's proteins to its genetic constitution.

7. Some people fear that the United States' human gene pool is slowly changing, with the proportion of deleterious genes increasing. We will discuss this possibility in more detail in later chapters. For now, can you think of any situation that might prompt such fears?

PROBLEMS

1. In 1962, G. R. Thompson reported that in Accra, Ghana, children had the following genotypes and phenotypes in the specified proportions:

   AA (normal hemoglobin): 593
   AS (sickle cell trait): 123
   SS (sickle cell anemia): 4

   What is the frequency of the sickle cell allele (*S*) in this sample of the population?

2. Assuming Hardy-Weinberg equilibrium, give the expected genotype frequencies in each of the following cases.

103

Gene frequencies:
a. $D = 0.64, d = 0.36$
b. $I = 0.19, i = 0.81$
c. $G = 0.81, g = 0.19$
d. $K = 0.99, k = 0.01$

3. Consider a hypothetical human trait controlled by a recessive allele, r. People without the trait have at least one dominant allele, R. The frequency of the r allele in the population is 0.2. Calculate the probabilities that a person in the population will have the genotypes rr, Rr, or RR.

4. Consider a hypothetical trait that can take two forms. The F phenotype appears in 2999 individuals. The f phenotype appears in 1001 individuals. Do these numbers tend to show that this is a simple Mendelian trait with F as the dominant allele and f as the recessive allele? Why or why not?

5. If the frequency of a certain homozygous recessive trait in a human population is one in 10,000, and if mating is completely random:
a. What proportion of all matings will be between two persons who both show the trait?
b. What proportion of all matings will be between two persons who show the dominant form of the trait?
c. What proportion of all matings will be between two heterozygotes?

6. The basic symbols of population genetics are p and q. What do they mean when combined as follows?

$$(2pq)/(p^2 + 2pq)$$

7. Consider a recessive allele whose frequency in the general population is one in 100. You do not have the recessive trait, but you do not know whether you—or your spouse—are a heterozygous carrier for the allele.
a. What are the chances that you are a carrier of this allele?
b. What are the chances that both you and your spouse are carriers?
c. Since you don't know your genotype or your spouse's genotype, what can you say about your child's chances of actually having the recessive trait?

8. One kind of blood group is the MN system. It is a strictly Mendelian trait, and people have the genotype MM, MN, or NN. If a population of 1000 people has 220 MM, 20 NN, and 760 MN genotypes, is this population in Hardy-Weinberg equilibrium?

9. Select any gene frequencies and genotype frequencies that are not in Hardy-Weinberg equilibrium and show that it takes only one generation of random mating for the progeny to reach equilibrium.
Hint: Set up all the possible matings; apply the genotype frequencies; assign the resulting progeny according to Mendelian probabilities; calculate the gene frequencies for the total progeny; and finally, compare these gene frequencies with those of the parental population.

104

REFERENCES

Bodmer, W. F., and L. L. Cavalli-Sforza. 1976. *Genetics, evolution, and man.* San Francisco: W. H. Freeman & Co. Publishers.

Glass, B., and C. C. Li. 1953. The dynamics of racial intermixture—An analysis based on the American Negro. *American Journal of Human Genetics* 5:1–20.

Hardy, G. E. 1908. Mendelian proportions in a mixed population. *Science* 28:49–50.

Nei, M., and A. K. Roychoudhury. 1972. Gene differences between Caucasian, Negro and Japanese populations. *Science* 177:434–436.

Nei, M., and A. K. Roychoudhury. 1974. Genetic variation within and between the three major races of man, Caucasoids, Negroids, and Mongoloids. *American Journal of Human Genetics* 26:421–443.

Race, R. R., and R. Sanger. 1968. *Blood groups in man.* Philadelphia: F. A. Davis Co.

Stern, C. 1960. *Principles of human genetics.* San Francisco: W. H. Freeman & Co. Publishers.

Thompson, G. R. 1962. Significance of haemoglobins S and C in Ghana. *British Medical Journal* 1:682–685.

Weinberg, W. 1908. Uber den nachweis der Vererbung beim menschen. *Naturkunde in Wurttemberg, Stuttgart* 64:368–382. Translated in English in Boyer, S. H. 1963. *Papers on human genetics.* Englewood Cliffs, NJ: Prentice-Hall, Inc.

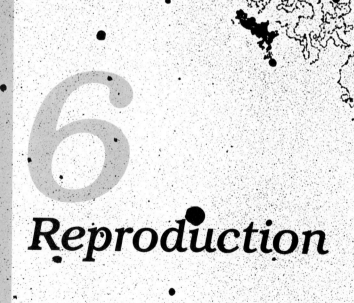

6

Reproduction

Reproduction provides the means for us to study genetics, the science of heredity. Members of one generation produce the next by the process of reproduction. Human reproduction, the means by which society biologically perpetuates itself, is complicated by social issues surrounding population growth, birth control, congenital disorders, and biological technology. We will address many of these issues in this chapter, after we have discussed the reproductive process itself.

There are two major kinds of reproduction. The first is **asexual**, which we see in strawberry runners, potato tubers, plant cuttings, the fission of bacteria and protozoa, and the budding of simple animals such as *Hydra*. The progeny resulting from asexual reproduction usually do not show any variation and are identical to their parent; they are clones grown from a few cells of the parent by mitosis.

Humans reproduce in a different way—**sexual reproduction**. In most sexual reproduction, two parents each contribute half of their chromosome complement to a special cell, a **gamete**. Human gametes are **haploid** cells, each bearing 23 chromosomes. The male gamete is a **spermatozoon**, or sperm cell. The female's is an **ovum**, or egg cell. The reproductive act, mating, brings one sperm and one ovum together. The two cells fuse to produce a single **diploid** cell, a fertilized egg or **zygote**, that can then develop into a new individual.

As you know, humans are not the only living things that reproduce sexually. Most plants and animals use this process, although some plants also rely heavily on asexual reproduction. Even bacteria and protozoa depend on a similar process at certain stages in their life cycles.

GENETIC DIVERSITY

Sexual reproduction is so widespread because it offers at least one compelling evolutionary benefit: it increases the genetic diversity of the population. The fusion of egg and sperm, each with half the chromosomes and genes of one parent, results in a **genetic recombination**, or a shuffling of genes. New combinations of genes appear in the offspring, and the offspring may or may not prove better fitted to survive than their parents.

In general, genetic diversity helps to ensure the survival of the species. Low diversity hampers survival, as when southern leaf blight attacked the corn crop in the United States in 1969 and 1970. The nation's farmers had planted their corn fields almost entirely with a variety of corn that lacked the genetic make-up that provides resistance to the blight. The disease spread rapidly and cut the nation's corn yield by about 15 percent—and by as much as 50 percent in some areas. The problem was solved only when new seed corn became available, incorporating the appropriate genetic material.

The case of the blighted corn tells us more about the value of high diversity than about how sexual reproduction increases diversity. But consider an individual with a dihybrid genotype for two genes, *DdEe*. Asexual reproduction can only produce *DdEe* progeny, for it depends on mitosis (*equational* cell division). However, if two *DdEe* heterozygotes mate and reproduce sexually, their offspring can have a variety of genotypes. Figure 6-1 shows how this typical dihybrid cross gives rise to nine progeny genotypes and four phenotypes (in the classic 9:3:3:1 ratio).

The key to this way of increasing diversity by sexual reproduction is the independent assortment and segregation of chromosomes in

FIGURE 6–1
Sexual reproduction allows offspring of two *DdEe* heterozygous parents to have nine genotypes and four phenotypes.

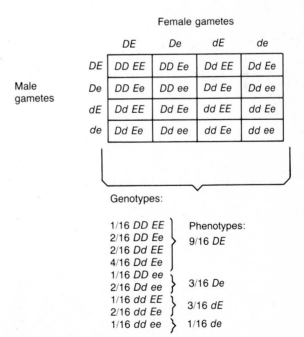

the meioses that produce the gametes. Humans have 23 pairs of chromosomes in each of their somatic cells, and meiosis can give rise to 2^{23} or 8,388,608 distinct chromosome combinations. The union of gametes then gives $2^{23} \times 2^{23}$ (2^{46}, or 70,368,744,000,000) possible combinations in the progeny.

Adding to this diversity, meiosis can scramble the genes even more through the phenomenon of **crossing over**, discussed in Chapter 3. During the prophase of the first meiotic division, the chromosomes undergo **synapsis**. While they are paired to form **tetrads**, the duplicated chromatids of homologous chromosomes can exchange segments one or more times as shown in Figure 3-13.

The result is an increased recombination of the genes. Humans have an estimated 50,000 to 100,000 genes. If we assume the number is 75,000 for simplicity, then we can calculate that the genetic recombination that results from crossing over and chromosome segregation can theoretically generate an astronomical $2^{75,000}$ different gene combinations. In fact, this number of combinations is higher than one would really expect to observe because genes that are closely linked on the same chromosome rarely undergo crossing over. With this vast reservoir of human variation, it is no wonder that **monozygotic (identical) twins**, derived from the division and separation of a fertilized egg, are the only

FIGURE 6-2
Schematic diagram of oogenesis in the female.

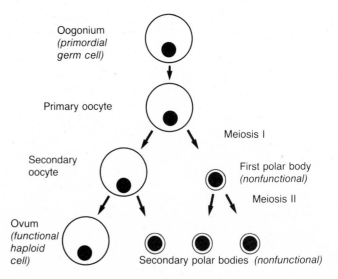

Oogonium
(primordial germ cell)

Primary oocyte

Meiosis I

Secondary oocyte

First polar body *(nonfunctional)*

Meiosis II

Ovum *(functional haploid cell)*

Secondary polar bodies *(nonfunctional)*

110

humans to share the same genotype. The rest of us are all genetically unique.

THE FEMALE REPRODUCTIVE SYSTEM

Although genetically the female and the male are each responsible for half the process of sexual reproduction, physiologically, the female has more than half the work. She must shelter within her body the developing **embryo**. Only very recently has technology begun to forecast a day when she can leave this job to others, or even to machines.

Generally, the human female is fertile from about 13 to 49 years of age. During these years, the **ovary**, the primary reproductive organ, produces an egg approximately each month, on schedule. In each cycle, an ovum develops to the point where it can be fertilized by a sperm cell. The ovary sheds this cell in the process of **ovulation**.

When the ovum is discharged, it has not yet completed meiosis. Technically, it is a **secondary oocyte**, as shown in Figure 6-2. It arose from a **primary oocyte** by means of the first meiotic division. The primary oocyte itself was formed from an **oogonium** before the woman's birth. It remained suspended in mid-meiosis for from 12 or 13 years (until puberty) up to 49 or 50 years (until the last ovulation). The secondary oocyte does not complete meiosis until after fertilization by a sperm cell. The steps in this process are shown in Figure 6-3.

The woman's monthly, or **menstrual**, cycle is regulated by hormones. The cycle begins at puberty when the **pituitary gland** begins to secrete **gonadotropic** hormones that stimulate the monthly ripening of one ovum in one ovary. The ovum enlarges, and with its surrounding ovary cells, forms a **Graafian follicle**, illustrated in Figure 6-4. The follicular cells secrete the hormone **estrogen**, causing the lining of the **uterus** to thicken. At ovulation, the follicle ruptures, releasing the ovum into the **Fallopian tube** or **oviduct**, through which it moves to the uterus. After ovulation, the follicle becomes the **corpus luteum** (yellow body), which secretes both estrogen and **progesterone**.

The ovarian hormones cause the uterine lining or **endometrium** to thicken and secrete nutritive substances, preparing it to host a fertilized egg. If the egg is not fertilized soon after ovulation, it dies, the corpus luteum degenerates, and the hormone secretion ceases. The thickened endometrium is discharged in the process of **menstruation**. It regenerates when the next Graafian follicle begins to form.

111

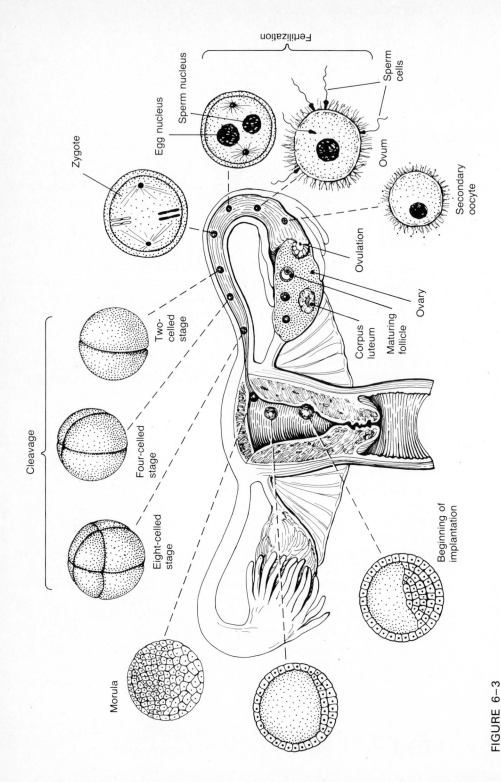

FIGURE 6-3
The release of the secondary oocyte, fertilization, early embryonic development, and implantation. (From Hole, Jr., John W., HUMAN ANATOMY AND PHYSIOLOGY, 2nd ed. © 1978, 1981 Wm. C. Brown Company Publishers, Dubuque, IA. All Rights Reserved. Reprinted by permission.)

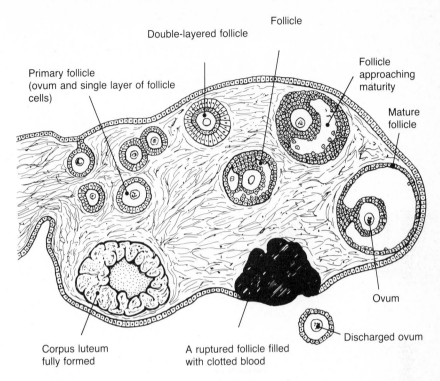

FIGURE 6–4
The development of the Graafian follicle and ovulation. (Reproduced with permission from P. Rhodes, *Carolina Biology Reader 4*, 1976. Published and copyright by Carolina Biological Supply Company, Burlington, North Carolina, USA.)

THE MALE REPRODUCTIVE SYSTEM

Unlike the female, the male is not subject to a reproductive cycle. Beginning at puberty, under the influence of the pituitary gonadotropins, he produces haploid spermatozoa continuously. The sperm cells are produced in the convoluted **seminiferous tubules** of the **testis** from parent cells called spermatogonia. These **stem cells** give rise by mitosis to **primary spermatocytes**, each of which undergoes meiosis to yield four haploid **spermatids**. The spermatids then develop into spermatozoa. The process of spermatogenesis is summarized in Figure 6-5.

The spermatozoa are stored in the **epididymis**. This structure connects to the **vas deferens**, a duct that delivers sperm to the **urethra** in the **penis** at the time of mating, or **copulation**. Both the epididymal

113

FIGURE 6-5
Schematic diagram of spermatogenesis in the male.

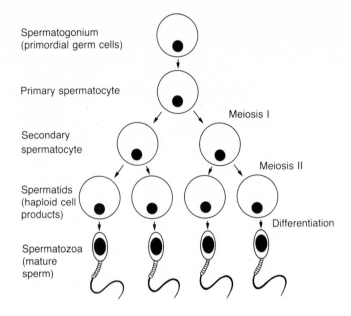

Spermatogonium (primordial germ cells)

Primary spermatocyte

Meiosis I

Secondary spermatocyte

Meiosis II

Spermatids (haploid cell products)

Differentiation

Spermatozoa (mature sperm)

FIGURE 6-6
The male reproductive system and genitalia. (Reproduced with permission from P. Rhodes, *Carolina Biology Reader 4*, 1976. Published and copyright by Carolina Biological Supply Company, Burlington, North Carolina, USA.)

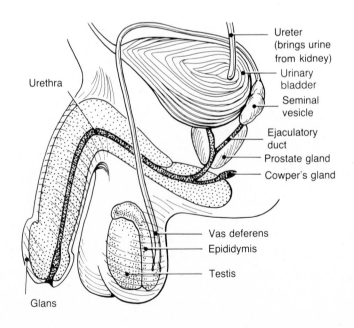

Urethra

Ureter (brings urine from kidney)

Urinary bladder

Seminal vesicle

Ejaculatory duct

Prostate gland

Cowper's gland

Vas deferens

Epididymis

Testis

Glans

tubules and the vas deferens have muscular walls and **cilia** to propel the sperm.

As sperm are being released and deposited in the female, they are mixed with the secretions of several glands in the male, the **prostate gland**, the **seminal vesicles**, and the small **Cowper's gland**, all illustrated in Figure 6-6. These secretions are rich in fructose, citrate, phosphate, and other nutritive substances; they make up the bulk of the 3 to 5 milliliters of **semen** discharged during **ejaculation.**

The testis produces more than sperm. In between the seminiferous tubules lie the **interstitial cells** secreting the male sex hormone, **testosterone**. This hormone is involved in the production of sperm, growth, and the development of male **secondary sex characteristics** such as facial hair, deepened voice, and heavier musculature.

FERTILIZATION

When a man inseminates a woman, he deposits millions of spermatozoa into her **vagina**. See Figure 6-7 for a magnified photograph of spermatozoa. Many of these sperm cells reach the uterus and begin to swim up the Fallopian tubes where fertilization occurs. Aided by contractions of the uterus and tubes, thousands may actually reach the egg, although only one penetrates and fertilizes it.

At the time of fertilization, the egg is surrounded by a clear **zona pellucida** and a layer of follicular cells, the **corona radiata**. Both are broken down by enzymes carried by the sperm cells, until one sperm can actually penetrate the egg's cell membrane. As soon as this happens, the egg changes and its surface is impenetrable to other sperm.

FIGURE 6-7
Human spermatozoa. (From A. M. Winchester and T. R. Mertens. 1983. *Human Genetics,* 4th ed. Charles E. Merrill Publishing Company, Columbus, Ohio.)

115

The sperm cell loses its whiplike tail as it penetrates the egg. At about the same time, the egg completes its second meiotic division, extruding one daughter nucleus as a **polar body**, which soon degenerates. The other daughter nucleus remains in the egg; it is haploid, with 23 chromosomes. The egg nucleus fuses with the sperm nucleus, also haploid, completing fertilization and producing a diploid zygote with 46 chromosomes, half from each parent. The zygote is a single cell, but it is also the beginning of a new organism, capable of growth, development, and eventually, of reproduction.

EARLY DEVELOPMENT

After fertilization, the new zygote grows by mitosis. By successive **cleavages**, shown in Figure 6-8, it becomes a two-celled, four-celled, and eight-celled embryo. It soon forms a ball of cells called a **morula**. When the morula develops a hollow interior, it is a blastocyst. One side of the **blastocyst** is thicker—this is the **embryonic cell mass** that will become the infant human being. The rest of the blastocyst will become the membranes that enfold and protect the growing embryo. Part of it, with the uterine lining, will become the **placenta**, a remarkable organ

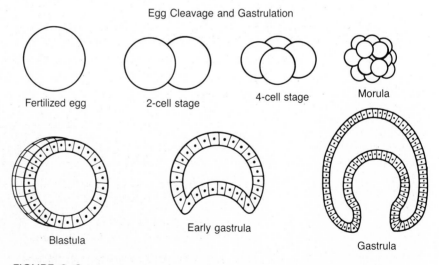

Egg Cleavage and Gastrulation

Fertilized egg   2-cell stage   4-cell stage   Morula

Blastula   Early gastrula   Gastrula

FIGURE 6-8
Two-celled and four-celled embryos, morula, and blastocyst with embryonic cell mass.

116

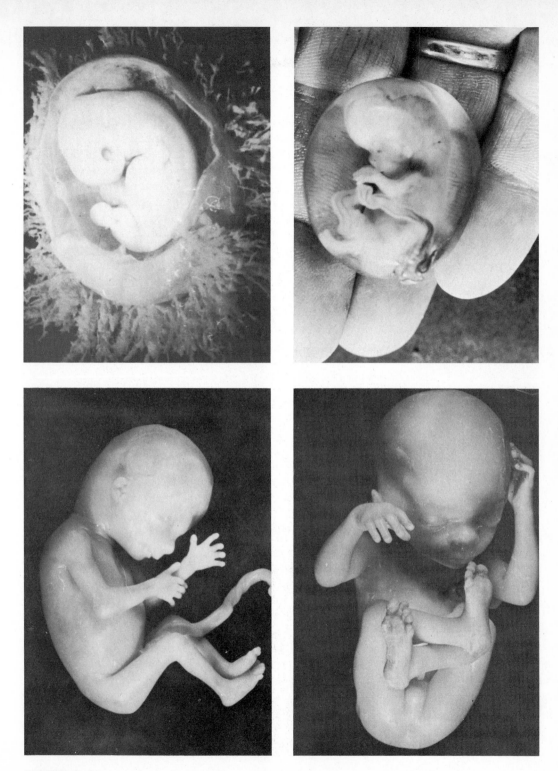

FIGURE 6–9
Human development in the uterus. (Upper and lower left and lower right photos courtesy of Landrum B. Shettles; upper right photo © Donald Yeager 1978.)

that supplies nutrients and oxygen to the embryo and removes wastes. It is the site of exchange between the embryo's blood and the mother's.

Five to eight days after fertilization, the developing embryo reaches the uterus and embeds in the endometrium; this is **implantation**. Two weeks later, the embryo has the beginnings of a skeleton and a nervous system. By the age of four weeks, it has a recognizable head and a beating heart. By six weeks, it is about an inch long and has eyes and limb buds. You can get a better idea of these stages of development by studying Figure 6-9. Shortly after six weeks, many of the internal organs begin to develop, and the embryo is properly called a **fetus**. The remaining months of development are largely devoted to growth and refinement of form. At birth, approximately nine months after fertilization, the infant will usually weigh five to ten pounds and be about 18 inches long.

THE TECHNOLOGY OF REPRODUCTION

The reproductive process is so complex that, even though it normally proceeds smoothly, it can go awry in many ways. Medical science has developed solutions for many of these problems. They include artificial insemination, *in vitro* fertilization, embryo transfer, surrogate mothers, birth control, and elective abortion. In addition, the future may lead us to parthenogenesis and cloning of individuals. However, each of these technological innovations can arouse strong feelings and conflicts in society.

Artificial Insemination

Artificial insemination (or "gamete transplantation") is the manual deposition into a female's vagina of semen collected from a male. The semen may come from the woman's husband (**AIH**) or from an anonymous donor (**AID**). Pregnancy results in 60 to 80 percent of the attempts, and many thousands of children are born each year as results of AIH or AID.

An estimated 12 to 15 percent of all married couples cannot, or choose not to, conceive a child. The reasons include male sterility (reduced number or quality of sperm in the semen), male ejaculation into the bladder instead of through the penis, and the desire to avoid passing on a hereditary disorder. In cases involving a male fertility problem, the couple may choose AID in order to have a child. An increasing number of single women are also choosing AID. Anonymous donors are often

118

medical students or others deemed to have high intelligence and appearing to be in good health. However, almost no information is collected regarding the presence of genetic disorders in the donors and their families. When couples are involved, the donors are chosen to match the husband's phenotype as closely as possible. The offspring resulting from AID are 58 percent male and 42 percent female, and the incidence of spontaneous abortions and genetic defects may be less than in the general population.

Though the process is similar, AIH is used for different reasons than AID. Some women have used AIH when their husbands were in another country for long periods of time. Other husbands with only a few sperm in their semen have had their sperm concentrated in the laboratory; then AIH has been used to inseminate their wives. Still others have had samples of their sperm frozen and stored before being sterilized, sent to war, or undergoing radiation treatment for cancer.

The freezing process—**cryopreservation** or **cryobanking**—is simple and quite reliable. Collected sperm are treated with glycerol, placed in plastic tubes, and stored in liquid nitrogen at $-196°$ Celsius. **Sperm banks** can keep sperm stored in this way for many years, and the sperm remain able to fertilize an egg. This has been demonstrated convincingly in the livestock industry. However, cryopreservation does require continuous monitoring of the equipment and stored sperm to ensure that the sperm do not warm and deteriorate.

Of the many sperm banks in the United States, the most controversial is surely the one that purports to store the sperm of Nobel Prize winners. Its aim is to inseminate carefully selected women, thereby increasing the number of exceptionally bright children in the population. This enterprise is based on extending the principles of plant and animal breeding to humans, without, perhaps, giving adequate weight to the importance of environment in human development. Not surprisingly, the existence of this sperm bank has generated vehement objections from scientists and the general public.

There is some fear that the freezing required for artificial insemination may damage genes, although no such effect has emerged in artificial insemination of livestock. There is also the possibility that through widespread use of artificial insemination, an AID offspring may inadvertently marry a close relative. The probability of an AID child marrying a first cousin or closer relative has been calculated as once in 4.5 years in the United States. Another potential problem is that donor anonymity may make it hard for an AID child to find a needed blood transfusion or organ transplant. Further, some couples may be disturbed by a connotation of adultery or illegitimacy, although the courts regard a

119

child conceived by AID with the consent of the husband as his legitimate and legal offspring. An AID child probably will not know his own, or his children's, chances of developing hereditary disorders. Many of the problems faced by AID children are not unique; similar concerns affect adopted children.

There are so many questions about AID that anyone who feels reluctant to use it should avoid the process. How should society regard AID without the consent of the husband? How should society regulate the process, and sperm banks? How does AID, especially as used by single women, correspond with religious and moral concerns? How much should we count the happiness of couples who could not have had children without AID?

In Vitro *Fertilization*

In vitro means "within glassware." *In vitro* fertilization (IVF) is fertilization *outside* the body, in a test tube or petri dish. Though it has been possible for many years, it is now practical. The procedure, outlined in Figure 6-10, begins at ovulation, when a physician uses a **laparoscope** (a device equipped with a viewing system and a suction tube that can be inserted through a small abdominal incision to reach the ovary) to collect a freshly released egg cell. Various drugs can be used to time the ovulation conveniently or to induce multiple ovulation. The captured egg (or eggs) is then put in a glass dish and properly treated sperm cells from the woman's husband or a donor are added. If this step is successful, fertilization occurs.

Twelve hours later, the fertilized egg is transferred to a second dish. There a nutrient medium supports the first few cleavages of the zygote. In two days, the zygote is an eight-celled embryo. Shortly thereafter, the physician collects the embryo with a little of its nutrient fluid and flushes it into the woman's uterus, where, if all goes well, the small embryo implants. The resulting offspring is often called a **test-tube baby** even though it is within glassware for only two or three days of its pre-birth development.

This process was first perfected by P. C. Steptoe, a gynecologist, and R. G. Edwards, a reproductive physiologist, and the first test-tube baby was born in England on 25 July 1978. Her name is Louise Brown. The first American test-tube baby, Elizabeth Carr, was born in late 1981. Since then, many successful *in vitro* fertilizations and pregnancies have been reported, although the success rate was initially fairly low.

120

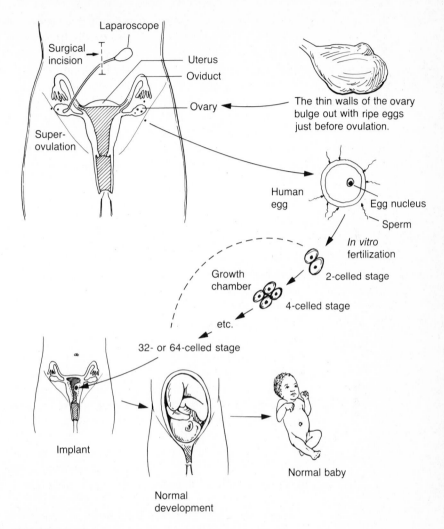

FIGURE 6–10
The Steptoe and Edwards procedure of *in vitro* fertilization. (From Kieffer, BIOETHICS: A TEXTBOOK OF ISSUES, © 1979. Addison-Wesley, Reading, MA. Figure 3.4. Reprinted with permission.)

In vitro fertilization has also been accomplished with many animals. The success rate for embryo implantation in cows has been about 50 percent. In humans, it was only 11 percent as of 1980, and the overall probability of actually giving birth after *in vitro* fertilization and

121

implantation was only 3 percent, but the success rate is increasing rapidly. Many of the failures are due to spontaneous abortion, and some of the aborted fetuses have had chromosomal abnormalities. However, the successfully born infants have seemed completely normal thus far.

No doubt, *in vitro* fertilization involves a large wastage of embryos. This has raised some ethical questions. Do the embryos represent human life and is the experimenter or physician morally and legally responsible for them? Such questions can only be answered on the basis of society's definition of human life. We must keep in mind that natural reproduction also includes a high rate of spontaneous abortion and fetal wastage. Furthermore, *in vitro* fertilization offers the chance to be parents to thousands of people who could not otherwise have children. Louise Brown's mother had suffered nine years of infertility before turning to Drs. Steptoe and Edwards. Because her Fallopian tubes were blocked, fertilization was impossible. Other women—20,000 in the United Kingdom and several times that many in the United States—experience the same problem or lack Fallopian tubes altogether. Still others who have been sterilized (their Fallopian tubes have been cut and/or blocked) decide they want children.

New, reversible sterilization techniques are being developed, but a reliable *in vitro* fertilization technique might prompt more women to choose sterilization as a birth control method. This is one of the important arguments in favor of the process. Other positive arguments come from reproductive physiologists who want to use *in vitro* fertilization experiments to study the occurrence of chromosome abnormalities in humans, the effects of contraceptives, and the process of freezing embryos for later use.

Society can influence research directly, by choosing what work to fund. The forum for the choice is the political process, where infertile parents, religious groups, and other concerned parties make their wishes felt. As a result of political pressure, a moratorium was imposed on federally funded *in vitro* fertilization research in the United States in 1976. Private funding has been available, however, and by 1981 the United States had two clinics for this kind of work. By 1983, there were several more, 11 countries had clinics as well, and more than 120 babies had been born as a result of the technique. The moral, ethical, and social questions, however, remain with us. Will *in vitro* fertilization make possible the screening of embryos to eliminate those carrying hereditary disorders, or to select the sex of the offspring? Will it become possible to grow a fetus *in vitro* from conception to birth? It is already possible for parents to hire surrogate mothers—wombs for rent—to bear their children. These questions will become more urgent as soon as *in vitro* fertilization is an efficient, economical, routine procedure.

Embryo Transfer and Surrogate Mothers

Just how far can we separate sexual activity from procreation by manipulating human reproduction? Both artificial insemination and *in vitro* fertilization can promote the use of **surrogate mothers**. In the first case, the surrogate mother is inseminated with sperm from the husband of an infertile couple; she also serves as the "egg donor." In the second case, the *in vitro* fertilized egg is implanted in the uterus not of the egg's source, but of a third party, who serves as an incubator for the fetus. Transfer of an embryo from the genetic mother's uterus to that of a host mother is another technique that has already been accomplished in human beings.

Embryo transfers in animals are now routine and impressively successful. With cattle, they are a big business, with about 17,000 transfers in 1979 in North America alone. They serve to increase the rate of reproduction of valuable cows, bypass infertility problems, transport embryos to other sites (frozen or in other animals such as rabbits), make genetic tests, and even increase the reproduction of endangered species.

The main reason for the use of human surrogate mothers will probably be the prevalence of married women who cannot bear children for one reason or another. Other factors could be the high demand and long waiting lists for adoptees in the United States, and the wish of some women not to interrupt careers with pregnancies. In addition, cryopreservation of eggs and/or embryos may allow careful planning and timing of births. A young couple can place a future baby in cold storage and rear it later in life. Cryopreservation of embryos has been accomplished, first in Australia and soon after in the United States. If the technique becomes cost effective, it could even lead to an embryo adoption system. Again, the need for ethical guidelines is evident.

It seems certain that surrogate mothering will be one of the future's novel professions. Although one motive is surely money, some women have expressed compassion for those who cannot have children. Still others say they simply enjoy being pregnant. But before these motivations can be fully explored, society must make a number of decisions and regulations. How often and under what circumstances might a surrogate mother decide to keep the child? Could an artificially inseminated "egg donor" sue a sperm donor for child support? Could the ge-

123

netic parents refuse to accept a hosted child who had a hereditary or other defect? Are the sperm donor, egg donor, hired surrogate, and the man and woman who pay the expenses to obtain a child all equally "parents"? What impact might mechanical wombs (**extracorporeal gestation**) have? What about the use of a chimpanzee as a surrogate mother? It just might be possible!

Parthenogenesis

Parthenogenesis is "virgin reproduction." It occurs when an unfertilized egg (or other cell) develops into a new individual. Since the cell usually has to be diploid if it is to develop, it cannot be the product of a normal meiosis. It must be produced either by a faulty meiosis or a mitosis.

Parthenogenesis occurs naturally in some lower animals, but it is quite rare in **vertebrates** (animals with backbones). It has been noted in certain lizards and in one variety of turkeys. However, researchers have found ways to induce it in the laboratory in several species.

Can parthenogenesis occur in humans? Thorough interviews, blood tests, and biochemical tests have disproved most claims of virgin birth, for a virgin's child must match the mother in every way. Certainly, no virgin can bear a son. The few remaining cases are still dubious because not all possible tests could be done. Good genetic matches between mother and daughter can arise from heredity and coincidence. Nevertheless, some scientists do believe human parthenogenesis is possible, though it must be rare.

Cloning and the Concept of Totipotency

Cloning is the growth of one or more individuals each from a single cell taken from an existing individual. Cloned progeny are identical to each other *and* to their "parent," for all have the same genotype. The only differences arise from mutation and other phenomena that can prevent genetic homogeneity.

Cloning is theoretically possible because every diploid cell of an organism has the same set of genes. However, during development of the organism from a fertilized egg to an adult, it acquires a number of different tissues by the process of **differentiation**. Groups of cells diverge from each other in structure and function, and though they keep all their genes, some become inactive. Normally, only the zygote is **totipotent**. That is, only in the zygote are all the appropriate genes available for use in normal development. But under the right conditions, inactive genes in nonzygotic cells can be reactivated; that is, totipotency

124

can be restored. This happens naturally in vegetative reproduction in plants, and to a lesser extent, when a salamander regenerates a lost leg or tail. In the laboratory, it has been demonstrated for single cells of carrots and other plants. It has also been demonstrated with frogs, toads, fish, and mice, and progress has been made with cattle.

Robert Briggs and Thomas J. King first cloned frogs in 1952. Because they believed that something in a fertilized egg's cytoplasm encouraged totipotency, they replaced a fertilized egg's nucleus with a nucleus removed from an embryo's cell. The enucleated egg then developed into a larva, a tadpole, and a normal adult frog. This process is outlined in Figure 6-11. Abnormal growth resulted if the transplanted nucleus came from a cell of an adult frog; by adulthood, apparently, too many genes have been permanently inactivated.

In 1962, J. B. Gurdon used ultraviolet light to inactivate the nuclei of unfertilized toad eggs. He then transplanted diploid nuclei from tadpole intestinal cells into the eggs. Some of these manipulated eggs then developed into adult toads, as indicated in Figure 6-12. In 1981, Karl Illmensee and Peter Hoppe successfully cloned mice in a similar way, using nuclei from the inner part of early embryos. The work with cattle aims to split eight-celled embryos into two- or four-celled embryos in order to accelerate reproduction with the aid of surrogate mothers.

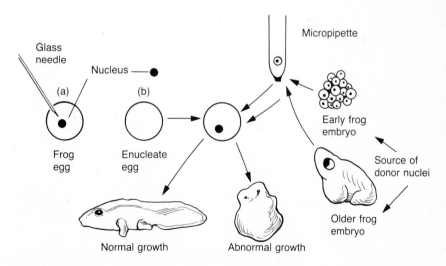

FIGURE 6-11
The basic format of the Briggs and King cloning experiments with frogs. (From Kieffer, BIOETHICS: A TEXTBOOK OF ISSUES, © 1979. Addison-Wesley, Reading, MA. Figure 3.1. Reprinted with permission.)

125

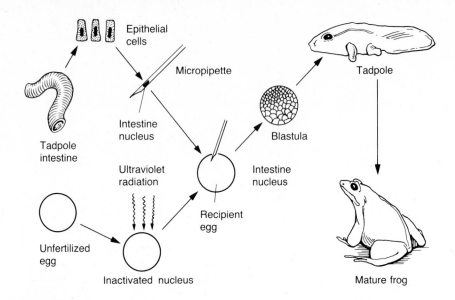

Epithelial cells

Micropipette

Tadpole

Intestine nucleus

Tadpole intestine

Blastula

Ultraviolet radiation

Intestine nucleus

Recipient egg

Unfertilized egg

Inactivated nucleus

Mature frog

FIGURE 6–12

The basic format of the Gurdon cloning experiments with toads. (From Kieffer, BIOETHICS: A TEXTBOOK OF ISSUES, © 1979. Addison-Wesley, Reading, MA. Figure 3.2. Reprinted with permission.)

Will it ever be possible to clone humans? In 1978, journalist David Rorvik claimed it had been done in his highly controversial book, *In His Image: The Cloning of Man*. Cloning researchers have called it a hoax, and even the publisher of the book now believes the story to be untrue. Human cell cloning, however, is achieved in cell biology laboratories. In 1951, George Gey succeeded in growing in glass bottles cells from a cancer of the cervix found in a 31-year-old Baltimore woman. Her name was Henrietta Lacks. The cells became known as **HeLa cells**, and they are still growing in laboratories around the world. Invaluable in many research studies, they grow so rapidly and aggressively that they will actually take over cultures of other human cells. They are relatively easy to grow if given the right nutrients, and they never die out.

Growing cells such as HeLa cells in culture is actually the cloning of cell masses. No one has been able to prompt such cells to grow into human beings. But every diploid human cell has the full complement of genes, and it may be possible to restore totipotency to a human cell. Human cloning might even resemble the technique developed for frogs

126

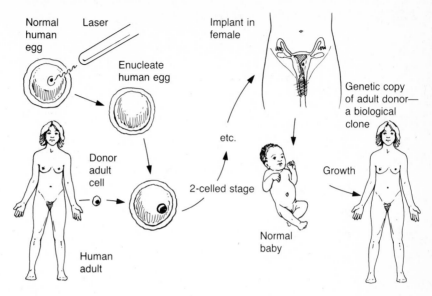

FIGURE 6–13

A theoretical method for cloning humans. (From Kieffer, BIOETHICS: A TEXT-BOOK OF ISSUES, © 1979. Addison-Wesley, Reading, MA. Figure 3.6. Reprinted with permission.)

and toads, but the process outlined in Figure 6-13 remains in the future.

Human cloning represents a much greater technological intervention in reproduction than artificial insemination or *in vitro* fertilization. It upsets many people and provokes many ethical questions. People worry about how to define ''parent'' for clones, about whether disposing of defective or unwanted clone embryos is murder, and about the objectives of the technology. Who should or should not be cloned? Might the process give tyrants new power? Might it depersonalize reproduction? Might it make males redundant? Many people believe there is no good reason for cloning humans; certainly, we should not do it simply because we have the capability.

On the other hand, the study of cell cloning and differentiation can provide valuable insights into cellular physiology, growth, and development. It may improve our ability to cope with cancer, other cellular diseases, wound healing, and the regeneration of lost organs and limbs. It may even aid our understanding of aging. And if we ever do grow adult clones, we will obviously learn a great deal about the interplay of heredity and environment in the developing phenotype.

127

POPULATION AND FERTILITY

The world's more than four and a half billion people increase by 250,000 every day, or 90 million per year. This population explosion, graphically presented in Figure 6-14, may not end until the world holds 10 to 16 billion people. Such numbers of inhabitants must inevitably affect the world economy, health, education, food supply, urbanization, and other aspects of the quality of life.

Many scientists and nonscientists alike believe that the only way to avoid disaster is to stop population growth. The most logical method is **contraception**, which controls fertility either by preventing the union of sperm and egg or by interfering with the result of their union.

The most effective contraceptive method, of course, is abstinence. Other preventive methods include the rhythm method, condoms, diaphragms, cervical caps, spermicides, penis withdrawal, contraceptive pills, and sterilization. Interference methods include elective abortion and **intrauterine devices** (IUDs), which seem to prevent the implantation of fertilized eggs in the uterine lining. All current methods of birth control offer some health risks or side effects and some unreliability or inconvenience. Sterilization is irreversible (though progress is being made on ways to reverse some kinds of sterilization), but it is still the most popular choice in the United States today.

For females, sterilization usually means a **tubal ligation**, in which the Fallopian tubes are cut and tied to prevent eggs from reaching the uterus and sperm from reaching the eggs. For males, sterilization means a **vasectomy**, in which the vas deferens is cut and tied where it passes under the skin of the scrotum. Since sperm may survive in the

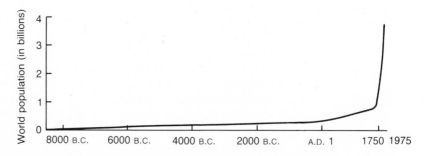

FIGURE 6–14
The human population between 8000 B.C. and 1975 A.D. (From D. L. Hartl. 1977. *Our Uncertain Heritage. Genetics and Human Diversity*. J. B. Lippincott Company, Philadelphia. Used with permission of the publisher.)

128

vas deferens for some time after their supply is cut off, the vasecto-mized male is not truly sterile until up to eight weeks after the operation.

Sterilization is the most effective contraceptive method (after ab-stinence). The least effective is the rhythm method, for even though ovulation occurs near the middle of the menstrual cycle and is indicated by changes in body temperature and in the viscosity of the cervical mu-cus, the timing can vary unpredictably. Also, there may be additional, **paracyclic ovulations** later in the cycle. Other methods have interme-diate effectivenesses, though their exact failure rates are still debated. The best of these may be the contraceptive pill, which supplies preg-nancy-mimicking hormones to suppress ovulation.

Researchers are still seeking the ideal method of reducing fertil-ity. They are considering anti-pregnancy vaccines, and male contracep-tive pills, for example. Their goal is a reliable, safe, and easily reversed method of contraception.

Even lacking an ideal method, many people are regulating their fertility. Unfortunately, many are not; and they will produce a large pro-portion of the next generation. Some geneticists worry that the people who regulate their own fertility are being selected against. There is little evidence, if any, that regulators and nonregulators differ in their genetic make-up; but if they do, the regulators' proportion of genes will even-tually decrease in the population, and the species' gene pool will change.

ELECTIVE ABORTION

Abortion is the removal of the developing embryo or fetus from the uterus. It is often used when contraception fails and is the leading method of birth control in some countries. Many abortions are per-formed because tests have shown that the fetus has some genetic dis-order. Many more are performed because there is a strong possibility of a disorder, because the pregnancy resulted from incest or rape, or be-cause the parents simply do not want the child.

Abortion upon demand is controversial. Religion, culture, and personal conviction fuel the debate. Some people will abort a fetus at any time for any reason, including convenience. Others will never abort a fetus for any reason whatsoever. People whose feelings lie between the extremes may agonize long over the decision of whether or not to abort.

The decision may be easiest when the fetus is likely to have a genetic disorder. The parents must then consider the disorder's probability and severity. For instance, two parents who are heterozygous for **sickle cell anemia** have one chance in four of producing a child with the disease:

$$\text{parents:} \quad AS \times AS$$

$$\text{offspring:} \quad \frac{1}{4} \ AA \ \text{normal}$$

$$\frac{1}{2} \ AS \ \text{sickle cell trait}$$

$$\frac{1}{4} \ SS \ \text{sickle cell anemia}$$

Sickle cell anemia is a severe disease marked by anemia, jaundice, poor blood flow, pain, and weakness. It is due to a change of just one amino acid in the hemoglobin molecule. However, in this case the parents may well choose not to abort. After all, they have three chances in four of having a healthy child without the sickle cell disease. If they want more complete information on which to base their decision, they might resort to fetal diagnosis. In this case, it is still a difficult procedure, since fetal erythrocytes need to be removed for the sickle cell anemia test.

Next, analyze a situation in which one of the parents is concerned about **Huntington's disease**, a severe affliction of the nervous system. This disorder is dominantly inherited, and it does not show up in most individuals until about age 40, although the onset can be quite variable. Persons heterozygous and affected, therefore, do not usually know that they are carrying the dominant gene until well into their reproductive years. The genotype of parents and progeny are as follows:

$$\text{parents:} \quad Hh \times hh$$

$$\text{progeny:} \quad \frac{1}{2} \ Hh \ \text{Huntington's disease}$$

$$\frac{1}{2} \ hh \ \text{normal}$$

The progeny of an affected parent would face a complicated decision about whether or not to abort a fetus. Such prospective parents could not be sure whether their genotype is *Hh* or *hh*. If one of the parents expressed Huntington's disease, the probability of each of their progeny being *Hh* is a rather high 50 percent. On the other hand, the onset of the disorder is usually later in life, allowing affected persons to live many normal years. Persons involved in this dilemma necessarily have to weigh all of these factors.

130

In 1973, the United States Supreme Court decided that it is a woman's constitutional right to have an abortion performed. In 1977, the Court upheld this right, but ruled that the state is under no obligation to pay for abortions. In 1981, a United States Senate Subcommittee held hearings on the possibility of bills to outlaw abortion and concluded that a human being exists from the time of conception. During these hearings, a number of scientists were asked, "When does human life begin?" The answers were varied, though many scientists believe that the question is meaningless and unanswerable. Certainly, a human zygote, embryo, and fetus are alive and human, as are the ovum and spermatozoon. A better question would have been, "At what point should we regard this product of human development as an individual or person, defined in the same way as other human lives and accorded the same rights?" This question goes beyond science. Its answer must come from society as a whole.

SUMMARY

Almost all organisms have some means of sexual reproduction, which is essentially the union of genetic material from two different sources. Sexual reproduction increases genetic diversity through the segregation of chromosomes during meiosis and the exchange of chromosome segments in crossing over. As a result, it aids the survival of the species.

By means of oogenesis, the human female's ovaries produce ova (usually one at a time) on a cyclic basis. Through spermatogenesis, the male's testes produce large numbers of spermatozoa. During the mating act, the male releases spermatozoa into the female's vagina, from which they move to meet the egg in the Fallopian tube. If the egg is fertilized, it begins cell division (cleavage) and moves to the uterus. The small embryo implants in the uterine wall and develops into a fetus that is nourished through the placenta.

Artificial insemination is the manual deposit of spermatozoa into the vagina. The technique is now common as either AID (by donor) or AIH (by husband). AID is often chosen because of the husband's sterility or concerns about genetic defects. Cryopreservation, placing spermatozoa in a deep freeze for later use, is sometimes used in conjunction with artificial insemination. Many people are concerned about the possibilities of recessive mutations, damage during cryopreservation, inbreeding, psychological problems, and the use of artificial insemination by single women.

131

In vitro fertilization (IVF) was accomplished in humans in 1978, and the technique has generated much interest, excitement, and publicity. Ova removed from the female are fertilized by a male's spermatozoa in a glass vessel. The resulting embryo is implanted into the woman's uterus. So far, the use of IVF has been restricted to women unable to conceive because of distorted, blocked, or absent Fallopian tubes. Research into this area now focuses on fertilization and early development.

Embryo transfers and surrogate motherhood are related aspects of manipulation and control of human reproduction. Both are widely used in livestock breeding, as is cryopreservation of embryos. These techniques are now being developed for use with humans. People have many reasons for wanting to use—or be—a surrogate mother. Another real possibility is the use of incubating machines for out-of-body (extracorporeal) gestation.

Parthenogenesis is the birth of an offspring without any fusion of gametes. The phenomenon occurs in some lower forms of life, and even in a few vertebrate species. However, its occurrence has been impossible to prove in humans.

The possibility of cloning humans has raised much concern. The potential of a somatic cell, or a somatic nucleus placed in an enucleated ovum, to develop into a differentiated organism is called totipotency. It has been demonstrated in some plants and animals. Certain human cells can be grown in cell culture, but they do not develop into an embryo. The famous HeLa cell line appears to be immortal in culture. No human has been cloned, and there is a tremendous amount of opposition to the idea. Scientists who work with cloning point out that their primary objectives are to learn more about cancer, other cellular diseases, wound healing, regeneration, development, and aging.

The rising world population growth has made birth control an urgent priority. Researchers continue to seek better and more convenient ways of controlling fertility. Many different methods are already available. When they fail, many people opt for elective abortion, even though other individuals and organized groups oppose it, often armed with legislative proposals to ban it.

DISCUSSION QUESTIONS

1. Give several reasons why sexual reproduction is a widespread phenomenon in the biological world.

2. Point out the differences between the structures of the spermatozoon and the ovum, and consider how these differences might relate to function.

3. Oogenesis and spermatogenesis differ markedly. In the first, a meiotic event yields a single ovum once a month. In the second, which is a continuous process, each meiotic event yields four functional spermatozoa. Explain why none of these male/female differences affect the genetic probabilities.

4. Most plant, animal, and bacterial species have two sexes, usually called male and female. How would you define *male* and *female*? How would sexual reproduction be different if there were three sexes?

5. You have heard, "Mary got all her genes from her father." Is this possible? Could Mary get all her genes from her mother? From the standpoint of heredity, what might prompt such a statement? Use your knowledge of genetics to discuss any factors that might relate to this situation.

6. What regulations should be imposed on artificial insemination in general, and on sperm banks specifically?

7. A nationwide Gallup Poll showed that 60 percent of the people were in favor of *in vitro* fertilization to relieve infertility, 27 percent were against it, and 13 percent had no opinion. Do you think the results would differ for men and women? College students and high school students? Westerners and southerners? People under 30 years old and people over 30 years old? Explain your answers.

8. What are the advantages of cloning for the cattle industry?

9. Discuss the advantages (if any) and disadvantages (if any) of a national army consisting of two million cloned individuals of the same genotype.

10. Explain why human cloning may actually work against long-term human survival as a species.

11. Suppose the unlikely event that someone has brought HeLa cells into the form of an embryo and placed the embryo in a woman's uterus. Suppose further that the embryo developed into a normal fetus and newborn child. Would we have Henrietta Lacks back with us? Discuss your thinking from all pertinent standpoints.

12. People have speculated that human cloning procedures might some day aid human organ transplantation. How might this be? Discuss the possible advantages of such a connection, and add any other relevant considerations.

13. How might the techniques of *in vitro* fertilization, embryo transfer, artificial insemination, and cloning be combined in various ways?

133

PROBLEMS

1. Approximately how many ripened ova does a human female produce in a normal lifetime?

2. Many genes are located on each chromosome. Think of three specific heterozygous gene pairs located on a particular pair of homologous chromosomes.
    a. How many different gene combinations could occur in the gametes if crossing over were not possible?
    b. How many different combinations could occur in the gametes with crossing over?

3. To demonstrate the vast diversity among humans, a genetics instructor asks 25 students to list their phenotypes for the hereditary traits:
    a. male or female;
    b. can or cannot taste phenylthiocarbamide (PTC);
    c. free vs. attached ear lobe;
    d. A, B, AB, or O blood type;
    e. Rh+ or Rh− blood type;
    f. M, N, or MN blood type; and
    g. normal or albino pigmentation.

    Considering only these seven traits, how many different phenotypes are possible? Assess the chances that any two students in the class will have the same phenotype.

4. Most of the alternative traits above do not occur with equal probabilities. For example, 70 percent of the population are PTC tasters and 30 percent are nontasters; approximately 85 percent are Rh+ and 15 percent are Rh−; and albinism occurs in only 1 person in 20,000. How do these data affect your calculation of the odds that two students will have the same phenotype?

5. How do you react to a woman's claim that her infant son was born by parthenogenesis? How do you react to a woman's claim that her daughter was born by parthenogenesis, when 12 tests show the mother and child to be identical for 12 biochemical traits?

6. In cattle, embryo transfer can be used to tell whether a particular outstanding cow carries a detrimental recessive gene. Using superovulation, breeders can conduct a larger number of genetic tests in less time. Assume that a bull homozygous for the recessive gene in question is used to fertilize superovulated eggs, producing eight embryos that are transferred to eight surrogate mother cows. All eight resulting calves are normal. What, then, is the probability that the outstanding cow is *not* a carrier? How long would it take to gather the same data using normal reproduction, if each calf takes a year to obtain?

7. The Jackson Laboratory in Bar Harbor, Maine, has produced some "half-cloned" mice by removing a fertilized egg from a female mouse, treating it with chemicals to destroy the genetic material from the male but not that from the female, treating the egg to double the female's genetic material, reimplanting the now-diploid egg, and allowing it to develop into an adult mouse.

   a. Does such a mouse have all the genetic make-up of the female parent? Explain.

   b. If more than one mouse were produced from a female in this way, would they be identical to each other? Why or why not?

   c. Are the results true cloning? Explain.

   d. A big problem with this technique could arise if the parent were carrying any lethal recessives. Explain this potential problem, using a diagram if necessary.

8. The effectiveness of birth control methods is sometimes given as the number of pregnancies that occur per 100 woman fertile years. (One pregnancy per 100 woman years equals one pregnancy per 100 women for one year.) For example, data for the contraceptive pill show less than one pregnancy per 100 woman years. Assume that a particular birth control method is practiced by 400 couples for 10 years, and these couples average 2 children apiece over this period. What is the pregnancy rate per 100 woman years? Your answer, if correct, is the actual rate for the use of spermicides for birth control. Compare this result with "no attempt to control births," which has a rate of about 40 pregnancies per 100 woman years, and with the rhythm method, which has a rate of about 24 pregnancies per 100 woman years.

REFERENCES

Briggs, R., and T. J. King. 1952. Transplantation of living nuclei from blastula cells into enucleated frogs' eggs. *Proceedings of the National Academy of Sciences* 38:455–463.

Briggs R., and T. J. King. 1953. Factors affecting the transplantability of nuclei of frog embryonic cells. *Journal of Experimental Zoology* 122:485–506.

Briggs, R., and T. J. King. 1959. Nuclear cytoplasmic interactions in eggs and embryos. In *The cell*, ed. J. Brachet and A. E. Mirsky, Vol. 1:537–617. New York: Academic Press Inc.

Curie-Cohen, M. 1980. The frequency of consanguineous matings due to multiple use of donors in artificial insemination. *American Journal of Human Genetics* 32:589–600.

Fraser, F. C., and A. Forse. 1981. On genetic screening of donors for artificial insemination. *American Journal of Medical Genetics* 10:399–405.

Gurdon, J. B. 1962. The developmental capacity of nuclei taken from intestinal cells of feeding tadpoles. *Journal of Embryology and Experimental Morphology* 10:622–640.

Hartl, D. L. 1977. *Our uncertain heritage: Genetics and human diversity*. Philadelphia: J. B. Lippincott Co.

Hole, J. W., Jr. 1978. *Human anatomy and physiology*. Dubuque, IA: Wm. C. Brown Group.

Huxley, A. 1946. *Brave new world*. New York: Harper and Brothers.

Illmensee, K., and P. C. Hoppe. 1981. Nuclear transplantation in *Mus musculus*: Developmental potential of nuclei from preimplantation embryos. *Cell* 23:9–18.

Kieffer, G. H. 1979. *Bioethics: A textbook of issues*. Reading, MA: Addison-Wesley Publishing Co. Inc.

Lopata, A. 1980. Successes and failures in human *in vitro* fertilization. *Nature* 288:642–643.

Rhodes, P. 1976. *Birth control*. Carolina Biology Reader 4. Burlington, NC: Carolina Biological Supply Co.

Rorvik, D. M. 1978. *In his image: The cloning of man*. Philadelphia: J. B. Lippincott Co.

Seidel, G. E., Jr. 1981. Superovulation and embryo transfer in cattle. *Science* 211:351–357.

Steptoe, P. C., and R. G. Edwards. 1978. Birth after the reimplantation of a human embryo. *Lancet* ii:366.

Steward, F. C., M. O. Mapes, and K. Mears. 1958. Growth and organized development of cultured cells. II. Organization in cultures grown from freely suspended cells. *American Journal of Botany* 45:705–708.

Trouson, A. O., J. F. Leeton, C. Wood, J. Webb, and J. Wood. 1981. Pregnancies in humans by fertilization *in vitro* and embryo transfer in the controlled ovulatory cycle. *Science* 212:681–682.

7

Mendelian Genetic Disorders

Genetic disorders are the leading cause of infant mortality in the United States. Between 3 and 5 percent of all live births have some kind of major genetic defect. The figure reaches 14 percent if we include the multitude of minor defects believed to be hereditary. Thus, 20 to 30 million Americans have a minor or major genetic defect, and approximately 100,000 persons with genetic defects are born in the United States each year. In about 20,000 people, the defect is serious. It has been estimated that 200,000 Americans have **muscular dystrophy** alone. Mental retardation affects 3 percent of the population of the United States, and 80 percent of the cases may involve a genetic component.

More than 3000 different genetic disorders and defects have already been identified, but only about 150 are well described medically, biochemically, and genetically. Many are **Mendelian** variants, caused by a single gene or gene pair, depending upon dominance and recessiveness. Because new disorders are continually being recognized, the few we know may be only the "tip of the iceberg."

Most genetic disorders are extremely rare, but because there are so many different ones, their overall frequency is rather high. This overall frequency depends, in part, on the concept of **genetic load**. One aspect of genetic load is based on the number of unexpressed deleterious genes maintained in a population or gene pool. Genetic load is measured as the average number of unexpressed deleterious genes per individual; in humans, the genetic load is estimated to be three to eight very deleterious alleles per person. Because most such alleles are recessive and rare, most of us are heterozygous for them. Therefore, most of us have a normal phenotype. Our children suffer a genetic defect only when they receive the same recessive allele from both parents. Genetic defects are as rare as they are because it is unlikely that any two randomly mating parents would have the same alleles among their three to eight deleterious alleles. Furthermore, even if both parents are heterozygous for the same deleterious allele, there is still only a 25 percent chance that their offspring will be homozygous for it. However, that 3 to 5 percent chance of a serious genetic defect for every child remains, no matter how healthy the parents appear to be.

The general public is interested in genetic health, but few lay people understand the role of genetics in disease. In fact, many medical curricula in the United States still do not offer training in genetics, in spite of the fact that the first course in medical genetics appeared in 1932. In a basic sense, all disorders result from an incompatibility between a gentoype and its environment. Overall, much medical attention centers upon infectious diseases, heart ailments, and cancer, but

138

human genetics is also making a significant impact there and will certainly become important in preventive medicine.

Genetic diseases vary in severity but many are physically and mentally debilitating. Since affected persons often cannot care for themselves, the social and economic costs can be enormous. The financial impact strikes family, neighborhood, community, and society in the form of rehabilitation services, medical insurance, special institutions, and taxes. The psychological impact affects mental and emotional health, relationships between husband and wife, biomedical ethics, and other complex issues. Coming to grips with genetic disorders is a major responsibility of society.

EPIDEMIOLOGY AND PEDIGREES

Epidemiology is the study of diseases, their distributions, causes, and control. More specialized, **genetic epidemiology** is concerned with the distribution, inherited causes, and control of genetic disorders in populations. Genetic defects can be caused by small changes—**mutations**—in genes, or by gross alterations in chromosomes. The chromosome variations can be changes in number or structure. Chromosome numbers, in turn, can be abnormal because of the presence of one or more additional chromosomes, the absence of one or more chromosomes, or the presence of one or more additional *sets* of chromosomes. The basic set of chromosomes in an organism is its **genome**.

Genetic defects resulting from mutations may be controlled by one gene or gene pair (**Mendelian model**) or by many genes or gene pairs (**multigenic model**). The Mendelian model, which accounts for about one fourth of all known genetic diseases, often produces recognizable patterns of inheritance in families. If one pair of genes affects several different phenotypic traits, the inheritance is called **pleiotropic**. Figure 7-1 diagrams these different models.

FIGURE 7–1
Genetic models.

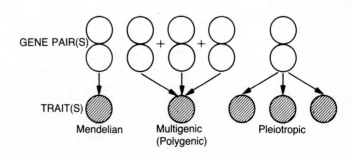

GENE PAIR(S)

TRAIT(S)

Mendelian   Multigenic   Pleiotropic
                 (Polygenic)

139

TABLE 7–1
Modes of inheritance for deleterious
traits in humans, and their incidences

Mode of Inheritance	Incidence (%)
Recessive gene pair	0.20
Dominant gene	0.08
Sex-linked (dominant and recessive)	0.05
Chromosome abnormalities	0.54
Multiple genes	2.00
Other congenital malformations (assuming half of them to have a genetic cause)	1.20
Total	4.07%

Many genetic disorders show **genetic heterogeneity**. That is, the genetic evidence indicates that the disorder may be produced by two or more different mechanisms. For example, **diabetes mellitus** may have more than one mode of inheritance.

Mendelian disorders resulting from mutations can be dominant or recessive. The responsible genes can reside on chromosomes involved in sex determination (**sex chromosomes**) or on chromosomes not directly involved in sex determination (**autosomes**). Genetic diseases can be **congenital** (expressed at birth), or they can occur at any time after birth, even late in adulthood. Table 7-1 summarizes the incidence of various modes of inheritance of deleterious traits in humans.

One very useful way to recognize modes of inheritance—especially Mendelian transmission—in humans is through the use of **pedigrees**. A pedigree is a diagram representing one person's ancestors and relatives. Figure 7-2 lists symbols commonly used in constructing pedigrees. Each generation in a pedigree is given a Roman numeral in sequence, and each individual within a generation is assigned an Arabic number in sequence. An example of a pedigree for an autosomal recessive trait (albinism) appears in Figure 7-3.

To assemble a pedigree, one starts with one person, the **index case**, also called the **proband**, or **propositus**. Through interviews and records, one then identifies siblings, parents, grandparents, and so on, and learns which of these people display the index case's particular trait or hereditary disorder. Once the pedigree has been constructed, it can be analyzed for patterns that reveal the various modes of inheritance. In **autosomal recessive inheritance**, the trait usually appears in the offspring of parents who do not show the trait, and the children of affected persons do not usually show the trait (especially if the trait is rare in the population). When many affected families are studied, one in

140

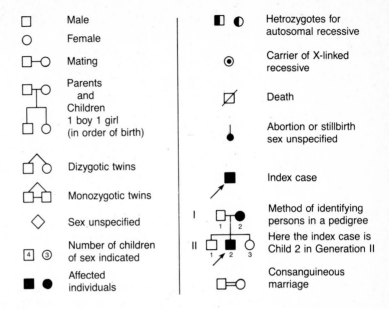

FIGURE 7–2
Symbols commonly used in constructing pedigree charts. (From J. S. Thompson and M. W. Thompson. 1980. *Genetics in Medicine*. W. B. Saunders Co., Philadelphia. Used with the permission of the publisher.)

four siblings of the index case will show the trait on the average, and males and females are affected equally often.

Autosomal dominant inheritance shows a different pattern. As one works back from the index case, the trait appears in every generation, in half the children of every affected person (on the average). It does not usually appear in the children of unaffected persons. The trait affects males and females in equal proportions, as shown in Figure 7-4.

With **X-linked recessive inheritance**, the trait is usually much more common in males than in females; the trait is passed from an

FIGURE 7–3
A pedigree of an autosomal recessive trait, albinism. (Reprinted with permission of Macmillan Publishing Company from *Genetics* by Monroe W. Strickberger. Copyright © 1968 by Monroe W. Strickberger.)

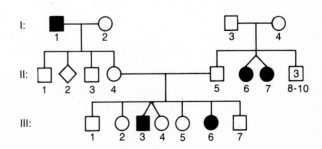

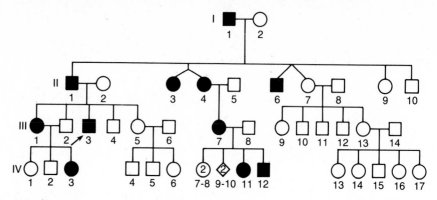

FIGURE 7–4

A pedigree showing autosomal dominant inheritance. (Reprinted with permission of Macmillan Publishing Company from *Human Genetics* by Edward Novitski. Copyright © 1982 by Edward Novitski.)

affected male through all his unaffected daughters to half the daughters' sons. The trait is never directly transmitted from a father to his son; and it can be transmitted through carrier females. Figure 7-5 shows an example of this mode of inheritance. With **X-linked dominant inheritance**, an affected male transmits the trait to all his daughters, but to none of his sons. More females are generally affected than males. Affected females, if heterozygous, transmit the trait to half their male and

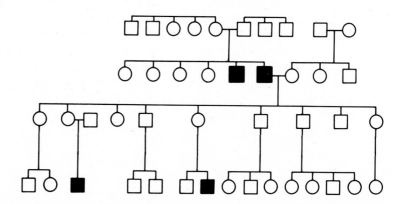

FIGURE 7–5

A pedigree showing the pattern of ichthyosis, an X-linked recessive trait. (From H. Harris. 1947. A pedigree of sex-linked ichthyosis vulgaris. *Annals of Eugenics*, 14:9. Used with the permission of Cambridge University Press.)

142

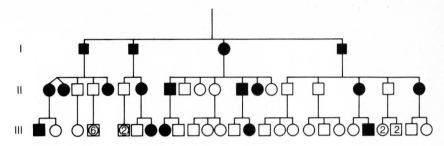

FIGURE 7–6

A pedigree pattern of vitamin D-resistant rickets, an X-linked dominant trait. The criterion for the phenotype diagrammed in the pedigree was a low serum phosphate. (After T. F. Williams, R. W. Winters, and C. H. Burnett. 1960. *The Metabolic Basis of Inherited Disease*, J. B. Stanbury, ed. McGraw-Hill Book Co., New York.)

female children, as shown in Figure 7-6, just as in autosomal dominant inheritance.

When pedigrees are inconclusive because more than one mode of inheritance can be derived from the available data, researchers seek additional information on the affected family or on other families with the same trait.

SICKLE CELL ANEMIA

Sickle cell anemia is an autosomal recessive disorder. Its fundamental symptom is an abnormality of the **hemoglobin** molecule. Hemoglobin is the molecule found in red blood cells that allows them to carry oxygen. Normal hemoglobin (A) has been thoroughly studied, and the peculiarities of sickle cell hemoglobin (S) have drawn a great deal of attention. In fact, this was the first disorder in which the exact molecular defect causing the problem was delineated.

People with sickle cell anemia suffer hemolytic crises. Chronically anemic, they often have mild to severe jaundice, fever, pain in the abdomen and joints, low resistance to infection, and damage to the kidneys, spleen, and lungs. They also have a shorter than normal life expectancy, although modern medical care allows many victims to live beyond age 50. Worldwide, the incidence of the disorder is highest among blacks; it is also high in Greece and Italy. In the United States,

143

one in 500 to 600 blacks has sickle cell anemia. More than 2,000,000 Americans carry the sickle cell allele.

Persons with sickle cell anemia are homozygous for the defective allele (*SS*) and their red blood cells are distorted. Figure 7-7 shows the shapes of normal and sickle red blood cells. Sickle cells do not live as long as normal ones. Because the bone marrow cannot replace the dying red blood cells quickly enough, anemia results. In addition, the distorted red blood cells tend to block small capillaries throughout the body, causing many of the symptoms. Heterozygotes or carriers (*AS*) have one allele for normal hemoglobin and one for sickle cell hemoglobin, and both types of hemoglobin are found in their red blood cells. These individuals have **sickle cell trait**. Most are clinically quite normal, but the sickling phenomenon that distorts the red blood cells can be triggered by reducing the amount of oxygen in the blood. The heterozygotes are more resistant to malaria than *AA* individuals. This relationship between malaria and sickle cell trait may be the evolutionary reason why the trait has persisted in certain geographic areas where malaria is endemic.

The molecular basis of sickle cell anemia was reported in 1949. Researchers using **electrophoresis** discovered that normal and sickle cell hemoglobin had different mobilities in an electric field. Their results appear in Figure 5-5, page 100. The difference is due to different electric charges on the two molecules, which indicate that they do not have identical structures.

The hemoglobin molecule consists of two alpha and two beta polypeptide chains. The gene or gene cluster for the alpha polypeptide

FIGURE 7–7
Normal and sickled red blood cells. (Photo courtesy Carolina Biological Supply Company.)

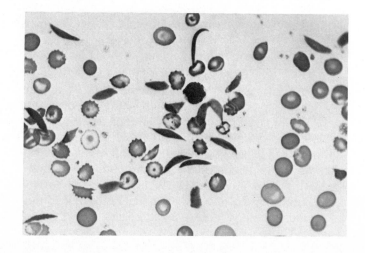

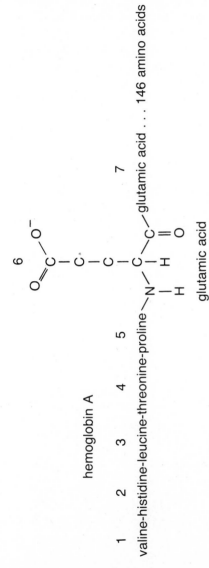

hemoglobin A

valine-histidine-leucine-threonine-proline

glutamic acid

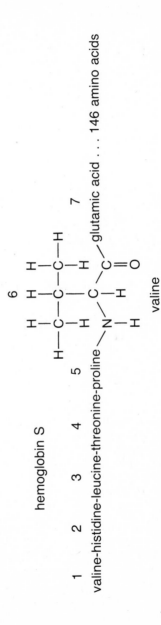

hemoglobin S

valine-histidine-leucine-threonine-proline

valine

FIGURE 7-8
The single difference between hemoglobin A and hemoglobin S. The amino acid in position 6 of A is glutamic acid and the amino acid in position 6 of S is valine.

145

is on chromosome 16; that for the beta polypeptide is on chromosome 11. Each alpha polypeptide consists of 141 amino acids; each beta consists of 146 amino acids. The hemoglobin molecule thus has 574 amino acids (2 × 141) + (2 × 146). In 1957, V. M. Ingram found that hemoglobins A and S differed only by one amino acid of the beta chain. Where hemoglobin A has a glutamic acid in the sixth place in the amino acid sequence, hemoglobin S has a valine. These differences are illustrated in Figure 7-8. Therefore, the difference between the two hemoglobins is just two amino acids out of 574 in the entire molecule.

How can such a slight molecular difference literally be the difference between life and death? Quite simply, the altered sequence changes the shape of the hemoglobin molecule from its normal configuration, shown in Figure 7-9, to one that forms cell-distorting crystals. The change in one amino acid is due to the replacement of a single nucleotide in the DNA of the gene. Sickle cell anemia provides a clear illustration of the one gene-one polypeptide concept. This is only one example of how a simple mutation can create a severe health problem.

So far, more than 275 variants—differing in one or more amino acids—from hemoglobin A have been discovered. Probably 80 percent of these variants are innocuous. The others cause sickle cell anemia, **thalassemia**, which is discussed later in this chapter, and other pathological conditions.

FIGURE 7–9

The hemoglobin molecule as it appears in three dimensions when viewed from the side. Two alpha and two beta subunits fit together to give the molecule its particular structure. The disks represent the heme groups. (Adapted from The hemoglobin molecule by M. F. Perutz. Copyright © 1964 by *Scientific American*, Inc. All rights reserved.)

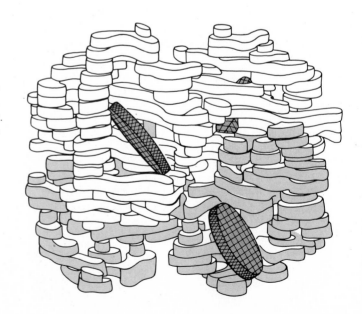

OTHER AUTOSOMAL RECESSIVE DISORDERS

Scientists understand only a few of the many genetic defects caused by autosomal recessive alleles. **Cystic fibrosis**, quite rare among blacks, affects one in 1600 Caucasians. The responsible allele is the most widespread recessive defect found among Americans, with more than 10,000,000 (one in 20) Americans believed to be carriers. Medically, only the phenotypic effects of the defect are well understood. The respiratory tract fills with thick mucus; the lungs can become severely infected; gastrointestinal problems occur; secondary infections are common; and much salt is lost in perspiration. Affected individuals require constant care, and death can strike at any time (usually before the affected person reaches age twenty). Nevertheless, some persons with cystic fibrosis have reproduced. The specific enzyme involved has not been identified, and an effective cure does not exist. Nor is there a reliable technique for identifying carriers.

One form of **albinism** is also an autosomal recessive defect. Albinism is not as severe as many other recessive disorders, but it still creates certain problems. Affected individuals, such as the one in Figure 7-10, lack the brown-black pigment **melanin** in skin, hair, and eyes. Albinos are extremely fair-skinned, and the iris of the eye is pinkish

FIGURE 7–10
An albino person among the natives of an island off the mainland of New Guinea. (Wide World Photos.)

147

since the blood vessels are not masked by melanin. Often albinism is accompanied by visual problems, extreme sensitivity to sunlight, and increased risk of skin cancer. The incidence of albinism is reportedly one in 10,000 to 20,000.

Phenylketonuria is routinely detected and treated in newborns. Although the condition is fairly rare, affecting about one in 14,000 newborns, the screening programs can be considered a significant medical advance. An individual with phenylketonuria appears in Figure 7-11.

FIGURE 7–11
A female with phenylketonuria. (Photo courtesy of Carl Larson.)

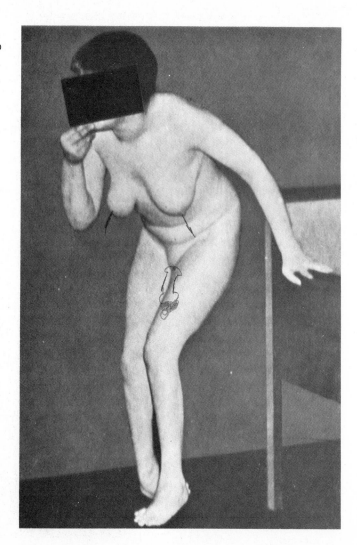

Phenylketonuria is due to a biochemical anomaly that causes mental retardation, which is often severe, and other neurological problems such as convulsions. A test performed shortly after birth reveals whether the blood contains an elevated level of the amino acid **phenylalanine**. A high level often indicates phenylketonuria, in which case a strict diet, low in phenylalanine, is immediately prescribed. The positive results of prompt treatment can be dramatic. Clear differences in the development of the mental capacities in treated and untreated cases can be easily seen. However, some problems are still associated with the disorder, as we will see when we discuss genetic screening and genetic counseling.

Tay-Sachs disease is rare except in certain Jewish populations. Among the Ashkenazic Jews of eastern Europe, its incidence is one in 3600, and one person in 30 is a carrier. In the general population, its incidence is only one in 360,000.

The effects of Tay-Sachs disease appear within one to six months after birth. These include mental retardation, with the possibilities of blindness, deafness, and paralysis. The child wastes away and usually dies by age three or four. There is no cure, and even with the best of care, the child dies by age five.

Tay-Sachs is a metabolic disorder in which lipids (**gangliosides**) accumulate in the tissues, especially in the central nervous system. One of the enzymes normally responsible for lipid breakdown is missing in the tissues of Tay-Sachs victims. The missing enzyme has been identified as **hexosaminidase-A**. Carriers can be detected by blood tests.

In **galactosemia**, which occurs in approximately one in 40,000 people, affected persons cannot convert the sugar galactose to glucose because of a deficiency of the enzyme **galactose-1-phosphate uridyl transferase** (GPUT). When it is not converted to glucose, galactose accumulates to toxic levels in body tissues. The brain and eyes are both affected, and the symptoms include mental retardation and cataracts, along with cirrhosis of the liver. A galactose-free diet is the only treatment.

Xeroderma pigmentosum is rare, but it has been intensively investigated because of its close association with cancers in skin areas repeatedly exposed to sunlight. Xeroderma pigmentosum is initially characterized by pigmentation abnormalities, dry and scaly skin, and possibly some scarring, as illustrated in Figure 7-12. A hypersensitivity to the ultraviolet component of sunlight causes the eruption of many skin tumors, largely because the victim's cells lack the ability of normal cells to repair the breaks that often occur in DNA, both spontaneously and in response to agents such as ultraviolet light.

149

FIGURE 7–12
Xeroderma pigmentosum. Note the
heavy freckling of the skin exposed to
light. (Photograph courtesy Guy's
Hospital, London, England.)

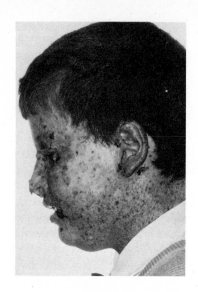

The thalassemias are characterized by impaired production or absence of either the alpha or the beta polypeptide subunit of hemoglobin. Because of the shortage or absence, not enough normal hemoglobin is synthesized, and the victim suffers severe anemia. Since beta thalassemic cells can make some beta subunits, but not enough, scientists think the beta thalassemias may result from problems with the regulation of the beta genes. The alpha thalassemias appear to be due to a reduced number of functional alpha genes.

Alkaptonuria is the genetic defect studied by A. E. Garrod very early in the development of medical genetics. It has sometimes been called the **black urine disease** because the urine of alkaptonurics tends to turn dark when exposed to light and air. This disease appears to be linked with the onset of arthritis later in life. Alkaptonuria is one of the classic metabolic disorders in which the cause is well defined biochemically. When the body metabolizes the amino acid tyrosine, it normally converts it to **homogentisic acid**, among other substances. Homogentisic acid is broken down by the enzyme **homogentisic acid oxidase**, but in persons with alkaptonuria this enzyme is absent. Consequently, homogentisic acid accumulates, and much of it ends up in body tissues and urine. One of the phenotypic indications of alkaptonuria is presented in Figure 7-13.

Some of the known metabolic pathways and genetic blocks that result in a few of the genetic diseases discussed so far are shown in Figure 7-14. **Hartnup's disease** causes a red, scaly rash, neuromuscu-

150

FIGURE 7–13

The ear of a person who has alkaptonuria. Homogentisic acid has accumulated in the cartilages of the ear and turned dark where it was exposed to light. (From A. M. Winchester and T. R. Mertens. 1983. *Human Genetics*, p. 145. Charles E. Merrill Publishing Company, Columbus, OH.)

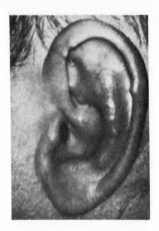

lar problems, and mild mental retardation in some affected persons. A deficiency of **alpha-1-antitrypsin** results in a high risk of cirrhosis of the liver, pulmonary emphysema, and enlargement of the right side of the heart. **Homocystinuria** causes mental retardation and dislocated lenses. **Hurler syndrome** occurs in late infancy, causing coarse features, stiff

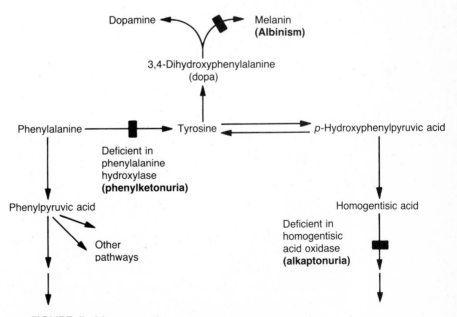

FIGURE 7–14

The metabolism of phenylalanine, showing the steps affected by the genetic defects, phenylketonuria, albinism, and alkaptonuria.

joints, mental retardation, hydrocephalus, and death by age ten. A few people have fructose intolerance, in which ordinary fructose sugar causes nausea, vomiting, excessive sweating, and tremors. Victims of **pentosuria** accumulate pentose sugars. The list of genetic disorders is very long, and it is constantly growing. Most of them are rare, but the overall frequency of genetic disorders is enough to concern most people.

SELECTED AUTOSOMAL DOMINANT DISORDERS

Disorders inherited as autosomal dominants occur even when only one copy of the allele is present in an individual. Since these genetic disorders are also rare, most victims are probably heterozygotes. As such, they have a 50 percent chance of transmitting the dominant trait to their offspring. Two unaffected parents can have an affected child only when one of their normal alleles mutates. Two unaffected parents could also have an affected child if one parent has the gene for the disorder but does not express the phenotype. This is the phenomenon of **penetrance**, and it complicates many theoretical discussions, as well as the practical interpretation of many pedigrees. An allele is not always expressed in the bearer's phenotype; to some extent, whether or not it is expressed depends upon the external environment. The expression of a penetrant allele may range from slight to severe. The *degree* of phenotypic expression of a particular genotype is referred to as **expressivity**, and it is also dependent upon the external environment.

If we assume that an autosomal dominant allele is expressed, we can easily determine mutation rates. We need only count the number of offspring having the dominant trait that are born to normal parents and the total number of offspring born during the same time in the same geographic area. Since a dominant trait requires a mutation in only one of the two involved gametes, the mutation rate is then: (affected offspring)$/2 \times$ (total offspring).

Neurofibromatosis is the "elephant man" disease that afflicts many Americans. Two forms of the genetic defect exist, both inherited as autosomal dominant disorders. The mutation rate for neurofibromatosis is 1×10^{-4} (one in 10,000), among the highest rates calculated for humans. However, there is much variability in the expressivity of the disorder. Symptoms may range from a few pale brown spots on the body to the severe disorder demonstrated in the Elephant Man. Comparable variation in phenotypic expression can be observed in many single-gene disorders.

Brachydactyly and **polydactyly** are two autosomal dominant disorders that affect the formation of fingers and toes. Brachydactyly means the affected person has very short fingers and/or toes, as illustrated in Figure 7-15. Several types of brachydactyly exist, with varying severities. In polydactyly the affected person has extra digits, as shown in Figure 7-16. The intensity of this malformation also varies greatly; extra digits may occur on both hands and both feet, or there may be only a slight abnormality on one hand.

Achondroplastic dwarfism occurs in one in 10,000 live births. A defect in bone and cartilage formation in the limbs results in a shortening of the arms and legs. A typical affected individual appears in Figure 7-17. Although the physical malformation is drastic, affected persons are mentally normal. Most affected persons are heterozygotes; homozygotes for the dominant allele are even more severely affected. Based on recent results, researchers believe that they have found the genetic defect responsible for this disorder. The gene under study directs the production of one type of **collagen**, a constituent of cartilage. Currently, researchers are working to develop a prenatal test to diagnose achondroplasia.

Some researchers believe that Abraham Lincoln may have had **Marfan syndrome**, an autosomal dominant condition marked by dispro-

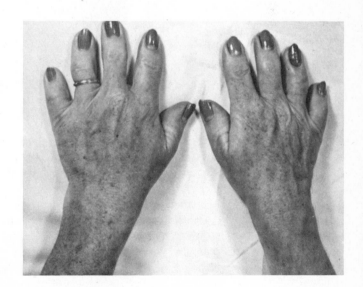

FIGURE 7–15
The hands of a person affected with the autosomal dominant disorder, brachydactyly. (From V. A. McKusick. 1969. *Human Genetics*. Prentice-Hall, Inc., Englewood Cliffs, NJ. Used with the permission of the publisher. Photo courtesy of Dr. Victor A. McKusick, Johns Hopkins Hospital.)

153

FIGURE 7–16
The hands and feet of persons affected with polydactyly, an autosomal dominant trait. (From A. M. Winchester and T. R. Mertens. 1983. *Human Genetics*, p. 236. Charles E. Merrill Publishing Company, Columbus, OH.)

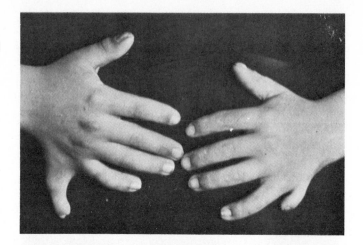

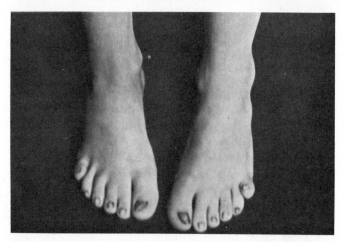

portionate elongation of the skeleton. Affected persons are generally tall, lean, long limbed, loose jointed, long fingered, and gaunt faced. A malformed breastbone and eye abnormalities may also appear. Progressive heart failure is the most serious aspect of the syndrome, and it is this malfunction that often causes death. Strangely enough, the second most common cause of death is suicide.

One of the better known dominant diseases is **Huntington's disease**, even though its incidence is only about one in 18,000 in the United States. The famous folk singer, Woody Guthrie, was stricken by this neurological disorder and eventually succumbed to it in 1967. Most

154

FIGURE 7–17
An achondroplastic boy. (Photo courtesy of Dr. J. Hall.)

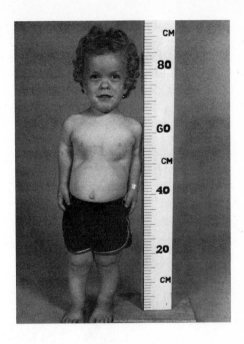

often, the disorder's onset is delayed until about 35 to 40 years of age, though it can strike before age 20 or after age 50. At the onset of the disease, most victims have already had children, so whenever a parent expresses Huntington's disease, the probability that his or her children will also be stricken is 50 percent. Woody Guthrie's children and all other such offspring face these formidable statistics. Their uncertainty cannot be relieved because there is no reliable test to determine whether a person is carrying the dominant allele before it is expressed. If such a test were available, affected individuals could avoid passing the disease to another generation by choosing not to have children.

Huntington's disease is a devastating degenerative disorder. The symptoms vary but commonly include loss of motor control, jerky movement, memory loss, and decrease in other mental capacities. Eventually, the victim may become prone to rages. The cellular basis for the disease seems to be a premature death of nerve cells, perhaps because of a defect in repair mechanisms for DNA.

Recent developments have been reported in which a segment of DNA that lies close to the Huntington's disease gene has been identified and located on chromosome 4. Recombinant DNA techniques were used to pinpoint this DNA. These powerful molecular techniques are described in Chapter 19. The presence of this particular DNA seg-

155

ment can be used as an indicator for the presence of Huntington's disease gene. This means that persons at risk for eventually having the disease might be able to have a pre-diagnostic test to indicate their chances of remaining healthy. The test is based upon the fact that a person would most likely inherit both DNA segments together because they are so closely linked. Researchers wonder whether, when the test becomes clinically available, persons at risk will want to know whether they have the Huntington's disease gene. Much remains to be learned about this disease before it can be effectively treated and/or reduced in the population.

Uninformed people often worry that dominant genetic disorders such as brachydactyly and Huntington's disease will spread throughout the human population. However, the Hardy-Weinberg principle illustrates that this will not happen with either dominant or recessive traits, *unless* a trait imparts a selective advantage to its bearers. Such an advantage would have to result in their having more offspring than unaffected people.

SEX-LINKED DISEASES

Sex linkage means that the gene for a particular trait is located on either the X or Y chromosome. Only males, of course, can have **Y-linked traits**. Human geneticists know of only a very few Y-linked traits. One is **hypertrichosis of the pinna**, or hairy ears; males with this trait have a pronounced growth of hair from the outer rim of the ear, shown in Figure 7-18. The gene for the **H-Y antigen**, which seems to be involved in initiating the development of the male primary sex characteristics, is also apparently located on the Y chromosome. Since Y-linked genes have no alleles on the X chromosome, they must be transmitted from male to male without skipping any generation. Figures 7-19 and 7-20 show several aspects of Y-linkage. This mode of transmission is called **holandric** inheritance.

Partial sex linkage can occur, at least theoretically, if a gene has alleles on both the X and Y chromosomes. No such cases have been documented in humans, but a small sector of the X chromosome does appear to be homologous to a small sector of the Y chromosome. It is at this point that the two chromosomes pair during meiosis. If such paired genes are ever found on the X and Y chromosomes, they will show inheritance patterns unlike either Y-linkage or X-linkage.

156

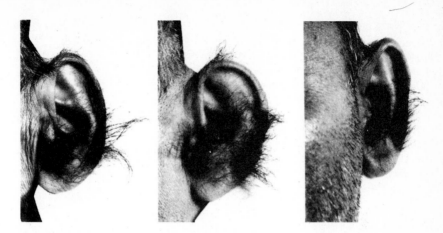

FIGURE 7–18
The hairy pinna trait, believed to be controlled by a gene located on the Y chromosome. (Reprinted from *The American Journal of Human Genetics*, 16:455–471, by C. Stern, W. R. Centerwall, and S. S. Sarker by permission of the University of Chicago Press. Copyright © 1964 by the University of Chicago Press.)

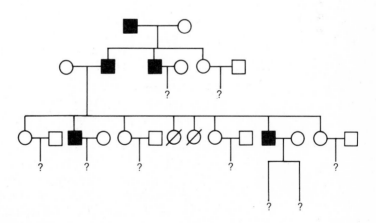

FIGURE 7–19
The pedigree of hypertrichosis of the pinna (hairy ears), a possible Y-linked trait. (Adapted from K. R. Dronamraju. 1960. *Journal of Genetics*, 57:230–243, with permission of the publisher.)

157

FIGURE 7–20

A pedigree depicting epidermolysis bullosa (a skin disease). Note the transmission of the trait from generation III to generation IV through the female III-9, ruling out complete Y-linkage. (Reprinted from *The American Journal of Human Genetics*, 9:147–166, by Curt Stern by permission of the University of Chicago Press. Copyright © 1957 by the University of Chicago Press.)

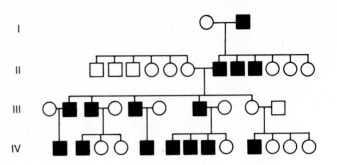

X-linkage is the usual form of sex linkage. Like autosomal genes, X-linked genes can have dominant and recessive alleles. Thus far, more than 150 genes have been assigned to the X chromosome. Because normal females have two X chromosomes, they can be homozygous or heterozygous for X-linked alleles. Normal males, however, can only be **hemizygous**, for they have but one X chromosome.

Vitamin D-resistant rickets is one of the few dominant disorders attributable to a gene on the X chromosome. It is a disorder of calcium metabolism that cannot be treated with vitamin D. It occurs in both males and females, but the condition is usually more extreme in males, who can never have a counterbalancing normal allele.

Hemophilia, one of the best known X-linked recessive disorders, is often called the bleeder's disease. Affected persons suffer uncontrollable bleeding, or at least very slow blood clotting. Certain drugs are now available to aid these people, but the cost of medical care is very high. Several forms of the disease exist; **hemophilia-A** and the rarer **hemophilia-B** are both X-linked. Although the severity of hemophilia varies, hemophiliac females (A^hA^h) are clinically affected much like hemophiliac males (A^hY). Hemophilia is known to be transmitted through the descendants of Queen Victoria of England, who was a carrier for the disorder. In fact, it was this "royal disease" and the pedigree shown in Figure 7-21 that allowed some of the first insights into sex linkage.

Lesch-Nyhan disease is a devastating X-linked recessive disorder. Affected children suffer from **cerebral palsy** (paralysis resulting from brain damage), **uric aciduria** (excess uric acid in the urine), and mental retardation. One of the most striking features of these children is the compulsive urge to self-mutilation. If not restrained, they will often gnaw their fingers away or gnaw their lips and mouths, producing hideous disfiguration. These children die relatively young as a result of

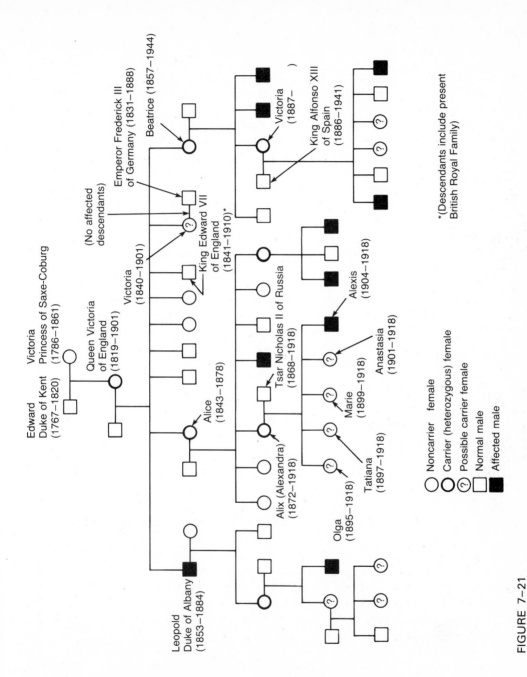

FIGURE 7-21

The pedigree showing the descendants of Queen Victoria of England. (From D. Hartl. 1977. *Our Uncertain Heritage. Genetics and Human Diversity.* J. B. Lippincott Co., Philadelphia. Used with the permission of the publisher.)

159

various physical problems. Because hemizygous males never live long enough to reproduce, there can be no homozygous females with the trait.

Duchenne muscular dystrophy is another X-linked disorder. It appears only in male children, usually by the age of six, and few if any affected individuals have ever reproduced. Other forms of muscular dystrophy exist, and in all forms the muscles literally waste away for reasons that are not known.

Among the other X-linked recessive disorders are **ocular albinism**; **Fabry's disease**, which appears as a skin rash, usually with pains in the extremities and death by age 40 because of renal failure; **glucose-6-phosphate dehydrogenase deficiency**, which causes hemolytic anemia as a result of the destruction of red blood cells; and **ichthyosis**, a skin condition in which the oil glands do not function properly, causing a scaly, fishlike skin surface. Part of the reason there are so many X-linked diseases is that X-linkage is relatively easy to detect. Affected males can transmit an allele on the X chromosome they contribute to their daughters, but never to their sons. Affected females always transmit the allele and the disorder to their sons. Since males are hemizygous for X-linked genes, recessive alleles on the X chromosome are always expressed. These criteria unique to the X chromosome have allowed us to assign many genes to it.

ADVANCES IN THE TREATMENT OF GENETIC DISEASES

The best way to reduce the frequency of any particular genetic abnormality in the population is to reduce the rate of reproduction by those individuals capable of having affected offspring. At the moment, there is no way to *cure* genetic disorders, although treatment is often possible. Genetic counselors can help people decide whether to have children, using family histories or pedigrees of parental disorders to calculate the probabilities of passing on the disorder. Prenatal diagnosis can give a mother the option of having an abortion, or it can alert parent and physician to the need for treatment. Some treatments can even begin before birth. True cures may only become possible years from now, when the highly publicized genetic engineering techniques become practical.

Some genetic disorders appear only under specific environmental conditions. If we knew more about genotype-environment interactions, perhaps we could institute more effective methods of health maintenance and treatment. We might attempt to avoid or produce specific environments. For instance, researchers have averted cleft lip and palate in mice by exposing the mother to high doses of oxygen during pregnancy. In humans, cystic fibrosis can be treated by regulating the diet, administering special enzymes and antibiotics, and using a mist tent. Without such measures, 95 percent of affected persons die before age five. If such treatments are implemented before severe lung damage occurs, 50 percent of affected persons live past age 20.

Treatments for genetic disorders take several forms. Dietary therapy can avert or ease some disorders. Excluding phenylalanine from the diets of phenylketonurics and lactose (milk sugar) from the diets of galactosemics is helpful. Occasionally, a missing cellular product can be supplied, as in the case of insulin used for diabetes mellitus. Drug therapy is most effective when treatment must reduce the accumulation of some toxic material, such as a missing enzyme's substrate. For instance, defects in uric acid metabolism, which cause ammonia to accumulate in the body, can be treated by administering sodium benzoate. The sodium benzoate reacts with glycine, the product is excreted, and the accumulated ammonia is used up in making more glycine. The basic concept underlying this type of treatment is to encourage chemical side reactions that use up the accumulated toxic substance. Of course, individual reactions to therapeutic drugs may be affected by hereditary factors, and this is the subject area of the increasingly important field of **pharmacogenetics**.

Treating the fetus as a medical patient is a relatively new idea. Although there have not been many success stories, the fetus is slowly becoming more accessible to therapies. One of the first successful attempts at fetal therapy was the treatment of **methylmalonic acidemia**, in which acid accumulates in the body and causes vomiting, mental retardation, and usually a very early death. Affected individuals lack an enzyme that activates vitamin B_{12}. Without active vitamin B_{12}, methylmalonic acid cannot be degraded, and the symptoms follow. The treatment—*not* a cure—is massive doses of vitamin B_{12} (as much as 1000 times normal) to the pregnant mother to avert the symptoms in the newborn.

Most genetic disorders are due to the lack of a critical enzyme. In a few cases, when a toxic substance builds up in the blood, the enzyme has been injected directly into the bloodstream. The enzymes can also be packaged in empty red blood cells (red blood cell "ghosts") and injected into the bloodstream where the enzymes are released. This technique has not proved highly effective.

161

Usually, the toxic buildup is within the body's cells. In efforts to reach this target area, researchers have synthesized artificial bodies called **liposomes**. These are small vesicles or bubbles that form when water-insoluble **phospholipids** (fatty substances) are mixed with water. As they form, liposomes can trap various substances in their surface membranes or interiors. When liposomes are injected into a patient, some of them are taken up by the body's cells. Absorption is more efficient if the liposomes are coated with antibodies. The cells then engulf the liposomes by **phagocytosis**, **pinocytosis**, or fusion of cell and liposome. Phagocytosis is the engulfment of solid matter by cells; pinocytosis is the taking-in of small liquid droplets. Once the liposomes are inside the cell, they fuse with cellular structures called **lysosomes** that digest the liposomes and release their contents.

The diseases under attack by these relatively new techniques are mostly the **lysosomal storage diseases**. There are more than 50 different lysosomal storage diseases in which the victim lacks an enzyme essential to normal cellular digestion. Without the enzyme, the cell becomes congested with some substance, which can cause grave problems. Some of the genetic disorders in this category are **Gaucher's disease**, Hurler syndrome, Fabry's disease, and Tay-Sachs disease. Gaucher's disease is a gradually progressive disorder characterized by enlarged spleen, bone pain, and pathological fractures. The other disorders were previously discussed.

Enzyme replacement with liposomes has succeeded most often *in vitro*, not in patients. The prospects are enticing, but actual clinical benefit to patients has been scarce. The most promising results have appeared in some trials with Gaucher's disease. The groundwork has been laid, though numerous problems remain to be conquered before these procedures can become practical. For example, researchers must learn ways to direct the enzymes to specific cells and tissues, and to get the enzymes past the blood-brain barrier to treat those disorders that involve the central nervous system.

Breakthroughs in enzyme replacement could lead to an alternative to abortion. Still, a choice would have to be made between fetal therapy and abortion; that is, between treatment and cancellation of the problem. Other ethical issues may also arise, for risks and benefits to the mother will also have to be assessed, and the appropriate conditions for fetal therapy will have to be determined. Because these treatments are *not* genetic cures, successfully treated persons would be able to pass their detrimental genes on to the next generation. Some people look upon such therapy as another systematic way to speed up the deterioration of the human gene pool. It is easy to find strong arguments on both sides of this issue.

SUMMARY

Genetic disorder is an incompatibility between a genotype and its environment. The study of genetic disorders, medical genetics, is becoming increasingly important in health management. More than 3000 genetic disorders are already known, and it is predicted that many more diseases will eventually be attributed to genetic causes. Many genetic disorders are recessive and rare, which means that many detrimental genes in a population are hidden because the bearers are heterozygous and appear to be normal. Nonetheless, considerable genetic disability afflicts the human population and causes serious problems for society.

Generally, epidemiology is the study of diseases and their causes. Genetic disorders are caused by small changes in the DNA (mutations) or by larger alterations in structure or number of whole chromosomes. They can be Mendelian, resulting from one gene pair, or multigenic, resulting from many gene pairs. Some disorders may be caused by different modes of inheritance in different individuals; this is genetic heterogeneity.

Genetic disorders resulting from gene mutations can be dominant or recessive, located on autosomes or sex chromosomes, and congenital or expressed later in life. Pedigrees offer one means of determining the mode of inheritance, for they reveal how certain disorders follow specific patterns of transmission from one generation to another. Sickle cell anemia is one genetic disorder for which we know the precise change in the gene and its product and the physiological consequences of this solitary change.

Dominant disorders afford a good way to determine mutation rates since normal parents should have only normal children, except when new mutations occur. The frequency of a dominant trait will not increase in a population that is subject to the Hardy-Weinberg principle. Sex-linked traits are those in which the gene or genes responsible reside on either the X or the Y chromosome. Hypertrichosis of the pinna (hairy ears) and the H-Y antigen are two traits strongly suspected of being Y-linked in humans. Many examples of X-linked traits have been identified. Although no examples are known in humans, partial sex linkage occurs when alleles for a trait exist on both the X and the Y chromosomes.

163

Some advances have been made in treating genetic disorders. Progress can be seen in dietary therapy, product therapy, and drug therapy. Also, the fetus is becoming more accessible for treatment; physicians have averted methylmalonic acidemia symptoms in the fetus by giving the pregnant mother massive doses of vitamin B_{12}. Enzyme replacement techniques involve injecting a missing enzyme directly into the bloodstream; enzymes packaged in synthetic liposomes can actually penetrate cells. These techniques are especially promising for the liposomal storage diseases, but so far they have been more successful *in vitro* than in patients.

DISCUSSION QUESTIONS

1. Some autosomal recessive disorders in humans result in death before sexual maturity. What factors prevent such a disorder from quickly disappearing from the population?

2. Some people insist that no parents have the right to have children if those children will suffer and/or be a burden to society. What is your opinion on this matter?

3. What are some of the molecular genetic concepts we have learned from the study of human hemoglobin?

4. What are the genetic consequences of the successful medical treatment of serious X-linked diseases such as hemophilia?

5. In simple terms, explain why disorders due to dominant alleles will not increase in a population that is in Hardy-Weinberg equilibrium.

6. In what way or ways would partial sex-linkage differ from X-linkage as a mode of inheritance?

7. What is the genetic consequence of successful medical treatment of an autosomal dominant disorder such as retinoblastoma? Retinoblastoma is a cancer of the retina; it was often fatal in childhood but can now be corrected by ophthalmectomy.

8. Do you think the magnitude of the human genetic load is increasing? Should we be concerned?

PROBLEMS

1. Is the trait shown in the following pedigree dominantly inherited or recessively inherited? Explain your position.

164

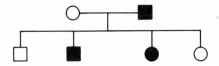

2. Consider a family in which both parents are heterozygous for sickle cell anemia (they have the sickle cell trait), the mother is colorblind, and the father has normal vision. Give the expected proportions of phenotypes (including sex) for the children in this family.

3. Analyze the following pedigree, and if possible determine the mode of inheritance. Discuss the bases for your conclusion.

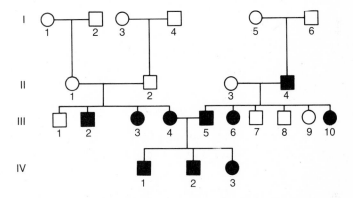

4. Consider a woman who is heterozygous for an allele that causes a dominant X-linked disorder. If she marries a man who is recessive for this gene and phenotypically normal, what kinds of children will they have, and in what proportions?

5. According to the pedigree below:
   a. Is the trait due to a dominant allele or a recessive allele?
   b. As far as possible, assign the appropriate genotype to each of the individuals in the pedigree using *E* and *e* as the two alleles. Remember that some individuals may not be assignable from the information given.

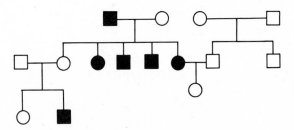

6. Phenylketonuria (PKU) is an autosomal recessive disorder with a frequency of about one in 14,000 in the human population. A couple who are both normal have a child with PKU.
   a. What is the probability that their second child will have PKU?
   b. What is the probability that the second child will be a carrier?
   c. What is the probability that both parents are carriers?
   d. What is the probability that any particular person in the general population is a carrier?

7. The partial pedigree below describes a family with cystic fibrosis.
   a. What is the probability that both of the two offspring denoted with question marks are normal?
   b. What is the probability that one is normal and one has cystic fibrosis?
   c. What is the probabiliity that both have cystic fibrosis?

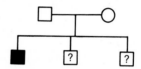

8. About one in 40,000 newborns has the autosomal recessive disorder, galactosemia. Therefore, the frequency of carriers for this gene is one in 100. Use the appropriate calculations to show how the two incidences relate to each other mathematically.

9. Consider a 16-year-old person, one of whose parents was stricken with Huntington's disease at age 39.
   a. What is the probability that the 16-year-old will also have Huntington's disease?
   b. Because of this probability, assume that this person and others in the same predicament abort all the pregnancies they have with normal spouses. What percent of those aborted fetuses would actually be genetically normal?

10. Galactosemia is an autosomal recessive disorder. What is the probability that two normal parents will produce a galactosemic child *if* it is known that all four grandparents were carriers?

REFERENCES

Childs, B., C. A. Huether, and E. A. Murphy. 1981. Human genetics teaching in U.S. medical schools. *American Journal of Human Genetics* 33:1–10.

Delhanty, J. D. A., J. M. Parrington, G. Casey, J. Attwood, L. West, D. Kirk, and G. Corney. 1981. Growth, DNA repair, sister chromatid exchange and chromosome studies in fibroblasts from Huntington's disease patients. *Annals of Human Genetics* 45:181–198.

Dronamraju, K. R. 1960. Hypertrichosis of the pinna of the human ear, Y-linked pedigrees. *Journal of Genetics* 57:230–243.

Farnsworth, M. W. 1978. *Genetics*. New York: Harper & Row, Publishers Inc.

Harris, H. 1947. A pedigree of sex-linked ichthyosis vulgaris. *Annals of Eugenics* 14:9.

Hartl, D. 1977. *Our uncertain heritage: Genetics and human diversity*. Philadelphia: J. B. Lippincott Co.

Ingram, V. M. 1957. Gene mutations in human haemoglobin: The chemical difference between normal and sickle cell haemoglobin. *Nature* 180:326–328.

Mabry, C. C., J. C. Denniston, T. L. Nelson, and C. D. Son. 1963. Maternal phenylketonuria: A cause of mental retardation in children without the metabolic defect. *The New England Journal of Medicine* 269:1404–1408.

McKusick, V. A. 1969. *Human genetics*. Englewood Cliffs, NJ: Prentice-Hall, Inc.

McKusick, V. A. 1978. *Mendelian inheritance in man*. Baltimore: The Johns Hopkins University Press.

Novitski, E. 1982. *Human genetics*. New York: Macmillan Publishing Co. Inc.

Okada, S., and J. S. O'Brien. 1969. Tay-Sachs disease: Generalized absence of a beta-D-N-acetylhexosaminidase component. *Science* 165:698–700.

Pauling, L., H. A. Itano, S. J. Singer, and I. C. Wells. 1949. Sickle cell anemia, a molecular disease. *Science* 110:543–548.

Perutz, M. F. 1964. The hemoglobin molecule. *Scientific American* 211:64–76.

Stern, C. 1957. The problem of complete Y-linkage in man. *American Journal of Human Genetics* 9:147–166.

Stern, C., W. R. Centerwall, and S. S. Sarkar. 1964. New data on the problem of Y-linkage of hairy pinnae. *American Journal of Human Genetics* 16:455–471.

Strickberger, M. W. 1968. *Genetics*. New York: Macmillan Publishing Co. Inc.

Thompson, J. S., and M. W. Thompson. 1980. *Genetics in medicine*. Philadelphia: W. B. Saunders Co.

Volpe, E. P. 1981. *Understanding evolution*. Dubuque, IA: Wm. C. Brown Group.

Winchester, A. M., and T. R. Mertens. 1983. *Human genetics*. Columbus: Charles E. Merrill Publishing Co.

Wyngaarden, J. B., and D. S. Fredrickson, ed. 1960. *The metabolic basis of inherited disease*. New York: McGraw-Hill Book Co.

167

8

Variations in Human Chromosomes

G roups of genes, together with protein molecules, form the structures we call the **chromosomes**. Chromosomes are present throughout the cell's life cycle, but they are easiest to see during mitosis and meiosis. At these times, they form dense bodies (somes) that are easily colored (chromo) by certain dyes. Scientists first observed these "colored bodies" over a century ago. The first diagrams of mammalian chromosomes, from the testes of a cat, were published in 1865. The first drawings of human chromosomes appeared in 1879 and were improved in 1884. They showed tangled clumps of overlapping structures that were so hard to count that everyone thought humans had 48 chromosomes. Finally, in 1955, Albert Levan and Jo Hin Tjio published convincing evidence that human beings had 46 chromosomes. In order to draw this conclusion, they analyzed 265 preparations from 22 different cell cultures made from the lung tissues of four different legally aborted embryos. One of their microscope preparations appears in Figure 8-1.

How could Levan and Tjio see the correct number of chromosomes when others could not? The biggest difference was in their technique. Levan and Tjio cultured human cells *in vitro* and treated them with **colchicine**, a drug that halts mitosis at metaphase and causes the chromosomes to become more compact. As a result of this treatment, the mitoses of a large proportion of the cells in culture were arrested at an ideal stage for analysis. Then the researchers ex-

FIGURE 8–1
One of the early human chromosome preparations, showing 46 chromosomes. (From Tjio, J. H., and Levan, A. 1956. *Hereditas*, 42: 1–6. Used with the permission of the publisher and the authors.)

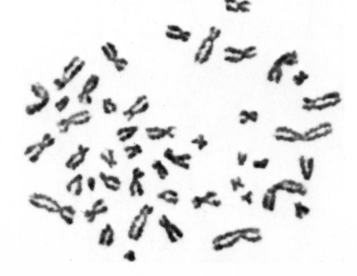

posed the cells to a **hypotonic** solution, which causes cells to absorb water and expand. As the cells filled with water, the chromosomes moved farther apart. Levan and Tjio then put the cells on a microscope slide and applied slight pressure to the cover slip to flatten the chromosomes into a single thin layer for easy and accurate counting.

Knowing how many chromosomes normal humans had was a milestone for researchers in human genetics. For one thing, it made it possible to recognize chromosome abnormalities. By 1959, Jerome Lejeune had reported that individuals with **Down syndrome** did not have 46 chromosomes, but 47, because of an extra chromosome No. 21. This discovery prompted extensive work on the association of human illness with chromosomal abnormalities.

Geneticists still do not fully understand how a human chromosome, which in a nondividing cell exists as a DNA-protein (**chromatin**) strand averaging some 3.77 cm or 1.5 inches long, condenses into the short, rod-shaped structure visible during mitosis. Nonetheless, these condensed chromosomes are essential to many genetic studies. Using photographs of spread-out chromosomes from the same cell, researchers routinely sort and arrange the chromosomes by size and centromere position into a **karyotype**. Figure 8-2 shows an example of a chromosome spread on a microscope slide and the finished karyotype. Each karyotype is composed of 22 pairs of **autosomes** and one pair of **sex chromosomes**. Their various structural features and landmarks al-

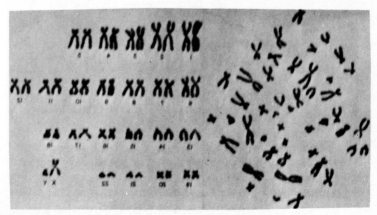

FIGURE 8–2
From the chromosome spread shown at the left, cytogeneticists have constructed the karyotype at the right. The chromosomes are organized into homologous pairs. Each chromosome consists of two sister chromatids because replication has taken place, but not division. (From Winchester and Mertens. 1983. *Human Genetics,* 4th ed. Charles E. Merrill Publishing Company, Columbus, Ohio.)

171

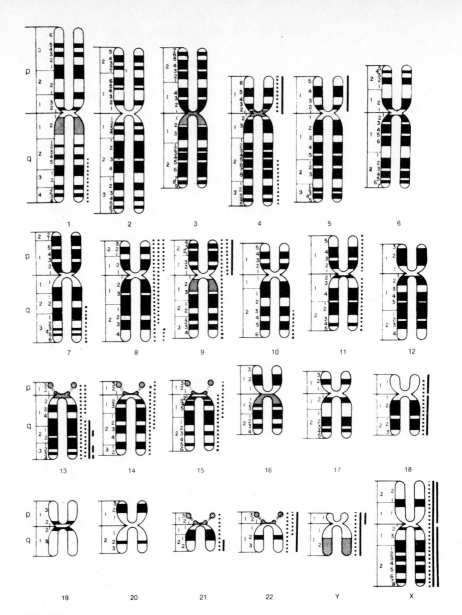

FIGURE 8–3

A diagram of the human chromosome banding patterns revealed by various types of staining methods. The letter *p* designates the shorter of the two arms, and *q* designates the longer arm. A two-digit numbering system specifies the bands. (From Paris Conference, 1971: *Standardization in Human Cytogenetics*. Birth Defects: Original Article Series, VIII: 7, 1972. The National Foundation, New York.)

172

low us to divide the autosomes into seven *groups*, marked A through G. Recent technical advances have revealed that each chromosome has a unique pattern of bands, so now *individual* chromosomes can be identified and numbered. A diagram of a human karyotype, showing characteristic banding patterns for each chromosome, appears in Figure 8-3. The use of chromosome bands has made the analysis of chromosome abnormalities much easier and more accurate. It has also helped to make **cytogenetics**, the combination of genetics and cell biology, the exciting field it is today.

— stop

CHROMOSOME VARIATIONS

Chromosome number and structure are relatively constant, but they can change. The effects of these changes on the phenotype range from none at all to very serious, and sometimes lethal. The severity of the phenotypic effect depends on the amount of genetic material involved and its function. In humans, most chromosomal variations are detrimental. We are diploid organisms, and we function best with exactly two doses of every gene on the autosomes. Any deviation from this dosage is a **genic imbalance**. Unfortunately, such detrimental imbalances are not at all rare.

Table 8-1 classifies chromosomal variations as deviations in structure and number. For example, a **euploid** cell, tissue, or individual has one or more complete sets of chromosomes. A cell with one set (n) of chromosomes, such as a gamete, is **haploid**; for humans, the haploid number is 23. A cell with two sets ($2n$) is **diploid**; for humans, the diploid number is 46. Similarly, **triploid** ($3n$) means three sets, **tetraploid** ($4n$) means four, and so on. Any multiple above diploid is called a **polyploid**.

An **aneuploid** cell, tissue, or individual has one or more whole chromosomes missing from or added to the basic complement, as shown in Figure 8-4. **Nullisomics** are missing both members of a chromosome pair, and **monosomics** are missing only one. A **trisomic** has three copies of a specific chromosome, a **tetrasomic** has four, and so on. A **double trisomic** is trisomic for two different chromosomes.

Structural variations often occur as the result of a chromosome break. Sometimes, the break simply results in chromosome fragments. Other times—probably more often—cellular enzymes repair the breaks. The repair may restore the original configuration or, especially when two or more breaks must be repaired simultaneously, it may rearrange the pieces. Only the broken ends of chromosomes can be joined by the

173

TABLE 8–1
Classification of some basic types of
chromosome variation

I. Variation in chromosome number
   A. Euploidy
      1. haploidy ($1n$)
      2. diploidy ($2n$)
      3. polyploidy
         a. triploidy ($3n$)
         b. tetraploidy ($4n$)
         c. pentaploidy ($5n$) etc.
   B. Aneuploidy
      1. nullisomy ($2n - 2$)
      2. monosomy ($2n - 1$)
      3. trisomy ($2n + 1$)
      4. tetrasomy ($2n + 2$)
      5. double trisomy ($2n + 1 + 1$) etc.
II. Variation in chromosome structure
   A. Deletions
   B. Duplications
   C. Inversions
   D. Translocations
      1. reciprocal
      2. nonreciprocal
   E. Isochromosomes
   F. Ring chromosomes

enzymes; a broken end cannot be attached to an unbroken end. This control mechanism protects the structural integrity of the chromosomes.

Breaks are not uncommon in mitosis and meiosis, before and after replication. The resulting fragments can remain aligned with the main chromosome or be displaced from it, as shown in Figure 8-5. When they remain aligned, we can see a **gap** in metaphase chromosomes; it is not clear whether such gaps represent complete or partial breakage or partial repair.

When break fragments do not remain aligned, they can give rise to various structural rearrangements, some of which are shown in Figure 8-6. A **deletion** or **deficiency** occurs when part of a chromosome is lost; the lost piece may come from one end (terminal deletion) or from the interior (intercalary deletion). A **duplication** occurs when a chromosome segment is doubled. An **inversion** is a reversal in the sequence of genes in part of a chromosome; it happens when two near-simultaneous breaks free a chromosome piece, the piece turns end over end, and the breaks are repaired. **Reciprocal translocations** occur when breaks in two nonhomologous chromosomes result in an exchange of

174

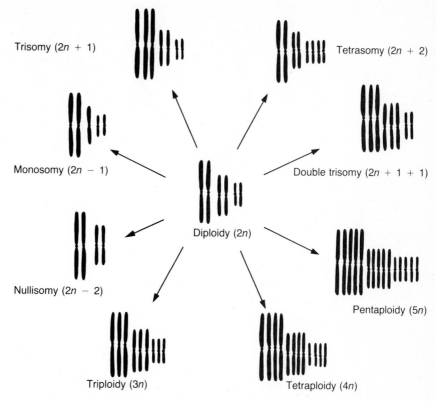

FIGURE 8–4
Some of the variations in chromosome number that can occur in organisms.
For clarity, the normal diploid number in this example is three pairs of chromosomes ($2n = 6$).

fragments. In a **nonreciprocal** translocation, a broken fragment from one chromosome attaches to another chromosome.

Chromosomes can also undergo large-scale changes. An **isochromosome** results when a centromere breaks perpendicularly to the two sister chromatids it joins. The resulting chromosomes each have two identical arms, unlike the original arrangement. **Ring chromosomes** are unusual structures that sometimes occur when a chromosome loses both its ends and the repair process joins the broken ends to form a ring.

Most of the many different structural changes result in genic imbalance for part of a chromosome. At the very least, they form new gene arrangements, and in every case they can have grave conse-

175

A gap occurred in a chromatid *after* replication.

A gap occurred in a chromosome *before* replication.

A break and displacement occurred in a chromatid *after* replication.

A break and displacement occurred in a chromosome *before* replication.

FIGURE 8–5
The different consequences of gaps and breaks occurring in chromosomes *before* and *after* replication. The affected chromosomes are subsequently viewed in metaphase.

quences. Large-scale cytogenetic surveys including chromosome analyses of aborted fetuses, newborns, children, and adults have aided in epidemiological studies, the detection of new chromosomal syndromes, and the collection of basic medical data. Chromosomal banding techniques have made it possible for researchers to recognize previously undetectable abnormalities. And in the future, automated instruments will further increase the cytogeneticist's capabilities.

SEX DETERMINATION AND SEXUAL DIFFERENTIATION

As we have already noted, human cells contain 22 pairs of autosomes and one pair of sex chromosomes. There are two kinds of sex chromosomes, X and Y, and they provide the chromosomal basis of sex determination. Normal females have two X chromosomes; and because they can generate only X-bearing gametes (eggs), we call them **homogametic**. Normal males have one X and one Y chromosome and can make two kinds of gametes (spermatozoa), X-bearing and Y-bearing; we call them **heterogametic**. Female offspring result when sperm and egg both contribute an X to the zygote. Male offspring arise when the sperm contributes a Y. This process is summarized in Figure 8-7.

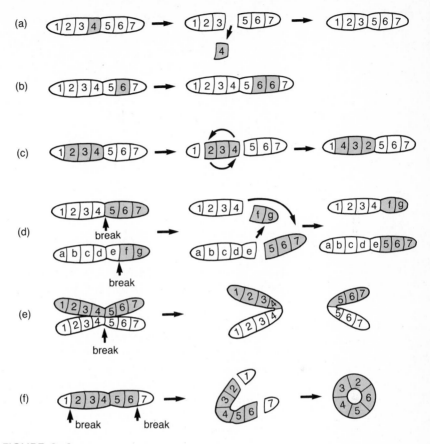

FIGURE 8-6

Chromosome aberrations: (a) intercalary deletion, (b) duplication, (c) inversion, (d) reciprocal translocation, (e), isochromosome, and (f) ring chromosome.

During meiosis, the male's X and Y chromosomes pair at their ends. This helps to ensure normal separation of the chromosomes into gametes and, as a result, equal probabilities of male and female offspring. It appears that the Y chromosome carries very few genes. The X, on the other hand, has hundreds or even thousands. The many X-linked hereditary disorders include ichthyosis, ocular albinism, congenital deafness, Lesch-Nyhan syndrome, hemophilia, Hunter syndrome, Duchenne muscular dystrophy, Fabry's disease, and two types of color blindness. The X chromosome also carries the genes for the Xg blood group, **glucose-6-phosphate dehydrogenase**, and many other traits.

177

FIGURE 8–7

The chromosomal basis of sex determination.

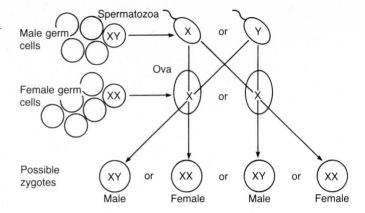

Sex determination itself seems to be associated with a particular cell membrane component, the **H-Y antigen**, that plays a role in controlling gonad formation. The H-Y gene seems to be on the Y chromosome, for almost all men have the antigen and almost all women do not. However, it may occasionally—perhaps because of a translocation—appear on the X chromosome. A very few XY females who lack the antigen and XX males who have it do exist. A few researchers argue that the H-Y gene is on an autosome, but is controlled by a gene normally on the Y chromosome.

Male and female embryos initially develop in the same way, both forming an "indifferent" gonad. If the H-Y antigen is absent, the gonad develops into an ovary. If it is present, it triggers testicular development. Once the gonads begin to differentiate early in development, they secrete hormones that induce the differentiation of other sex-specific structures. The levels of these hormones in the fetus can be influenced by the levels of hormones in the mother's body, which can in turn be affected by stress. Although medical scientists observe many anomalies of sexual development in newborns, adolescents, and adults, only some of these are genetic in origin.

SEX RATIOS AND SEX SELECTION

Theoretically, since spermatozoa are 50 percent Y-bearing and 50 percent X-bearing, there should be a 1:1 ratio of male and female newborns. Although this 1:1 ratio may be accurate if we consider all conceptions (including abortions and stillbirths), there are more male than

178

female newborns. Ratios such as 106:100 are common, especially in industrialized nations.

Traditionally, most cultures have preferred sons, at least for the first child, for reasons often related to financial inheritance and education. This preference exists in the United States today, where 80 percent of men polled wanted their first child to be a son, and so did a surprisingly high percentage of women. The sex ratio is thus a social issue, and will become a potentially controversial one as soon as some method of **sex preselection** is perfected. Present methods are only partially effective. Some are based on the fact that sperm containing the larger X chromosome have 3.4 percent more DNA and are heavier than sperm with the Y chromosome. Researchers can use **differential centrifugation** to spin a suspension of sperm at high speed so the X-bearers tend to settle to the bottom. But sharp separations are not possible yet, and the process can damage the sperm. Similar results can be obtained by putting the sperm into a dense medium that inhibits their swimming ability. The fastest swimmers then tend to move to the bottom, and 65 to 90 percent of them are Y-bearers. Electrophoresis, which exploits differences in both mass and electrical charge, can also be useful for differentiating between sperm. All three of these methods must be used in conjunction with artificial insemination.

Artificial insemination is unnecessary with some methods. For instance, douches can be used to make the vagina alkaline, a condition favoring Y-bearing sperm, or acidic, which favors X-bearers. Insemination 48 hours before ovulation favors fertilization by an X-bearing sperm, while insemination at the time of ovulation favors fertilization by a Y-bearer. Other methods under study include the use of antibodies to prevent one kind of sperm from functioning effectively.

If a completely effective method of sex preselection is ever devised, there may be serious social effects. Many couples are already using prenatal diagnosis and selective abortion to enforce their preference. If they could make their choice in an easier and cheaper way, so many couples might choose sons that the population would become predominantly male. Some people are concerned that this would lead to more violence, crime, and alcoholism (and fewer church-goers). If sex preselection is expensive, that might mean that more girls would be born to less favored socioeconomic groups, so that women would tend to be more disadvantaged than ever before. On the other hand, many people would find it immensely satisfying to be able to plan their families. The individual couple's freedom from continuing to have babies until a son or daughter is born could work in favor of population control.

179

BARR BODIES AND THE LYON HYPOTHESIS

The chromosomal difference between mammalian males and females is visible. Scientists can distinguish male interphase cells from female ones by using a good microscope and the appropriate stain. M. L. Barr and E. G. Bertram first discovered this in cat nerve cells in 1949, but it is true in most mammalian cells. In humans, it is conveniently clear in cells scraped from the lining of the cheek (**buccal cells**), transferred to a drop of water on a microscope slide, and suitably stained. In cells from a female, a small, deeply stained body is visible just inside the nuclear envelope. This is the **Barr body** or **sex chromatin**. The Barr body is rarely seen in male cells. Very occasionally a cell will have two or more Barr bodies, as shown in Figure 8-8.

There is another visible difference between the cells of males and females. **Polymorphonuclear leukocytes** are white blood cells (infection fighters) with multilobed nuclei. The leukocytes of females often have a small, stalked projection, or **drumstick**, attached to one nuclear lobe, as illustrated in Figure 8-9.

What are the Barr bodies and drumsticks? In the early 1960s, Mary Lyon formulated the **Lyon hypothesis**, which proposes an answer to this question. Although female cells have two X chromosomes, the

FIGURE 8-8
Photomicrograph of cheek (buccal) cells showing Barr bodies. (From Redding, A., Hirshhorn, K.: *Guide to Human Chromosome Defects.* Bergsma, D. (ed.): White Plains: The National Foundation—March of Dimes, BD: OAS IV (4), 1968.)

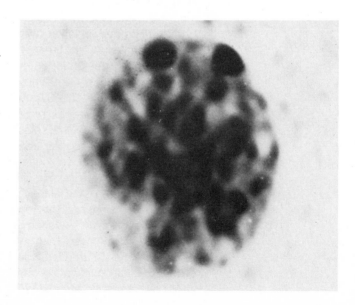

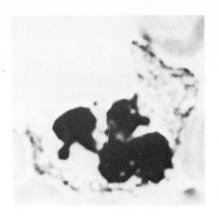 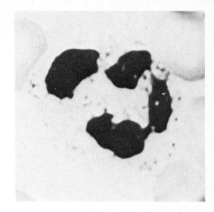

FIGURE 8–9
Photomicrograph (left) shows the drumstick of a polymorphonuclear leukocyte from a female. A polymorphonuclear leukocyte from a male is shown on the right. (Photographs by Richard Kowles.)

evidence shows that in any one cell only one X chromosome is active, with its genes being transcribed. The other is inactive, and its chromatin substance is condensed into **heterochromatin**. This condensed chromosome appears as a small, discrete structure. Males lack Barr bodies and drumsticks. The one X chromosome they have is active in the cell; they do not have an extra one to be inactivated and condensed.

The determination of which X chromosome will be active in a female cell occurs very early in the embryo. Parts of one X are inactivated as early as 16 days after conception, and most of the chromosome is inactive by five weeks. In some cells, the inactivated X is the paternal one, which came from the father; in others, it is the maternal one. The inactivation is thought to be random and is not the same for all cells. But once made, the inactivation decision remains the same for all mitotic descendants of the cell.

Generally, no more gene product is synthesized from the two X chromosomes in a female's cells than from the one X chromosome in a male's cells. A female who is homozygous for an X-linked allele is no more affected than a hemizygous male (who has only one copy of the allele). Apparently, females receive the benefits of what we call **dosage compensation**, by means of X-inactivation. Generally speaking, every person has one *functioning* X chromosome per cell, regardless of how many X chromosomes are actually present.

181

As you might expect, females heterozygous for one or more genes located on the X chromosomes are genetic **mosaics**. Some of their cells will express one allele, while others express the other. The mosaicism results from the random X-inactivation early in embryonic development. This mechanism is summarized in Figure 8-10. Regardless of mosaicism, a heterozygous female generally expresses the dominant allele overall. Evidently, enough cells expressing the dominant trait are usually present in the body to manifest the dominant phenotype. For example, a female heterozygous for hemophilia, an X-linked gene, would have a normal blood-clotting system. Rare exceptions do exist, for it is remotely possible for a heterozygous female to have mostly cells with the normal X chromosome inactivated. Such females do express the recessive condition even though they are heterozygotes; they are called **manifesting heterozygotes**. Figure 8-11 illustrates the phenotype associated with an X-linked skin condition, **anhidrotic dysplasia**, in three generations of heterozygous women. The random X-inactivation can be detected by the presence or near absence of sweat glands in the skin.

FIGURE 8–10

A diagrammatic explanation of X-inactivation and the resulting mosaicism. The symbols *p* and *m* indicate the paternal and maternal origin of the X chromosomes. The dark and light cells signify different phenotypes in the two types of cells. (a) A small group of cells is depicted for clarity. Researchers have estimated that the actual number of cells present at the time of X-inactivation is 1000 to 2000; (b) the X-inactivation event; (c) only one X chromosome remains active in each cell of a mosaic embryo and, eventually, a full-grown mosaic individual.

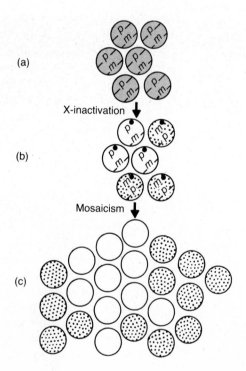

FIGURE 8–11

Three generations of females hetero-zygous for anhidrotic dysplasia, an X-linked allele that reduces the number of sweat glands in the skin: (I) the heterozygous mother, (II) a heterozy-gous daughter, and (III) heterozygous identical twin granddaughters. Shad-ing (abnormal) and clear (normal) areas represent the phenotypic expression in the skin, depending upon which X chromosome is active in the area. (Redrawn from Kline, A. H., Sidbury, Jr., J. B., and Richter, C. P. 1959. The occurrence of ecto-dermal dysplasia and corneal dysplasia in one family. *Journal of Pediatrics*, 55: 355–366. Used with the permis-sion of C. V. Mosby Co. and the au-thor.)

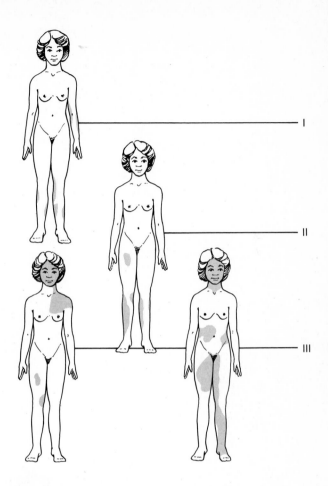

ANEUPLOIDY OF SEX CHROMOSOMES

The normal human chromosome number is 46, but exceptions are fairly common. Variations in the number of sex chromosomes are among the most common. The first example of sex chromosome aneuploidy was found in 1959, when it was discovered that some females lacked Barr bodies. They had only 45 chromosomes and were missing an X chro-mosome; their genotype was described as 45,X or 45,XO. Such women were **sex-chromatin negative**, and they suffered from **Turner syndrome**. They had female genitalia, but underdeveloped ovaries; they did not menstruate and were sterile; they tended to have sparse body

183

hair, small breasts, a wide (webbed) neck as a result of fluid accumulation in the neck, and other physical anomalies. A characteristic patient appears in Figure 8-12. The incidence of this XO monosomy has been estimated to be from one in 1200 to one in 2500 female live births, although 98 percent of all XO conceptions appear to end in spontaneous abortions.

At about the same time, researchers found males who had one Barr body in each of their cell nuclei. They were **sex-chromatin positive** and had 47 chromosomes, with an XXY genotype. Their condition, **Klinefelter syndrome**, does not exhibit pronounced abnormalities but includes testicular atrophy, beardlessness, a feminine body contour, a high-pitched voice, occasional mild to moderate mental retardation, and larger than normal breasts. Some of these characteristics are shown in Figure 8-13. Most of these people are also tall, thin, long legged, and sterile. The incidence of Klinefelter syndrome is one in 500 to one in

FIGURE 8–12
A child with Turner syndrome, whose sex organs will never mature. (Photo courtesy of Dr. Stella B. Kontras, Children's Hospital, Columbus, Ohio.)

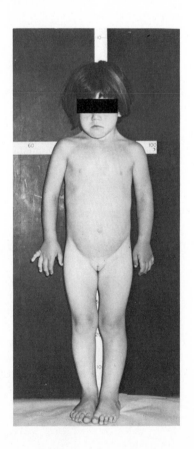

FIGURE 8–13
A man with Klinefelter syndrome, expressing some of the physical features associated with the 47,XXY genotype. (Photo courtesy of Povl Riis.)

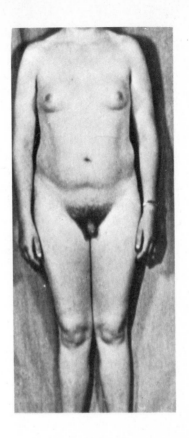

1000. We should note that despite the extra X chromosome, the presence of the Y chromosome ensures male genitalia in accordance with what we understand of sex determination.

Some aneuploid males do not have Barr bodies, but have an extra Y chromosome. Males with the **XYY** karyotype often have above average height, but many such persons have average intelligence and fertility. Curiously, the children of XYY men are usually normal (XX and XY). The frequency of XYY men is probably one in 700 to one in 1000.

Both XXY and XYY individuals, some researchers have claimed, have a tendency to be more aggressive and antisocial than normal, and hence are more likely to end up in penal and mental institutions. True or not, the claims have prompted argument over the existence of the "bad seed" and whether such people can be truly guilty of crimes. We will consider this issue further in Chapter 17 on human behavior.

Many other sex chromosome aneuploidies can also exist. The **triple-X syndrome** (XXX) is sometimes inaccurately called the "super-

185

female." Such women can show a phenotype ranging from normal to one resembling Turner syndrome; XXX women are usually fertile and have normal children. Triple-X females are often mentally retarded. The frequency of XXX women ranges from one in 500 to one in 1200 in different populations. Much rarer are XXYY, XXXY, XXXXY, and XXXYY males, and XXXX and XXXXX females. The XXYY frequency is about one in 10,000. Medical researchers have reported as many as five X and three Y chromosomes in live births, and as expected the detrimental symptoms are more pronounced in these syndromes. For example, the testes of the 49,XXXXY male are very minute.

Table 8-2 summarizes the sex chromosome aneuploidies, including the number of Barr bodies associated with each, and the primary sex of the affected individuals.

Mechanisms Underlying Aneuploidy

The most frequent cause of aneuploidy is probably chromatid and/or chromosome **nondisjunction**—failure to separate—during meiosis, or chromatid nondisjunction during mitosis of somatic cells. As a result, one daughter cell acquires two chromosomes or chromatids while the other has none. **Anaphase lag**—failure of a chromosome to move fast enough to be included in a newly formed nucleus—can have a similar result.

Nondisjunction can occur in either or both parents, in mitosis or meiosis, in the first, second, or both meiotic divisions. These possibil-

TABLE 8–2
Sex chromosomal genotypes and numbers of associated Barr bodies

Genotype	Primary Sex	Chromosome Number	Barr Bodies
XO	female	45	0
XY	male	46	0
XYY	male	47	0
XX	female	46	1
XXY	male	47	1
XXYY	male	48	1
XXX	female	47	2
XXXY	male	48	2
XXXYY	male	49	2
XXXX	female	48	3
XXXXY	male	49	3
XXXXX	female	49	4

FIGURE 8–14

The possible zygotic results of nondis-junction in the second division of meiosis in the male.

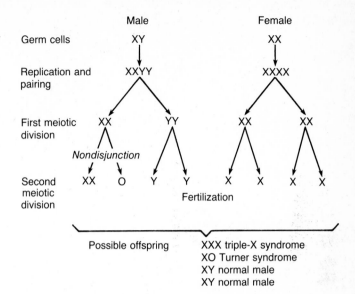

ities and combinations readily account for all the aneuploidies discussed thus far, as well as others. Figure 8-14 shows nondisjunction in a male's second meiotic division. The resulting sperm are two normal Y bearers and two abnormal ones—an XX-bearer and an O bearer. If the abnormal sperm fertilize a normal egg, they will give rise to XXX and XO offspring.

Chromosomal Mosaicism and Chimerism

When nondisjunction occurs after fertilization, during a mitosis in the embryo, it results in **mosaicism**. A mosaic individual has cells of two or more different genotypes or chromosomal constitutions; nondis-junction mosaics are sometimes called **mixoploid**. Mosaics that result from the fusion of two separate embryos are **chimeras**; when the two fused embryos are male and female, the resulting individual is an XX/XY chimera who may develop into a **hermaphrodite**.

Chromosome nondisjunction and anaphase lag seem to be the main causes of mosaicism, producing both two-genotype (XX/XY, XO/XX, XO/XY, X/XYY, XO/XXX, etc.) and three-genotype (XO/XX/XXX, XO/XY/XYY, etc.) mosaics, as shown in Figure 8-15. Together, mosaicism and chimerism can cause many unusual genotypes and phenotypes. Mosaicisim is often the basis for abnormal sexual development. Because of mosaicism, researchers can sometimes make mistakes in chromosomal analysis of aneuploid syndromes and in fetal diagnosis.

187

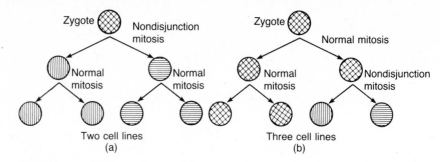

FIGURE 8–15
Depending upon when it occurs, mitotic nondisjunction in the zygote or very early embryo forms the basis for mosaicism with (a) two cell lines, or (b) three cell lines.

Particularly worrisome are the errors in diagnosis that may result from mosaicism occurring in tissue culture.

Other Sex Anomalies

Because sex determination is a complex process, it can go wrong in many ways. Among the simplest of these ways are the anomalies we have already discussed. There are also various **intersexes**, people with ambiguous external genitalia and/or abnormal internal sexual organs. Some of these cases show sex chromosome abnormalities, and they have helped researchers to clarify the mechanisms of sexual development. More bewildering are cases in which a phenotypic abnormality cannot be associated with a chromosomal abnormality. For example, there are a few XX males and XY females. Such people may have an undetected mosaicism, a minuscule fragment of the Y chromosome may have been translocated to a new site, or they may have a single-gene mutation.

True hermaphrodites, who have both ovarian and testicular tissues, are rare (one in 24,000 newborn males). They usually have an ovary on one side, with at least a few follicles, and a testis on the other, with at least a few seminiferous tubules, but they are not reproductively functional. They are best described as intersexes showing sexual ambiguity. They may lack some sex characteristic criteria, and they may have symptoms resembling Klinefelter syndrome. Most have XX genotypes; XY genotypes are rarer. Some are XX/XY mosaics. The mechanisms that produce hermaphroditism are not well understood.

Pseudohermaphrodites have only testicular or only ovarian tissue but nevertheless resemble the other sex in some ways. Female pseudohermaphrodites may have a penis or scrotal remnants. Some persons may suffer **testicular feminization** because of a single gene (the **tfm** gene). Most such persons have female external genitalia and well-developed female secondary sex characteristics, but no pubic hair. Internally, these people have a blind vagina, a rudimentary or no uterus, and undescended testes. They are sterile, but they can often enjoy adequate marital relations.

Homosexuality may seem a kind of intersexuality, but it has not yet been linked to any chromosomal anomaly. **Transsexualism** refers to voluntary sex changes by people with ambiguous external genitalia or by those who feel more comfortable as the opposite sex. The reasons are usually psychological, and the operation is usually chosen by men who wish to become women. It involves both surgery and the application of sex hormone.

Many concerned observers feel that the medical profession is indifferent to sex reversals. One's perception of one's sexual identity is an important issue that deserves careful analysis. It raises serious questions about maleness, femaleness, and the difference between the sexes, as well as about the proper treatment for psychological and physical difficulties. Indeed, proper treatment is essential, for abnormal sexual development poses difficult problems for individuals, families, schools, and society. Sexual anomalies, as a group, are not rare.

AUTOSOMAL ANEUPLOIDY

Aneuploidy can also affect the autosomes, though autosomal aneuploidies are rare among live births. Most are trisomics, and most are observed in fetuses. The human body tolerates extra autosomes less well than it does excess sex chromosomes. Monosomies are even less likely to survive.

Most trisomics that survive birth are seriously afflicted both physically and mentally, and few live very long. You will notice that only trisomics for very small chromosomes are able to survive. The best known examples are **trisomy-13**, **trisomy-18**, and **trisomy-21**, in which the cells have three chromosome 13s, 18s, and 21s, respectively. Other trisomies are much rarer.

Trisomy-18 (**Edwards syndrome**) has a 90 percent mortality within one year of birth and 99 percent within ten years; its victims do

189

not live to reproduce. Symptoms include a high-pitched cry, low-set ears, mental retardation, congenital heart disease, and finger overlap in the clenched fist. An infant with trisomy-18 appears in Figure 8-16. The frequency of this syndrome is one in 4500 live births.

Trisomy-13 (**Patau syndrome**) has 100 percent mortality soon after birth. In general, the severities of the trisomies are loosely related to the size of the chromosomes involved. Chromosome 13 is larger than chromosome 18, therefore its effects are greater. Symptoms include eye defects, congenital heart disease, severe mental retardation, **microcephaly** (small head), and cleft palate and lip, as shown in Figure 8-17. Its frequency is about one in 5000 live births.

Trisomy-21 (**Down syndrome**, or mongolism) is one of the most common human chromosomal abnormalities. It is responsible for about 30 percent of all mentally retarded children in the United States. Children with Down syndrome have slanted eyes, short broad hands, poor muscle tone, and other physical anomalies. A typical patient is shown in Figure 8-18. They are also prone to heart trouble, gastrointestinal disorders, and acute leukemia. They are easily identified by appearance, but Down—and other—syndromes should always be confirmed by observing the actual chromosome arrangement (karyotype analysis). Karyotyping will distinguish the 95 percent of individuals with Down syndrome whose trisomy is due to nondisjunction in a parental meiosis, shown in Figure 8-19, from the 4 percent whose third chromosome 21 is translocated onto a nonhomologous chromosome, and the 1 percent whose condition arises from normal/trisomic mosaicism. As you might

FIGURE 8–16
An infant with trisomy-18, or Edwards syndrome. Note the low-set ears and the clenched fist with the fingers overlapping. (Photo courtesy Richard C. Juberg, M.D., Ph.D., The Children's Medical Center, Dayton, Ohio.)

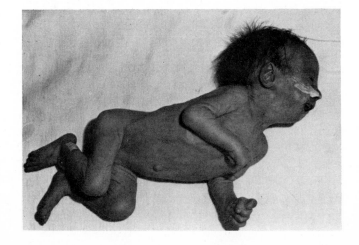

FIGURE 8–17
An infant with trisomy-13, or Patau syndrome. (From Thompson, J., and Thompson, M. W. 1980. *Genetics in Medicine*. W. B. Saunders Co., Philadelphia. Used with the permission of the publisher.)

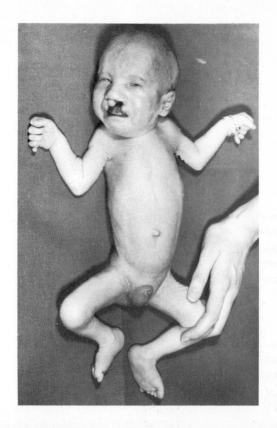

FIGURE 8–18
A child with Down syndrome. (Photo by Ted Shenenberger.)

FIGURE 8–19

The karyotype of a male with Down syndrome. Note the three chromosome 21s. (Courtesy of Dr. Catherine G. Palmer, Indiana University School of Medicine.)

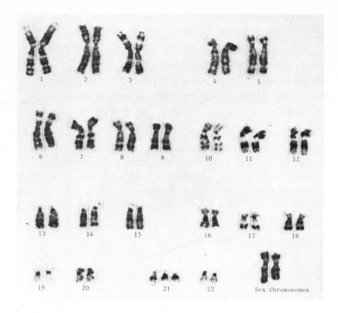

expect, symptoms may vary in mosaics, since their severity depends on the proportions of the two cell lines in the body.

The overall frequency of Down syndrome is between one in 600 and one in 800. However, the frequency goes up sharply with maternal age, from one in 1000 at age 20 to 25 to one in 10 at ages over 49. The frequency starts to rise steeply after the mother reaches age 30–34. Statistics showing this increase appear in Table 8-3. The frequency of Down syndrome in the population depends on average child-bearing age, and there is some concern that as more women delay child-bearing to build careers, more infants with Down syndrome will be born in the United States.

Nondisjunction in the female is not the only cause of Down syndrome. The extra chromosome 21 comes from the father 20 to 30 percent of the time. Reports that the frequency goes up with the father's age have not, however, been documented.

Translocation Down syndrome occurs when a chromosome 21 is translocated to one of several different chromosomes, such as chromosome 15. The translocation can occur in meiosis or in the zygote, or it can exist in one parent (the carrier), where it seriously affects the types of children he or she can have, and the proportions of affected to normal children. Figure 8-20 shows the latter situation. One parent has the normal amount of DNA, but only 45 chromosomes; one is the struc-

TABLE 8-3

The estimates of Down syndrome rates per 1000 live births given by single year maternal age intervals from three different studies

Maternal Age (years)	Massachusetts Study[a]	New York Study[b]	Swedish Study[c]
20	0.57	0.52	0.64
21	0.60	0.59	0.67
22	0.64	0.65	0.71
23	0.67	0.71	0.75
24	0.70	0.77	0.79
25	0.74	0.83	0.83
26	0.77	0.89	0.87
27	0.80	0.95	0.92
28	0.84	1.01	0.97
29	0.87	1.07	1.02
30	0.90	1.13	1.08
31	0.93	1.21	1.14
32	1.15	1.38	1.25
33	1.55	1.69	1.47
34	1.98	2.15	1.92
35	2.53	2.74	2.51
36	3.22	3.49	3.28
37	4.11	4.45	4.28
38	5.24	5.66	5.60
39	6.68	7.21	7.32
40	8.52	9.19	9.57
41	10.86	11.71	12.51
42	13.85	14.91	16.36
43	17.66	19.00	21.39
44	22.51	24.20	27.96
45	28.71	30.84	36.55
46	36.61	39.28	47.79
47	46.68	50.04	62.47
48	59.52	63.75	81.67
49	75.89	81.21	106.76

Source: [a]Hook, E. B., and Fabia, J. J. 1978. *Teratology*, 17: 223-228.
[b]Hook, E. B., and Chambers, G. M. 1977. *In* Bergsma, D., Lowry, R. B., Trimble, B. K., and Feingold, M. (eds.). *Birth Defects*. Original Article Series 13 (3A): 123–141. New York: Alan R. Liss, Inc.
[c]Hook, E. B., and Lindsjo, A. 1978. *American Journal of Human Genetics*, 30: 19–27.

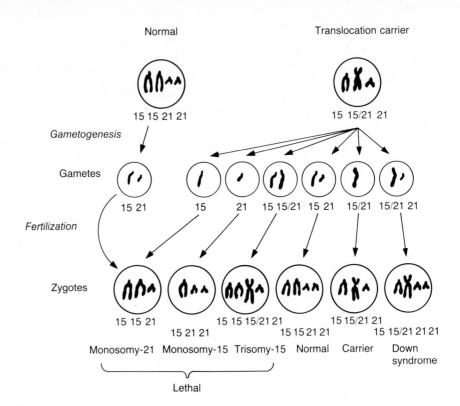

FIGURE 8-20

All the possible results of matings between a 15/21 translocation carrier and a normal parent. Only the three zygotes on the right result in live births: theoretically, each should occur one third of the time.

ture formed by combining one chromosome 15 and one 21. Because of the unusual behavior of the translocation chromosome, the carrier can produce six different gametes, and six different zygotes, accordingly. Three of the zygotic combinations are lethal. The other three, all equally probable, include one normal, one carrier, and one child with Down syndrome. For reasons not yet understood, such cases produce only 10 percent, not 33 percent, children with Down syndrome; however, even 10 percent is a dismaying probability, especially compared with the overall Down syndrome frequency of one in 700.

Only a few matings between females with Down syndrome and normal males are known to have produced children. Theoretically, such matings should produce Down syndrome and normal children in a 1:1 ratio, as shown in Figure 8-21. The actual data, though limited, come

194

FIGURE 8–21
The consequences of matings between a parent with Down syndrome and a normal parent.

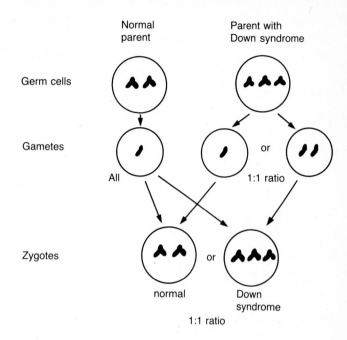

very close to this ratio. It is uncertain whether any males with Down syndrome have ever reproduced, but they would be expected to show the same statistics.

One of the most serious questions in human cytogenetics is how an excess of normal genetic information can be so damaging. Researchers do not know how the extra chromosome 21 actually causes Down syndrome. Nor do they know much about the genetic content of chromosome 21. Investigations of these problems are underway.

STRUCTURAL ABERRATIONS IN HUMAN CHROMOSOMES

Structural aberrations—deletions, duplications, inversions, and translocations—have been seen in all 23 human chromosomes. Many have been revealed only by the newer cytological techniques, for the associated phenotypes are quite varied and there are few well-defined sets of symptoms, or syndromes. Some aberrations are proving to be associated with cancers, so that there can be no doubt that chromosomal aberrations are a significant threat to human health.

Deletions are common, and one of the better known is responsible for **cri du chat** or **cat-cry syndrome.** Only a small piece of the short

195

FIGURE 8-22
An infant with the cat-cry syndrome, also known as cri du chat. Note the small head, rounded face, and slanted eyes. (Photo courtesy G. H. Valentine, *The Chromosome Disorders,* 3rd edition, Heineman Medical Books.)

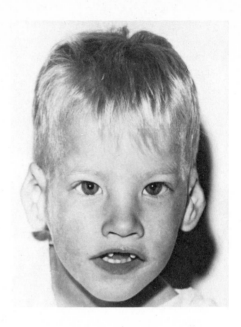

arm of chromosome 5 is missing, but afflicted children have a cry resembling a cat's, mental retardation, small round faces, microcephaly, and slanted eyes. An affected infant appears in Figure 8-22.

Translocations can be **balanced** or **unbalanced**. In a balanced translocation, the cell contains all the normal genetic material, but in a new arrangement. The individual does not always show a phenotypic effect, but he or she is a translocation carrier. His or her rearranged chromosomes will segregate at meiosis in such a way that many gametes—and offspring—will have too much or too little genetic material (an unbalanced genotype). The end result is embryonic or fetal abortion, stillbirth, infant lethality, or an abnormal phenotype such as mental retardation.

FREQUENCIES OF CHROMOSOMAL ABNORMALITIES

In the past decade, researchers have analyzed the karyotypes of thousands of live births. They have found that 5 to 8.34 per 1000 live births have chromosomal abnormalities. Because they cannot detect the very small abnormalities, and some have not used the most sensitive chromosomal banding techniques, the true figures are probably higher.

196

Most of the abnormalities are sex chromosome aneuploids. Other kinds of abnormalities, such as translocations, are more common in live births in a balanced form.

About 15 percent of all known pregnancies end in spontaneous abortion (many pregnancies are never recognized as such, because they end in very early spontaneous abortions). The resulting **abortuses** show a chromosomal abnormality rate of 40 to 50 percent (or more). About half the abnormalities are various trisomies, as indicated in Table 8-4. Table 8-5 shows that trisomy-X accounts for about 2 percent of all trisomies among abortuses. The rate for trisomy-16 is much higher (32.4 percent) than chance would suggest (100/23 = 4.3 percent), and we do not know the reason for this. Most others show less-than-chance rates, perhaps because the trisomy has such drastic effects in early development that the pregnancy is never recognized. In the case of trisomy-X, the low rate indicates that many affected infants are born and not aborted. The rate for trisomy-21 is remarkably high, despite the high percentage of Down syndrome births.

The frequency of chromosomal abnormalities among spontaneous abortions (50 percent) is 100 times greater than the frequency among live births (0.5 percent). Clearly, most chromosomal abnormalities are far more likely to result in abortion than in live birth. In addition, couples who have had multiple spontaneous abortions show frequencies of chromosomal abnormalities ranging from 3.0 percent to as high as 31.0 percent, compared with 0.5 percent for the general population.

Chromosomal abnormalities are definitely not rare. Half of all spontaneous abortions alone sets the frequency at 7.5 percent of conceptions. The 0.5 percent frequency among live births raises the minimum frequency to about 8 percent for all conceptuses, regardless of their fate. Detection of very small chromosome aberrations would surely

TABLE 8-4

Relative frequencies of chromosome abnormalities in human abortuses from spontaneous abortions

Total Number of Abortuses with a Chromosome Abnormality	Trisomic Condition %	XO Condition %	Triploid %	Tetraploid %	Other Abnormalities %
1863	52	18	17	6	7

Source: From the compilation by Carr, D. H., and Gedeon, M. 1977. *In* Hook, E. B., and Porter, I. H. (eds.). *Population Cytogenetics. Studies in Humans*, p. 4. New York: Academic Press.

TABLE 8–5
Trisomies among human spontaneous abortions

Chromosome Number	% of All Trisomies
1	0.0
2	5.6
3	0.8
4	2.5
5	0.1
6	0.3
7	4.5
8	3.7
9	2.8
10	1.9
11	0.2
12	0.8
13	5.7
14	4.2
15	7.3
16	32.4
17	0.7
18	5.1
19	0.1
20	2.7
21	8.3
22	10.1

Source: Data have been compiled from eight surveys by Chandley, A. C. 1981. *Annales de Genetique*, 24: 7.

raise the frequency even more, as would detection of very early abortions. When all factors are considered, scientists estimate that 20 to 40 percent, or even more, of all conceptuses carry a chromosomal abnormality. However, we should realize that these sobering statistics indicate that a rather effective genetic screen is at work. Most chromosomal aberrations are eliminated by spontaneous abortions long before they can be born.

SUMMARY

Humans carry their genes on 46 chromosomes, as can be shown by the process of karyotyping. Karyotypes reveal that the 46 chromosomes come in 23 pairs, differing in size, centromere position, and banding patterns. One pair, XX in females and XY in males, is the sex chromosomes. The rest are autosomes.

SUMMARY

Chromosomal abnormalities—variations in either number or structure—are common. Any change that causes more or less than two doses of genetic material results in genetic imbalance. The effects of these changes on the phenotype range from inconsequential to lethal. When cells contain an additional set or sets of chromosomes, they are called polyploid. When a cell has one or more chromosomes either absent or in addition to the basic set, it is aneuploid. Other abnormalities result from structural rearrangements following chromosome breakage. Some of the more common are deletions, duplications, inversions, translocations, isochromosomes, and ring chromosomes.

Sex determination in humans depends on the XX/XY system. The Y chromosome is necessary to initiate the development of male genitalia. A substance known as the H-Y antigen is fundamental in this development; it is produced as a result of the presence of the Y chromosome, though the H-Y gene may not be on the Y chromosome. Subsequent production of hormones contributes to the secondary sex characteristics. The many anomalies of sexual development include XX males, XY females, hermaphrodites, and pseudohermaphrodites, and others. Sometimes abnormal phenotypes are associated with abnormal genotypes, but clear associations do not always exist. More biological data are needed.

The theoretical 1:1 sex ratio is not observed. More males than females are generally born in large populations. People may soon be able to use reliable methods to choose the sexes of their children, and some unfavorable response to this prospect has already arisen from society.

Every X chromosome in the cell except one is inactivated very early in embryonic development. The inactivated chromosomes form compact masses called Barr bodies, which are easily visible with the microscope. The active X chromosome in each cell of the female may be of either maternal or paternal origin; the choice is initially random, although once made it is fixed for all of the cell's mitotic descendants. Hence females heterozygous for genes located on the X chromosome are actually mosaics for different alleles of the gene. This concept is the Lyon hypothesis.

Some of the sex chromosome aneuploidies are XO (Turner syndrome), XXX (trisomy-X), XXY (Klinefelter syndrome), and XYY. The XXXX, XXXXX, XXYY, XXXYY, and XXXXY individuals are rarer. Most of these sex chromosome anomalies are associated with a certain group of phenotypic traits, often including mental retardation. Generally, the more drastic the genotypic deviation, the more severe the phenotypic deviation. The principal mechanism causing aneuploidy is nondisjunction of chromosomes or chromatids during mitosis or meiosis in either or both parents. Mosaics (from one zygote) can result from nondisjunction, and chimeras (from two or more zygotes) can occur following the fusion of zygotes.

Autosomal aneuploidy also exists, but most types abort. Among live births, Down syndrome (trisomy-21) is the most common. Trisomy-18 and trisomy-13 are also observed among live births, but these children usually die very young. Down syndrome can be due to three separate copies of chromosome 21, a translocation involving chromosome 21, or mosaicism. The incidence of the syndrome increases with maternal age, but the nondisjunction event that supplies the extra chromosome 21 can occur in either parent.

Other chromosomal anomalies, such as deletions, inversions, and translocations, are often detectable with karyotyping, and many different ones are known. Their effects vary greatly; cat cry syndrome, which is due to a small deletion in chromosome 5, is fairly well known. The frequency of chromosomal anomalies is about one in 200 live births, and one in two spontaneous abortions. Most analyzed abortuses are trisomic for one of the chromosomes. The spontaneous abortion rate is at least 15 percent of all pregnancies, and—since we fail to detect many early abortions and small chromosomal abnormalities—the true rate for chromosomal abnormalities is probably 20 to 40 percent, or more.

DISCUSSION QUESTIONS

1. What made it difficult to count the exact number of human chromosomes? How were these difficulties overcome?

2. Regarding the possibility of someday being able to choose the sex of our children:
   a. Discuss one or more of the serious concerns.
   b. Discuss a possible medical advantage.
   c. Discuss the possibility that such a practice could help control population numbers.

3. Normal XX persons have only one active X chromosome in each cell. Persons having Turner (XO) and Klinefelter (XXY) syndromes also have only one active X chromosome in each cell. Yet the XX, XO, and XXY phenotypes show marked differences. Offer some possible explanations for the differences.

4. What might be some of the societal issues surrounding transsexualism (medical sex reversals)?

200

5. Why do some XO (Turner syndrome) fetuses abort while others do not?

6. Females wishing to enter the Olympic games must have their chromosomes checked. Why (or why not) is this check warranted? What should be checked specifically?

7. A young couple whose first child has Down syndrome is told that at their ages the chance of having a second child with Down syndrome is only 1 in 1000. Can this be correct?

8. Is it biologically possible for a female with Down syndrome to have a child with an I.Q. of 148? Why or why not?

9. If you had a brother or sister with Down syndrome, what further information would you seek, and why? How could you obtain this information?

10. If one parent carries a chromosome 21 translocation, probability figures indicate that one-third of the couple's children will have Down syndrome. However, only one-tenth of such parents' children have the syndrome. Give one or more possible explanations.

11. A friend of yours and his wife have been trying for several years to have a child. So far, four pregnancies have ended in spontaneous abortions. Your friend would like to have his wife's chromosomes karyotyped to find out what her problem might be. Comment.

12. Most chromosomal aberrations, aneuploid or structural, are associated with mental deficiencies. Discuss any significant points related to this observation.

PROBLEMS

1. Chromosome banding techniques (b) reveal a chromosomal aberration where standard microscopy (a) does not. What type of chromosomal aberration shows in (b)? Is the aberration homozygous or heterozygous? In this particular cell, did the aberration exist before replication or did it occur after replication?

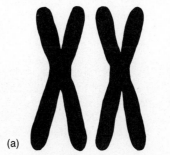
(a)

(b)

2. What is the sex chromosome constitution of a person with:
   a. male genitalia and two Barr bodies?
   b. female genitalia and no Barr bodies?
   c. female genitalia and three Barr bodies?

3. Give at least two sex chromosome constitutions that suggest that human maleness depends on the presence of the Y chromosome.

4. In the following chromosomal karyotypes, did nondisjunction occur in oogenesis or spermatogenesis? Could it have occurred in either of the parents? Support your answers with appropriate diagrams.
   a. XO
   b. XXY
   c. XYY

5. A child with an XXY karyotype has a normal XY father with a blood type of Xg^a and a normal XX mother with a blood type of Xg. The locus for the Xg^a and Xg alleles is on the X chromosome. What genotype must the child have for this blood type if the nondisjunction event occurred in the father? In the mother? Support your answers with simple diagrams.

REFERENCES

Altchek, A., ed. 1972. *The pediatrics clinics of North America*, vol. 9, no. 3. Philadelphia: W. B. Saunders Co.

Barr, M. L., and E. G. Bertram. 1949. A morphological distinction between neurons of the male and female, and the behavior of nucleolar satellite during accelerated nucleoprotein synthesis. *Nature* 163:676–677.

Carr, D. H., and M. Gedeon. 1977. Population cytogenetics of human abortuses. In *Population cytogenetics: Studies in humans*, ed. E. B. Hook and I. H. Porter. New York: Academic Press Inc.

Cavalli-Sforza, L. L. 1977. *Elements of human genetics*. Menlo Park, CA: W. A. Benjamin, Inc.

Chandley, A. C. 1981. The origin of chromosomal abberations in man and their potential for survival and reproduction in the adult human population. *Annales de Genetique* 24:5–11.

Degenhardt, A., P. Tholey, and H. Michaelis. 1980. Primary sex ratio of 125 males to 100 females? Analysis of an artifact. *Journal of Human Evolution* 9:651–654.

Hook, E. B., and G. M. Chambers. 1977. Estimated rates of Down's syndrome in livebirths by one year maternal age intervals for mothers aged 20 to

49 in a New York state study: Implications of the risk figures for genetic counseling and cost analysis of prenatal diagnosis programs. In *Numerical taxonomy on birth defects and polygenic disorders*, ed. D. Bergsma, R. B. Lowry, B. K. Trimble, and M. Feingold. Birth Defects Original Article Series 13(3A). New York: Alan R. Liss Inc.

Hook, E. B., and J. J. Fabia. 1978. Frequency of Down syndrome by single-year maternal age interval: Results of a Massachusetts study. *Teratology* 17:223–228.

Hook, E. B., and J. L. Hamerton. 1977. The frequency of chromosome abnormalities detected in consecutive newborn studies—differences between studies—results by sex and by severity of phenotypic involvement. In *Population cytogenetics: Studies in humans*, ed. E. B. Hook and I. H. Porter. New York: Academic Press Inc.

Hook, E. B., and A. Lindsjo. 1978. Down's syndrome in livebirths by a single year maternal age interval in a Swedish study: Comparison with results from a New York state study. *American Journal of Human Genetics* 30:19–27.

Jacobs, P. A. 1978. Population surveillance: A cytogenetic approach. In *Genetic epidemiology*, ed. N. E. Morton and C. S. Chung. New York: Academic Press Inc.

Jacobs, P. A., M. Melville, S. Ratcliffe, A. J. Keay, and J. Syme. 1974. A cytogenetic survey of 11,680 newborn infants. *Annals of Human Genetics* 37:359–376.

Kline, A. H., J. B. Sidbury, and C. P. Richter. 1959. The occurrence of ectodermal dysplasia and corneal dysplasia in one family. *Journal of Pediatrics* 55:355–366.

Lejeune, J. 1964. The 21 trisomy—Current stage of chromosomal research. *Progress in Medical Genetics* 3:144–177.

Lejeune, J., M. Gautier, and R. Turpin. 1959. Etude des chromosomes somatiques de neuf enfants mongoliens. *Comptes Rendus des Seances de l'Academie des Sciences* 248:1721–1722.

Lyon, M. F. 1961. Gene action in the X-chromosome of the mouse (*Mus musculus* L.). *Nature* 190:372–373.

Lyon, M. F. 1962. Sex chromatin and gene action in the mammalian X-chromosome. *American Journal of Human Genetics* 14:135–148.

Neilsen, J., K. B. Hansen, I. Sillsen, and P. Videbech. 1981. Chromosome abnormalities in newborn children: Physical aspects. *Human Genetics* 59:194–200.

Paris Conference. 1972. *Standardization in human cytogenetics*. Birth Defects Original Article Series 7(7). New York: The National Foundation.

Redding, A., and K. Hirschhorn. 1968. *Guide to human chromosome defects*. Birth Defects Original Article Series 4, ed. D. Bergsma. New York: The National Foundation.

Thompson, J., and M. W. Thompson. 1980. *Genetics in medicine*. Philadelphia: W. B. Saunders Co.

Tjio, J. H., and A. Levan. 1956. The chromosome number of man. *Hereditas* 42:1–6.

Walzer, S., and P. S. Gerald. 1977. A chromosome survey of 13,751 male newborns. In *Population cytogenetics: Studies in humans*, ed. E. B. Hook and I. H. Porter. New York: Academic Press Inc.

Winchester, A. M., and T. R. Mertens. 1983. *Human genetics*. Columbus: Charles E. Merrill Publishing Co.

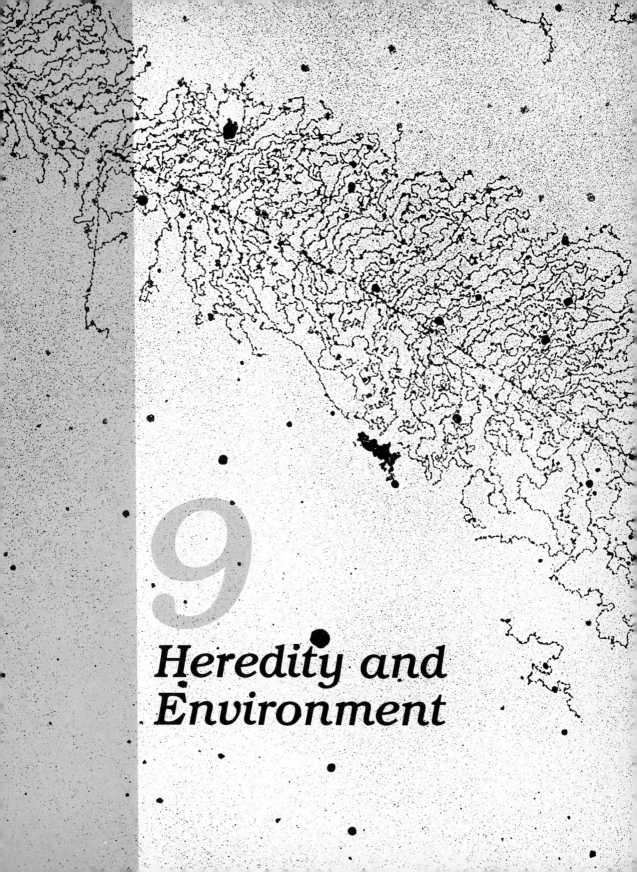

9

Heredity and Environment

What shapes us physically and mentally? Are we simply the products of our genes? Or are we instead the composites of influences of parents, teachers, nutrition, and climate? Are we the products of heredity or environment, of nature or nurture?

Perhaps the answer seems obvious to you, but it is a difficult question. The problem is that there is little agreement on just what that obvious answer is. The question has been debated since ancient times, and especially vigorously since Charles Darwin's extensive study of natural selection. Developments in the last decade have continued the controversy. The potent implications for society and politics have fueled the debate even though the data defining the roles of heredity and environment in humans are scarce.

The question has two opposing sides, represented by **hereditarians** and **environmentalists**. An extreme hereditarian advocates the position that certain traits are caused mostly or entirely by hereditary factors. Extreme environmentalists believe that traits are caused mostly or entirely by environmental influences. Since the most bitterly debated traits are directly or indirectly involved in sensitive social policies, the two positions are sometimes influenced by ideological bias, racial prejudice, and sexual stereotyping. Social prejudices are, of course, largely unrelated to the biological basis of human traits. In fact, many human traits result from an intricate combination of heredity, environment, and heredity-environment interactions.

Consider **scurvy**. Anyone can get this potentially fatal disease, which results from a vitamin C deficiency. Few people do get it because they ingest vitamin C (**ascorbic acid**) in fresh fruits and vegetables. However, in the seventeenth and eighteenth centuries, scurvy was common among sailors on long voyages, who had no access to fresh foods.

Is scurvy, therefore, an environmental disease? Many people would say so. However, the underlying cause is probably genetic. Unlike most other mammals, humans lack the enzyme L-**gulonolactone oxidase** and cannot synthesize their own vitamin C. Because of this metabolic block, humans must obtain the vitamin through their diet. We are all subject to a genetic deficiency that must be compensated for by an environmental supplement. The enzyme deficiency is a purely hereditary trait; however, scurvy is a disease with both hereditary and environmental components.

Internal and external factors mold our phenotypes, and the weight of their relative roles differs from trait to trait. The controversy arises because the roles are often difficult to separate, and because

206

some people place human beings—for whom learning is so important—in an evolutionary category separate from other animals. Perhaps the basic issue is one of definitions, as the example of scurvy shows.

DIFFICULTIES IN DETERMINING THE BASES OF TRAITS

The causes of many phenotypic traits remain mysterious because scientists have not yet sorted out all the complexities. This is true for many **congenital** malformations and diseases, which may have either hereditary or environmental causes. Furthermore, even when genetic epidemiologists have largely ruled out genetic causes, the specific environmental causes may remain elusive.

Researchers cannot do breeding experiments with humans, and they cannot readily control the human environment. They must rely on pedigrees to study related people, identifying phenotypic categories and attempting to discern patterns of heredity. Their work would be much easier if they had access to large populations of genetically identical humans, but only monozygotic twins (and triplets, etc.) share the same genotypes. It is rare when identical twins exhibit the particular traits that researchers wish to study. Frequently, the trait being studied is itself rare.

Pedigrees over many generations offer the best available data. Yet they are not perfect, for a trait may "run in a family" because of either heredity or a shared environment, or both. Child-rearing and dietary practices may persist through many generations in a family, and such factors may be just as important as heredity in explaining the differences between families such as the Jukes and Edwards.

In an early study of 2096 members of six generations of Jukes, researchers found 300 sexually immoral persons, 310 paupers, 207 criminals and thieves including 7 murderers, and 130 members convicted of various lesser crimes. Another study focused on Jonathan Edwards, a man of high intelligence and a past president of Princeton University. His family tree included 1300 people, among whom were 295 college graduates, 100 clergymen, 95 lawyers, 65 college professors, 60 medical doctors, 60 authors, 30 judges, and 80 public officials, and no convicted criminals.

Since in both familes all members were related to each other and shared at least some genes, it is easy to attribute their good or bad records to heredity. Yet family members tend to live together too, eat similar food, pass on their ways of coping with the world, and so on. It is just as easy to credit environmental factors for these family accomplishments. We must, therefore, be cautious in how we interpret repeating family tendencies, particularly in areas of behavior.

POLYGENIC INHERITANCE AND QUANTITATIVE TRAITS

Most human traits do not follow a simple mode of inheritance. Height, skin color, blood pressure, and intelligence are not *either/or* situations. Instead, such traits are continuous or **quantitative**; that is, their degree or magnitude can fall anywhere within a certain range. The best explanation for this variability in expression is **multiple gene** or **polygenic** inheritance combined with variable environmental influence.

Polygenic inheritance is more complicated than simple Mendelian inheritance. It involves more than one gene pair. Each gene pair behaves in Mendelian fashion with respect to its transmission, but usually the effect of each gene is small and **additive**. The genes' individual effects add together to produce a particular quantitative level of the characteristic. The effects of the various genes are sometimes assumed to be equal, but this may not be true for some traits. Certain genes may have greater effects than others. In addition, truly additive genes do not show dominance and recessiveness. The slight effects exerted by various alleles are simply summed.

Many human traits are probably determined by polygenic systems. Many congenital diseases may also be caused by the interaction of a fairly large number of gene pairs with environmental factors. Some good candidates are the neural tube defects, **anencephaly** and **spina bifida**, **cleft lip** and **palate**, and **clubfoot**, which all show great variation in expression. Scientists have hypothesized that polygenic traits can also be expressed discontinuously, as we will see later.

The inheritance of eye color has not been completely worked out, but it does not follow a simple Mendelian pattern. Many people believe dark eyes are dominant over light eyes, but the inheritance is more complex. The amount of **melanin** pigment produced in the eyes determines whether they will be dark (the browns) or light (the blues). Melanin production, in turn, seems to be controlled by several genes. Therefore, it is entirely possible for two light-eyed parents to pool all their genes for melanin production to produce a child with eyes that are darker than those of either parent.

208

How does polygenic inheritance work? Consider a simple hypothetical situation involving three gene pairs. Each capital-letter allele exerts a certain effect above a minimum on the phenotype; lowercase alleles do not. An *AABBCC* × *aabbcc* cross has *AaBbCc* F_1 progeny. The F_2 progeny of an *AaBbCc* × *AaBbCc* cross, however, show 27 different genotypes and 7 different phenotypes, as shown in Figure 9-1. The most common phenotype occurs 20 times in 64 offspring and is produced by 7 distinct genotypes, all of which have 3 capital-letter alleles.

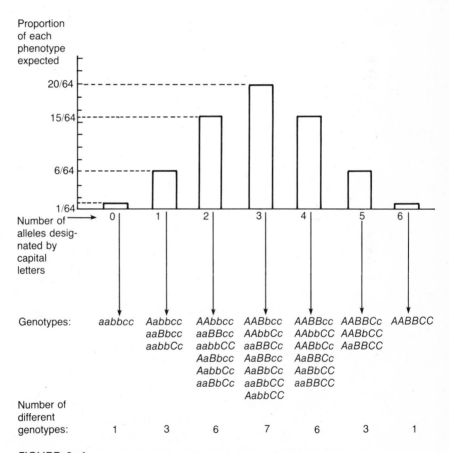

FIGURE 9–1

The proportion of different F_2 progeny expected from the cross *AaBbCc* × *AaBbCc* if the three gene pairs are involved in the expression of a polygenic trait.

209

In this basic example, we are assuming that all the genes contribute equally to the phenotype and that the genes are completely additive. This is not always the case. Furthermore, the environment usually plays some role in determining the phenotype. The expression of specific traits in a group of genetically identical individuals can vary considerably as a result of environmental differences. Identical twins are rarely identical in all physical respects; their heights, for instance, are often slightly different.

In Figure 9-1, three gene pairs produced seven phenotypes. With more gene pairs, the number of phenotypic classes increases, and the differences among them diminish. If the environment also affects the trait, phenotypes will tend to overlap. At this point, distinct classes will no longer be recognizable. The histogram (bar chart) becomes a continuous curve (a **normal distribution** curve), and it is difficult to distinguish hereditary from environmental effects.

HUMAN SKIN COLOR

Skin comes in many shades, depending on the amount of melanin present in skin cells. The amount of melanin, in turn, is controlled by three to six gene pairs and, to some extent, by the environment (exposure to sunlight). The trait serves as an excellent example to demonstrate polygenic inheritance with an environmental influence.

As early as 1913, C. B. Davenport concluded that two gene pairs govern human skin coloration. Although his methods were crude by today's standards, they do reflect considerable ingenuity in attempting to quantify skin color. Davenport studied the offspring of marriages between whites and blacks in Jamaica and Bermuda. He used paper disks with colored sectors of varying size, shown in Figure 9-2, to measure the shade of the skin of each experimental subject. He spun the disk like a top on a central axle so that the colors would appear to blend. The resulting whirling color could be made to match an individual's skin color simply by changing the proportions of the colors—especially black—in the disk sectors. The proportion of the black sector was reported as the measure of skin pigmentation. That crude measurement was taken on the upper arm, an area of the body Davenport assumed to be usually protected from sunlight by clothing. Although not very sophisticated, Davenport's results are informative. For example, one family gave the following readings: the white male parent matched a 5 percent black disk, and the black female parent matched a 71 percent black

210

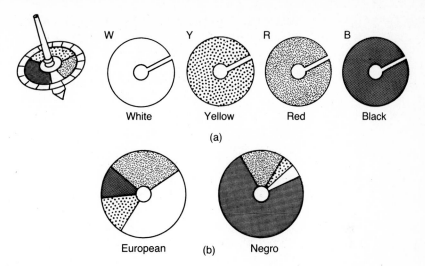

FIGURE 9–2
(a) The color top and the disks used by Davenport in his investigations of human skin coloration. (b) Proportions of colors set to match the skin of two persons in the investigation. (Redrawn from C. B. Davenport, The skin colors of the races of mankind. With permission from *Natural History*, Vol. 26, No. 1; Copyright the American Museum of Natural History, 1926.)

disk. Their seven children registered 37, 35, 35, 43, 37, 35, and 35, giving a fairly uniform cluster of readings. The mean, 36.7, comes very close to the **mid-parent value** (mean of the two parents) of 38.0. The tight clustering is a good indication that the two parental genotypes generate the extreme phenotypes. Apparently, the skin color genes of the two parents are all or mostly homozygous for their respective black and white phenotypes. In such a case, the phenotype of the F_1 progeny should be intermediate with respect to the parents if the system is truly additive.

Davenport calculated the number of gene pairs involved in the trait from data on F_2 progeny. By measuring skin shades in 32 children of parents whose shades fell between the extremes, he found a range of values. He recorded 2 of the 32 (or one in 16) as 0 to 10 percent black (the extreme white phenotype), 2 (or one in 16) as 51 to 60 percent (the extreme black phenotype), and the remainder between these extremes. He then concluded that two gene pairs controlled skin shade. His reasoning considered the genotype *AABB* as two gene pairs for black color and *aabb* as two gene pairs for white color, with the two gene pairs

211

behaving as a polygenic system. Then his observations can be explained as follows:

$$
\begin{array}{rl}
\text{parents:} & AABB \times aabb \\
\text{F}_1 \text{ intermediate progeny:} & AaBb \\
\text{F}_1 \text{ parents:} & AaBb \times AaBb \\
\text{Extreme F}_2 \text{ phenotypes:} & 1/16\ AABB \\
& 1/16\ aabb
\end{array}
$$

The other 14 out of 16 of the progeny show dark intermediate, true intermediate, and light intermediate phenotypes, as a Punnett square analysis would indicate. Davenport's rationale was precisely that of the Mendelian dihybrid cross, appropriately modified for additivity, and it extends to cases involving three or more gene pairs. In each case, the proportion of the F$_2$ progeny showing an extreme phenotype will be close to $(\frac{1}{4})^n$, where n is the number of gene pairs controlling the trait. The rationale can, however, fail if one or both parents do not represent the extreme phenotypes, if the gene pairs do not have equal weight in determining the trait, or if the environment greatly affects the trait.

Although Davenport's data seem to fit very well with the hypothesis that two gene pairs control skin color, subsequent research has used better instrumentation to measure skin shades and has shown that three to six gene pairs fit the data better. Davenport used a very small sample of the population that may not have been truly representative. Also, the environment certainly played a role regardless of the precautions taken, because sunlight causes the synthesis of melanin in skin cells. Nonetheless, Davenport's investigation is generally regarded as a creative piece of work, and the basic conclusion that multiple genes control human skin coloration was correct.

MEASURING THE HERITABILITY OF TRAITS

Variability and Variance

As you have just learned, today's researchers believe that three to six gene pairs control skin color intensity. If we consider the actual number to be five gene pairs, we can see that a single person can have 10, 9, 8, 7, 6, 5, 4, 3, 2, 1, or 0 capital-letter alleles and 11 different genotypes and phenotypes. This represents a large amount of variation in the trait, and the amount is still greater if we consider subtle (or drastic) environmental effects. It is not surprising that traits such as skin color show a continuous range of phenotypes rather than distinct, separable classes.

If we graph measurements of a continuous trait against frequency, we often obtain a normal distribution curve, as shown in Figure 9-3. This curve is symmetrical and bell-shaped, with its peak marking the **mean** or average of the measurements. However, it is not enough to know the mean to describe a distribution of values or to compare two distributions. It also helps to know the variability or spread of the data.

For example, consider the three distributions:

A: 10, 10, 10, 10, 10, 10   mean = 60/6 = 10
B: 5, 15, 5, 15, 5, 15   mean = 60/6 = 10
C: 1, 1, 1, 1, 1, 55   mean = 60/6 = 10

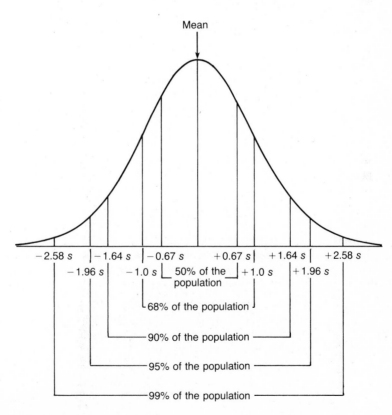

FIGURE 9–3
Curve of normal distribution.

FIGURE 9-4
Fundamental statistical formulas.

Mean: $\bar{x} = \dfrac{\sum\limits_{i=1}^{n}}{n} x_1$ where n is the number of observations

Standard deviation: $s = \sqrt{s^2}$

Variance: $s^2 = \dfrac{\sum\limits_{i=1}^{n} (\bar{x} - x_i)^2}{n - 1}$

Degrees of freedom: $n - 1$

All three groups of numbers have the same mean, but they are much more different from each other than this simple figure indicates. The differences show in the measure of variability called **variance**, and its square root, the **standard deviation**. Both calculations indicate how the data are distributed around the mean. Equations for these fundamental statistical measurements are shown in Figure 9-4. For our three sample distributions, the variances are 0, 30, and 486. For an example of the necessary calculations, see Table 9-1. The standard deviations are 0, 5.5, and 22.

In a normal distribution, about two-thirds (68 percent) of all measurements fall within one standard deviation ($\pm 1s$) of the mean. Slightly more than 95 percent of the measurements fall within two standard deviations of the mean, and over 99 percent fall within three (see Figure 9-3).

The Heritability Concept

Our brief excursion into basic statistics is essential if we are to understand how geneticists use the concept of variance to unravel the effects of heredity and environment on phenotype. The key lies in the fact that measurements of any quantitative trait show some variation that can be expressed as variance. This variance, the total **phenotypic variance** (V_t), can be viewed as having two parts, one due to genetic effects, V_g, and one due to environmental effects, V_e. The trait's **heritability** is:

$$h^2 = \frac{V_g}{V_t} \text{ or } \frac{V_g}{V_g + V_e}$$

Depending on the relative sizes of V_g and V_e, the heritability can have any value between 0 and 100 percent.

214

TABLE 9–1
Variances of three groups of hypothetical data

	X_i (Values)	$\bar{X} - X_i$ (Deviations)	$(\bar{X} - X_i)^2$ (Squared Deviations)	(Calculation of Variance)
	10	0	0	
	10	0	0	
Example 1	10	0	0	$s^2 = 0/5 = 0$
	10	0	0	
	10	0	0	
	10	0	0	
	$\Sigma = 60$	$\Sigma = 0$	$\Sigma = 0$	
	5	5	25	
	15	−5	25	
Example 2	5	5	25	$s^2 = 150/5 = 30$
	15	−5	25	
	5	5	25	
	15	−5	25	
	$\Sigma = 60$	$\Sigma = 0$	$\Sigma = 150$	
	1	9	81	
	1	9	81	
Example 3	1	9	81	$s^2 = 2430/5 = 486$
	1	9	81	
	1	9	81	
	55	−45	2025	
	$\Sigma = 60$	$\Sigma = 0$	$\Sigma = 2430$	

As a simple numerical example, suppose that we have measured a trait in seven individuals of different genotypes, all subject to the same environment. The trait has the values 1, 2, 3, 6, 10, 10, and 10, with a mean of 6 and a variance of 16.33. Since the environment is constant but the genotypes vary, all the variance is presumably due to heredity. That is:

$$s^2 = V_g = 16.33$$

Now suppose that we measure the same trait in seven individuals of the same genotype, but in different environments. The values are 4, 5, 5, 6, 6, 7, and 9, with a mean of 6 and a variance of 2.67. Now the variance is presumably all due to environment, so that:

215

$$s^2 = V_e = 2.67$$

We can calculate the heritability of this trait as:

$$h^2 = \frac{16.33}{16.33 + 2.67} = 0.86$$

What this means is that 86 percent of the variation observable in the trait in a population of varied genotypes living in a varying environment is due to the genetic differences. Only 14 percent is due to the environment.

The concept of heritability has proved very useful to plant and animal breeders. They are able to raise large groups of organisms (**inbred** lines) with identical known genotypes, and they can often control the environment to suit their needs. These advantages have helped them calculate the heritability of back fat thickness in pigs as 0.5 to 0.7, of body weight in sheep as 0.2 to 0.4, and of the rate of egg production in poultry as 0.15 to 0.30. The higher the heritability, the greater is the potential pay-off for selective breeding efforts. The lower heritability suggests that breeders will have more success if they manipulate feed, temperature, lighting, and other environmental factors.

Heritability is hard to calculate for human traits, because aside from identical twins there are no groups of genetically identical individuals. In addition, it is impossible to control the environment as researchers can for plants and animals. However, scientists do estimate human heritabilities by comparing identical twins reared in the same and different families, parents and offspring, uncles and nephews, and other combinations of relatives. Using these methods, for example, researchers have placed the heritability of cleft lip at 0.70 and of clubfoot at 0.80.

Heritability is more complex than we have indicated in this presentation. This can make calculations of heritability even more difficult. **Genetic variance** (V_g) can include variation that is due to dominant alleles (V_d), interactions of genes (V_i), and multiple genes with additive effects (V_a). Heritability *in the broad sense* is thus:

$$h^2 = \frac{V_d + V_i + V_a}{V_t}$$

If we assume that dominance and gene interactions play no part, then heritability *in the narrow sense* is:

$$h^2 = \frac{V_a}{V_t}$$

In essence, this figure applies to traits governed by polygenic inheritance.

We often view genotypes as if certain ones survive or reproduce best in all environments, while others are least successful in all environments. However, many data indicate that this is not always so. It is easy to find examples, such as:

	environment 1	environment 2
genotype 1	10	10
genotype 2	10	50
genotype 3	50	10
genotype 4	50	50

where the numbers represent arbitrary measures of some trait. Clearly two of the four genotypes respond very differently to the two environments. **Genotype-environment interactions** of this sort definitely exist, and the variance that results from such interactions (V_{ge}) is a separate and distinct component of phenotypic variance. Heritability in the narrow sense thus becomes:

$$h^2 = \frac{V_a}{V_t} = \frac{V_a}{V_g + V_e + V_{ge}}$$

Genotype-environment interactions may account for a large part of total phenotypic variation in humans, especially for such traits as hypertension and schizophrenia. However, interaction is hard to specify or measure. Our inability to account adequately for genotype-environment interactions is one weakness of heritability as it is used in human studies. There are enough other problems that some scientists believe heritability is too biased and misleading a concept to be useful in human genetics. These researchers think that too much information is lost or ignored in the process of developing figures on heritability. Other scientists favor using the concept, but with caution. We must always remember that we can only observe particular genotypes in particular environments, never all genotypes in all environments. Of course better techniques would be useful, but the problems are not easily eliminated.

BIOLOGICAL DETERMINISM

The concept of **biological determinism** states that human beings are essentially the products of their genes. In this view, the intrinsic nature of humans is quite rigid. However, many people challenge this concept, saying that the idea of biological determinism can be dangerous to society, because it can lead to discrimination based on a genetic rationale for racism, sexism, and class superiority. Yet extreme environmentalism can be just as dangerous. It holds that because humans are infinitely malleable, it should be possible to change them at will, ideologically, psychologically, and behaviorally. Both sides of the question have potent implications for society.

These views present the essence of the "nature versus nurture" argument. It is by no means dead, even though most researchers agree that few, if any, traits are solely genetic or solely environmental. Today, much of the argument focuses upon the relative importance of each component. All too often, the answers ignore the importance of genotype-environment interactions.

It is probably most accurate to say that an organism's genome defines its *potential* phenotype, but the environment determines the degree of the phenotype attained. For example, wild rabbits have white fat beneath their skin, while certain domestic strains of rabbits have yellow fat. The latter lack a specific enzyme which in the former breaks down **xanthophylls** (yellowish pigments in the plants rabbits eat) to colorless substances. Rabbits that lack this enzyme store the yellow pigments in their fat. However, those same rabbits will have white fat if fed on a xanthophyll-free diet. Genes determine potentials. Yet we should not lose sight of the genotype-environment interaction. This is a separate component, part of neither the genetic nor the environmental component of phenotypic variation.

The nature-nurture controversy rouses strong feelings, but we must keep our minds open. Biological determinism, which lacks a good scientific basis, is not very popular, because it places rather strict limitations on people. We must remain concerned about whether congenital disorders are genetic or environmental, or whether they are the products of interactions, and direct our search for treatments and cures accordingly. Ideally, our research efforts will ensure that all will achieve the maximum development their genetic potentials allow.

SUMMARY

The question of the relative influence of nature versus nurture (heredity versus environment) has been a long-standing controversy among biologists, psychologists, and others interested in the causation of human traits. The extremists are hereditarians and environmentalists. The traits most often debated obviously result from both genetic and environmental causes, often with an additional genotype-environment interaction component. Unraveling nature and nurture requires several assumptions. Pedigrees can help, even though they do not strictly separate environment from heredity.

Polygenic, or multigenic, inheritance results from many gene pairs. Polygenic traits are called quantitative since they spread across a measurable spectrum. They show a continuous distribution. The genes involved in polygenic inheritance have additive effects, rather than dominance and recessiveness. Usually, the environment plays an important role in the expression of quantitative traits. Because of the involvement of many genes and the environment, such traits exhibit a continuous normal distribution. Human skin color serves as an excellent example. It seems to be governed by three to six gene pairs, and there is no doubt about the effects of the environment.

In addition to the mean, the variance and standard deviation are useful statistics for describing the variability of a trait in a population. Variance is often used to determine the heritability of a quantitative trait. Heritability, in turn, is the proportion of the total phenotypic variance of a trait that is due solely to genetic effects. The total genetic variance can consist of additive effects, dominance effects, and gene interaction effects. In addition, one must reckon with the variances that are due solely to the environment and to genotype-environment interactions. The interaction variance can be quite extensive for some traits. Although heritability calculations have been extremely useful in plant and animal breeding, their value in human genetics has been vigorously contested. Some scientists believe that a lack of caution in heritability studies can erroneously promote the concept of biological determinism. At this time, however, a sound scientific basis for rigid biological determinism does not exist.

DISCUSSION QUESTIONS

1. What are the principal differences between complex multigenic traits and simply inherited Mendelian traits?

2. Exactly what is meant by multiple genic inheritance? Distinguish between multiple genes and multiple alleles.

3. Do you think that the height of each person is different from that of everyone else? Discuss the problems of proving or disproving your speculation.

4. Why is it sometimes difficult to determine whether the origin of a characteristic in humans is largely genetic or environmental? What measurement might help to answer this question? How can this measurement be accomplished?

5. Succinctly discuss how two short parents can have children who are very much taller than themselves.

6. One often hears that two definitely white parents can have a child who is definitely black. What do you think about such an event?

7. Crosses between parents of opposite phenotypes yield intermediate F_1 progeny. F_1 crosses yield F_2 progeny. Explain why the F_1 generation shows less variability than the F_2 generation.

8. A particular trait in a human population follows a normal distribution. Discuss whether the trait could be due to genetic causes alone, environmental causes alone, or both genetic and environmental causes.

9. The concept of heritability has become an integral part of some genetic specialties. What is your concept of heritability? Discuss its applications and its limitations.

10. How do the ways of life of human populations make it difficult to obtain solid heritability estimates from studies of similarities between relatives?

11. The following hypothetical data are the mean measurements of a trait for four different genotypes in two different environments. Explain why this is a good example of genotype-environment interaction.

| | phenotypic measurements | |
genotype	environment 1	environment 2
A	38	31
B	32	20
C	42	42
D	22	48

12. Knowledgeable farmers have long taken advantage of the phenomenon of genotype-environment interaction. How?

13. Is there any compelling reason for us to determine whether our quantitative traits are mostly genetic or mostly environmental? Whatever your answer, discuss it by offering reasons and pertinent examples. What dangers might arise in the pursuit of this information?

PROBLEMS

1. In the cross below, the parents have the indicated genotypes for a polygenic trait. Capital-letter alleles affect the traits equally. Lowercase alleles do not affect the trait. The subscript numbers refer to different loci.

    $$R_1R_1R_2R_2r_3r_3R_4r_4 \times r_1r_1R_2R_2R_3r_3R_4r_4$$

    a. Which parent has the higher degree of this trait?
    b. What is the maximum number of capital-letter alleles possible in any child from this cross?
    c. What is the minimum number of capital-letter alleles possible in any child from this cross?
    d. What is the probability that this mating will produce an $R_1r_1R_2R_2R_3r_3r_4r_4$ child?

2. To analyze the progeny of a cross between parents who are both heterozygous for the same two gene pairs, you need a 16-block Punnett square.
    a. How many different phenotypes, based on heredity alone, are possible among the progeny portrayed in the Punnett square if both genes affect one trait additively?
    b. How many different phenotypes are possible if the genes have slightly different effects on the trait?

3. The distributions below describe a particular polygenic trait in a species. Using the bottom graph, draw in the distribution you would expect for F_2 progeny.

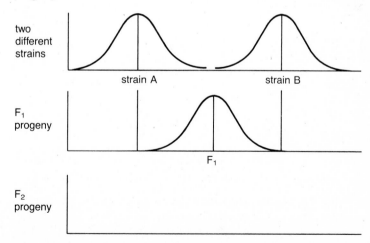

4. Assume that the normal curve below illustrates the expression of a polygenic trait in which many genes of equal importance are involved. Place

221

additional curves on the same axes in the appropriate positions to indicate the probable results of the following crosses. Do not forget that the environment can noticeably affect phenotypes, especially those due to a polygenic system.

   a.   (1) × (3)
   b.   (2) × (2)
   c.   (3) × (3)

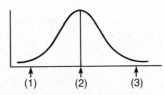

5. Assume that as a student of a particular quantitative trait in humans, you have been able to find many marriages between one person with the smallest possible measurement and one with the largest possible measurement. Their children all had measurements nearly intermediate to those of the parents. Among the children of matings in which both parents have this intermediate measurement, you count 102 very small, 412 fairly small, 601 intermediate, 405 fairly large, and 99 very large measurements. How many gene pairs are probably involved in the expression of this trait?

6. The mean intelligence quotient (I.Q.) of white school children in the United States is 100, with a variance of 225. Between what two I.Q. values would we find approximately two-thirds of the I.Q.'s for these children?

7. A particular trait in a population is studied to learn what contribution genes make to its expression. Researchers determine that environmental effects cause a variance of 52, genetic effects a variance of 26, and genotype-environment interaction a variance of 52. What is the heritability of this trait?

8. A study found that the heritability of trait A in a population was 80 percent. The heritability of trait B was 40 percent. Which trait has the greater environmental variance? Briefly explain the basis for your answer.

9. Given that in situation A the environmental effect on a trait is 50 percent, and in situation B the environmental effect is 40 percent, which situation shows the higher heritability? Briefly explain.

10. The environmental component of the total variance for a particular trait is four times as large as the genetic component of the total variance.
   a.   What is the heritability, assuming that there is no genotype-environment interaction variance?
   b.   What is the heritability if the genotype-environment interaction variance is 40 percent?

222

REFERENCES

Davenport, C. B. 1913. Heredity of skin color in Negro-White crosses. *Carnegie Institution of Washington* 188:1–106.

Davenport, C. B. 1926. The skin colors of the races of mankind. *Natural History* 26:44–49.

Harrison, G. A., and J. J. T. Owen. 1964. Studies on the inheritance of human skin colour. *Annals of Human Genetics* 28:27–37.

Stone, I. 1974. Humans, the mammalian mutants. *American Laboratory* 7:32–39.

223

10

Human Diseases with a Complex Genetic Basis

Many human diseases are due to complex **polygenic (multigenic** or **multifactorial)** inheritance. As we saw in Chapter 9, polygenic inheritance, when combined with environmental effects, leads to a continuous gradation of expression or symptoms. Researchers believe that polygenic inheritance may govern varicose veins, hypertension (high blood pressure), peptic ulcers, arteriosclerotic heart disease, migraines, cataracts, asthma, allergies, rheumatic fever, some forms of cleft palate, clubfoot, dislocation of the hip, and certain heart abnormalities, as well as some forms of cancer. Polygenic disorders may affect 1.7 to 2.6 percent of all live births, but polygenic inheritance affects us all—*polygenes* are probably associated with intelligence, emotional stability, and many other important human traits.

Polygenic inheritance usually involves many gene pairs, each with a small additive effect. Dominance and recessiveness do not apply to multigenic traits, and the **polygenes**—of which there can be hundreds—are not individually identifiable. The environment can significantly affect polygenic traits, and there can be strong genotype-environment interactions. Scientists question whether all the genes have equal weight in the expression of a polygenic trait or whether **major genes** exist that contribute disproportionately to the observed variability. Many researchers are convinced of the importance of major genes, but their existence is difficult to prove.

FIGURE 10–1
The presence or absence of three different genetic factors could be responsible for seven possible variations of a trait, independent of environmental influences.

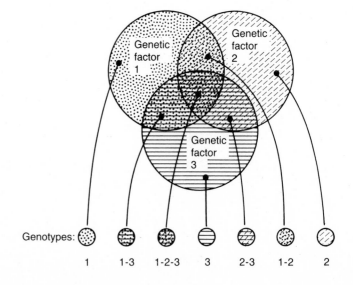

226

The hereditary bases for the tremendous variation in the expression of many traits, and in our susceptibility or resistance to some diseases, are probably very complex. For example, Figure 10-1 shows how the presence or absence of just three different hereditary factors can theoretically account for seven different phenotypes. Diverse powerful environmental influences on the system can increase the amount of variation immensely.

EPIDEMIOLOGICAL APPROACHES

Human disorders that affect more than one member of a family do not necessarily have a genetic origin. Family members share both a common environment and common genes, and the shared environment can be responsible for the disease. In fact, we can expect any trait to occur with a higher frequency among the siblings of affected persons than in the general population. Epidemiologists typically look first for a pattern of family resemblance in a given disorder, and only if that pattern is not obvious do they look elsewhere for the cause. However, if the risk to other members of the family is considerably higher than that of the general population, and no good evidence exists for a specific environmental cause, epidemiologists usually investigate a genetic cause. However, the genetic patterns usually deviate from the classic Mendelian ratios of 3:1 or 9:3:3:1. Often the risk to a patient's close relatives is only 2 to 5 percent, an observation that is best explained by polygenic inheritance—many genes and a strong environmental influence.

There are several criteria for deciding whether a disorder has any genetic basis. It should strike siblings, cousins, and other "blood" relatives in characteristic proportions. It should usually be absent among genetically unrelated family members, such as in-laws. If genetic, the disorder should have a characteristic age of onset and possibly a particular course of development. A researcher needs an adequate family history in order to identify the cause of the patient's disorder. Many physicians are now compiling such information routinely.

Geneticists have long used parent-child and sibling studies to evaluate the genetic bases of traits. Their favorite approach, however, is to study **dizygotic** (fraternal) and **monozygotic** (identical) twins. The key to the success of these studies is the concept of **concordance**. When both twins have the trait in question, they are said to be concordant for the trait. When one has the trait and the other does not, they are discordant. Researchers can calculate a **concordance rate** for any

227

TABLE 10-1
Concordances of certain diseases

	Monozygotic Concordance (%)	Dizygotic Concordance (%)
Cancer—same site	6.8	2.6
Cancer—all sites	15.9	12.9
Arterial hypertension	25.0	6.6
Manic depression	67.0	5.0
Asthma	47.0	24.0
Clubfoot	22.9	2.3
Measles	95.2	87.0

trait. They calculate this percentage for both monozygotic and dizygotic twins and compare them. Since monozygotic twins share 100 percent of their genes and dizygotic twins share, on the average, only 50 percent of their genes, a significantly higher concordance in monozygotic twins indicates a genetic cause. The relationship exists regardless of the magnitude of the concordances, as we can see in Table 10-1, which lists concordances for some human disorders. Since the high concordance for measles in monozygotic twins (95.2 percent) is almost as high in dizygotic twins (87.7 percent), we can be sure that measles must have an environmental cause, at least within those families. The concordance for clubfoot in monozygotic twins is relatively low (22.9 per-

TABLE 10-2
Heritability values for certain human disorders

Disorder	Heritability (%)
Asthma	80
Pyloric stenosis	75
Cleft lip and cleft palate	76
Late onset diabetes	35
Ankylosing spondylitis	70
Congenital clubfoot	68
Hypertension	62
Dislocation of the hip	60
Peptic ulcer	37
Congenital heart disease—all types	35

Source: Emery, A. E. H. 1976. *Methodology in Medical Genetics,* p. 54. New York: Churchill Livingstone.

cent), but in dizygotic twins it is only 2.3 percent, one-tenth of the value for monozygous twins. We can conclude that clubfoot has a large genetic component within those families.

Heritability expresses the degree to which a phenotype is genetically influenced, or the percent of the total variance of a trait that is due to genes. This measurement is of limited use when applied to human traits, and this creates an obstacle in the study of polygenic traits. Nonetheless, heritability is often calculated, reported, and discussed (or vehemently debated). Table 10-2 lists a few reported heritabilities for human disorders. Those for hypertension, asthma, and clubfoot reinforce the concordance rates in Table 10-1.

CANCER

It is not easy to see the effects of heredity in **cancer**, but they are there. A cancer (or a **neoplasm**) develops when for some reason a cell stops responding to the intercellular messages that normally regulate growth and development. The cell then multiplies without restraint, often forming a mass of cells called a **tumor**. The cell and its descendants may show various abnormalities of cell division and differentiation.

If a tumor lacks the ability to spread in the body, it is **benign**. If it can spread or **metastasize**, it is **malignant**; malignant cancers are often fatal, even when treated. There are about 200 distinct types of cancer known, and each one could have a different cause. Together, they account for almost 20 percent of deaths in the United States, and they offer biologists one of their most formidable challenges.

Etiology of Cancer

The **etiology** (cause) of the common forms of cancer is somewhat controversial. Epidemiological patterns have prompted some researchers to estimate that 60 to 90 percent of all human cancers are due to environmental causes, often **carcinogens** such as pollutants and food contaminants. However, there are also data that support genetic causes of cancer.

Cancer death rates vary greatly from one area to another. This may be due partly to genetic factors, but much of the variation could be due to differences in environments. For instance, the frequency of breast and prostate cancer is lower in Japan than in the United States, but when Japanese people emigrate to the United States, their frequencies for these cancers resemble the American frequencies within two

229

generations. Many other examples also show that the environment can play a decisive part in cancer etiology. On the other hand, some geographical differences in cancer rates can be linked to ethnicity to show that some genetic influence might also be at work. Here a good example might be **Ewing's tumor**, a cancer nearly absent in blacks no matter where they live. Studies of the distribution of cancers have given us some important clues to their general causes.

Twin studies have been less useful. Overall their evidence for a genetic influence on many cancers has not been impressive. For example, concordance rates for gastrointestinal and breast cancer have not been much higher for monozygotic than for dizygotic twins.

A better indication of genetic influence on cancer may come from inbred groups. Inbreeding increases the amount of homozygosity in an individual; that is, more gene pairs have identical alleles. Unfortunately, inbreeding has been difficult to study, mainly because of its rarity, especially in the United States. Certain groups known to be more homozygous than the general population, such as the Amish, do show higher susceptibilities to certain malignancies, but the cause may be either environmental or genetic.

When a cancer cell divides, both of its descendants are cancer cells. This suggests that genetic changes have occurred within the cell, and that we are confronting a genetic problem at the cellular level. What we want to know is whether individuals can have a genetic predisposition to the disease, regardless of the environment. The problem is not an easy one to solve, although there are many hypotheses. Figure 10-2 offers one simple hypothesis describing how a polygenic mode of inheritance might interact with various environmental influences. The more realistic **somatic mutation hypothesis** has received a great deal of attention. It suggests that an alteration of one or more genes in a somatic (body) cell might upset normal regulation of cell division and cause cancer. The most recent data strongly indicate that two or more such genetic changes are necessary to transform a normal cell into a cancerous cell. Such cellular changes can also be due to viruses. There is good evidence that specific viruses can cause cancer in animals and possibly also in humans.

Because any change in DNA structure can be harmful, it is not surprising that cells have an elaborate array of mechanisms for repairing damaged DNA. Research has shown that some rare genetic diseases, which are associated with a high incidence of cancer, are also associated with defective DNA repair mechanisms. These DNA repair diseases include **xeroderma pigmentosum**, **ataxia telangiectasia**, and **Fanconi's anemia**. Most evidence linking defective DNA repair to cancer

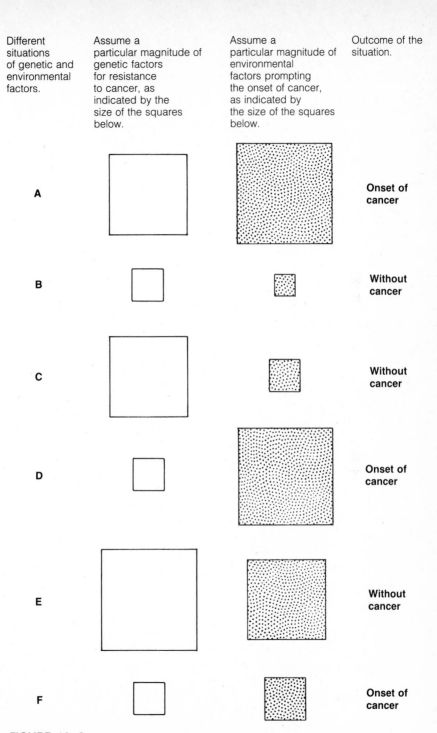

Different situations of genetic and environmental factors.	Assume a particular magnitude of genetic factors for resistance to cancer, as indicated by the size of the squares below.	Assume a particular magnitude of environmental factors prompting the onset of cancer, as indicated by the size of the squares below.	Outcome of the situation.
A			Onset of cancer
B			Without cancer
C			Without cancer
D			Onset of cancer
E			Without cancer
F			Onset of cancer

FIGURE 10–2

A hypothetical model for the occurrence or nonoccurrence of cancers as a result of polygenic inheritance in conjunction with environmental influences.

transformation is inferential, but some investigators are considering this possibility very seriously.

Cell growth is controlled by intercellular signals apparently mediated by molecules embedded in the cell membrane. These molecules include the **histocompatibility antigens**, which differ from individual to individual, and which can differ between normal and cancerous cells in a single individual. Since the body's immune system normally destroys cells with non-self histocompatibility antigens that differ from those that belong in the body, it may normally destroy the cancerous cells that continually arise. One hypothesis holds that cancer occurs when the immune system fails in this task. It is supported by observations that with age, the immune system loses vigor and cancer becomes more frequent.

Genes, Chromosomes, and Cancer

Many cancers run in families. Data indicate that 30 percent of all cancer patients have a close relative with cancer, 20 percent have two close relatives affected, and 7 percent have three or more. Some families have been completely devastated by the disease. For example, one woman with cancer of the cervix had a brother with cancer of the colon, a sister with breast cancer, and two nephews with rare blood cancers—and four of her six children had developed leukemia.

One study of 1142 traits controlled by single genes found 160 that were associated with increased risk of cancer. The best known of these traits is probably the recessive disease xeroderma pigmentosum (characterized by abnormal pigmentation, sensitivity to sunlight, freckling, and occasional mental retardation). As we have already discussed, xeroderma pigmentosum has been linked with malfunctioning DNA repair systems. Affected individuals almost always develop skin cancer before the age of 20. Two other recessive diseases that are strongly linked with cancer and thought to be due to a defect in DNA repair are ataxia telangiectasia and Fanconi's anemia. Even persons heterozygous for these diseases have three to five times the normal risk of dying of cancer. Although homozygotes for all three of these recessive diseases are fairly rare, their carriers are not uncommon.

In some cases, the genetic link to cancer is so strong that we can say one allele or a pair of alleles of a particular gene causes cancer. **Bilateral retinoblastoma**, marked by tumors of the retina in both eyes of children, acts as an autosomal dominant trait. **Polyposis** of the colon, marked by small, benign growths in the wall of the large intestine, also seems to be an autosomal dominant trait.

232

In still other cases, cancer seems to be linked with chromosomal aberrations. Down syndrome (trisomy-21) is accompanied by a greatly increased risk of leukemia. In **chronic myelogenous leukemia**, certain elements of the bone marrow proliferate, and 92 percent of the patients show a translocation of a small piece of chromosome 22 (the **Philadelphia chromosome**) to chromosome 9. Of the remaining 8 percent, most show a different translocation; very few patients have no detectable translocation. **Wilm's tumor** is associated with a third kind of chromosome defect, a small deletion.

Chromosomal abnormalities have been found in so many cancer cells, partly as a result of new high resolution and banding techniques, that some researchers are suggesting that almost all cancers must be associated with chromosome defects. We cannot definitively assert that the defects are causes or consequences of cancers, but recent research has shown that normal cells might contain **oncogenes**, genes with the potential to cause cancer. These genes may be activated by mutational events or by chromosomal rearrangements that change their relations to regulating genes. They can be activated by external events such as carcinogenic chemicals, ionizing radiation, and other factors. It seems wise for us to avoid such carcinogens if we wish to increase our chances to live long and healthy lives.

THE IDEA OF THRESHOLDS

Single-gene disorders such as xeroderma pigmentosum and diseases caused by chromosomal aberration, such as Down syndrome, are quite clearly all-or-none in their expression. A person either has the disease or does not. Polygenic diseases, on the other hand, appear as a range of conditions, their severities varying continuously from slight effect to fatal in some cases. Yet—obviously—even with such polygenic disorders as cleft palate, diabetes mellitus, peptic ulcer, and asthma, one either has the disease or not.

How can both statements be true? A possible explanation is the idea of a **threshold**, a level of severity below which the disease does not exist and above which it does. With a polygenic disease, this may mean that a particular number of specific alleles (polygenes) must be present in the genotype before the disease is expressed. This particular number of polygenes is the threshold. Figure 10-3 shows a threshold model designed to illustrate this concept.

A polygenically determined trait with a threshold does not seem to have a truly continuous range of severities. That is, it looks discontin-

233

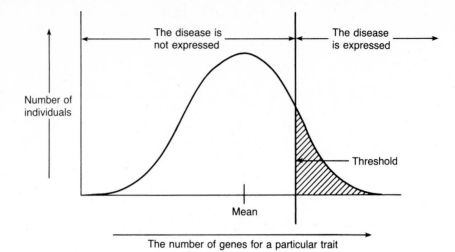

FIGURE 10-3
An interpretation of the threshold model to explain the occurrence of polygenic diseases on an all-or-none basis. Susceptibility to a particular trait is normally distributed, but the population is divided into two groups by a threshold: the normal and abnormal types. There is also likely to be an environmental factor contributing to the threshold.

uous or **quasicontinuous**. This is the case with **pyloric stenosis**, a constriction of the **pylorus** (the valve between stomach and small intestine), seen in 1.6 of every 1000 Caucasian newborns. This condition can be corrected by surgery. In this case, the threshold appears to be different for males and females. There is likely an environmental component to the threshold as well. More data are needed to support the concept, but already the idea of thresholds is helping to explain many of our observations of complex diseases.

DISEASE ASSOCIATIONS

Obviously, it would be immensely valuable if we could determine who carries a gene for a hereditary disease *before* the disease strikes. Then we might be able to begin treatment in time to do more good, for often any delay is damaging. The ideal early warning test would reveal the specific gene itself, but this is impossible when we do not know the gene's identity, location, or function.

234

A substitute for the ideal test is to associate a disorder with some other trait. If we can find a **marker** trait that accompanies the disorder more often than chance alone would allow, we can predict the occurrence of the disorder with some confidence. One of the first such markers was reported in 1953, when I. M. Aird and his colleagues associated blood group A with stomach cancer. Not every type A person gets stomach cancer, and not every stomach-cancer patient has type A blood, but a significantly higher proportion of stomach-cancer patients than of non-stomach-cancer patients has type A blood.

Since 1953, scientists have found many other associations between diseases and blood groups. Some of these associations are weak, but others are quite strong. Recently, researchers have emphasized associations between diseases and human **HLA antigens (human-leukocyte-associated antigens)**. This large and complex system of genetically controlled antigens affects the functioning of the immune system. The HLA antigen system is important in **transplantation** surgery, because when recipient and donor differ significantly in HLA antigen type, the recipient's body usually rejects the transplanted tissue or organ. HLA antigen compatibility can be important even in blood transfusions. Because there are so many different genes for HLA antigens, the HLA system has also been useful in paternity cases, twin diagnosis (monozygotic or dizygotic), and other identifications.

Some astonishing HLA antigen associations have been discovered. One of the most significant is between HLA antigen B27 and **ankylosing spondylitis**, a disease characterized by a stiff spine and arthritis. This disease occurs in 0.4 percent of the overall United States population, but in 6 percent of persons with HLA antigen B27. Even more convincing, 95 percent of all persons with this disease also have the B27 HLA type, and people with HLA B27 are 175 times more likely than others to develop the disease. This particular association is exceptionally strong, but researchers have also demonstrated other associations of varying degrees between more than 80 diseases and certain blood groups or HLA antigens. These associations include pernicious anemia with blood group A, gastric ulcer with blood group O, influenza with blood group O, multiple sclerosis with HLA antigen B7, diabetes with HLA antigen BW15, and Hodgkin's disease with HLA antigen B18.

A disease and a marker gene may be associated simply because the gene is located close to the marker gene on the same chromosome. On the other hand, the marker gene may actually be involved in the disease in some way we do not yet understand. This may be the case with the HLA markers, for the HLA antigens may play a part in the development of diseases or immunity to them. Clearly, people have

235

their personal repertoires of susceptibilities and resistances. At this time, however, the mechanisms that may link the HLA system, disease, and immunity are unknown.

Can known disease associations offer us any assistance in prevention or diagnosis? Detecting a marker can allow us to say that a person has a certain *probability* of developing a disease, but this is not the same as predicting the disease. At best, these associations will let us monitor high-risk individuals more closely in order to diagnose the disease early, when treatment is often more successful.

Markers may be most useful to genetic counselors, who can use them to refine their estimates of a person's probability of developing a disease or having a child with the disease. In extreme cases, having a particular HLA antigen might change a person's probability of contracting a disease from 1 percent (the value for the general population) to 9 percent, or more. Some aspects of genetic counseling are covered in Chapter 12.

In the future, it may be possible to determine HLA antigens and other markers very early in an individual's life. The individual could then be immunized against whatever disorders he or she is most likely to contract. Perhaps the information could be used to compile a list of the environmental factors to which the individual is most vulnerable. Before any such measures can become reality, we must know much more about the genetics and the environmental aspects of diseases, and about how the two groups of factors interact.

SUMMARY

Multigenic or polygenic inheritance provides the genetic basis of numerous human traits, including many diseases. Polygenic inheritance is characterized by many gene pairs with small additive effects, usually having a substantial response to environmental factors. Interaction between genetic and environmental factors can also be important. Much variation arises out of this complex web.

Family members generally have similar genotypes, but they also share similar environments. Separating the two components can be difficult. To assign a complex disease some genetic basis, epidemiologists look for definite proportions in the occurrence of a disease in a family, absence of the disease among genetically unrelated family members, and characteristic ages of onset. Detailed family histories are essential. Significantly higher concordance rates among monozygotic than among dizygotic twins indicate some genetic basis for a trait. When appropriate data are available, heritability can be calculated, but this measure has to be considered cautiously in human studies.

Cancer is a collective term for many diseases marked by abnormal cell growth and proliferation. Many medical investigators believe that most cancers are due to environmental agents, broadly called carcinogens. Correlations of cancer incidence with geography, in twins, and in inbred groups, among others, have suggested both environmental and genetic causes. Some of the popular hypotheses for the cause of cancers are somatic mutations, viruses, faulty DNA repair mechanisms, and a failing immune system. Some cancers are known to be caused by a single gene or gene pair. Recently, broken and abnormally rearranged chromosomes have attracted much attention as cancer causes.

The idea of thresholds in polygenic inheritance provides a plausible explanation for why polygenic traits do not always show a truly continuous distribution.

Many diseases are strongly associated with certain blood groups and HLA antigens (human-leukocyte-associated). The exact relationship involved is not known, but some medical geneticists are already contemplating the use of such associations for predictive purposes.

DISCUSSION QUESTIONS

1. What major characteristics of complex multigenic traits distinguish them from simply inherited monofactorial (Mendelian) traits?

2. How could a disease like cancer possibly be due to a multigenic inheritance when it is a discontinuous (either-or) trait?

3. As chief of a large, well-equipped research laboratory, you are given a budget to study the etiology of cancer. Your budget is just enough to employ three scientists with excellent research reputations. From which scientific discipline(s) would you select these researchers? Why?

4. What experiments would help you discover whether certain cancers are the result or the cause of chromosome abnormalities?

5. Study, draw appropriate conclusions, and comment on the following hypothetical data. Assume that a particular disease was investigated. The persons who eat food X live in a coastal region, and the persons who do not eat food X live inland. The table shows how many in each group get disease Y.

237

population studied	number of persons studied	number of persons who get disease Y by age 65
persons who eat food X 3 or more times per week	2460	86
persons who never or very rarely eat food X	9783	18

6. Millions of federal and private dollars have been spent on conquering cancer. Many people describe the situation as a huge "rip-off." What are some of the factors that have fostered this attitude?

7. In most countries, the incidence of stomach cancer rises with the mean amount of meat consumed per individual, as shown in the graph. Do these data mean that meat has a tendency to cause stomach cancer? Are alternative explanations possible? If so, what are they?

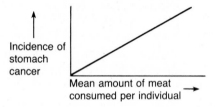

8. A very definite relationship exists between aging and the overall incidence of cancer. To which possible causes of cancer, if any, does this relationship lend support? Defend your choices.

9. If you were in complete control of the nation's cancer research budget, how would you apportion it (in percent) between (a) cure, and (b) prevention? Defend your position.

PROBLEMS

1. Many polygenic diseases show a 3 to 5 percent recurrent risk for the siblings of an index case. What would the recurrent risk be if the disease were caused by a recessive allele?

2. Often, the more severe a malformation, the greater the chance it will recur in another child born later into the same family. This observation supports the idea of polygenic inheritance as the cause for such malformations. Using letters for gene pairs, demonstrate why this conclusion makes sense genetically.

238

3. Certain strains of mice are 100 percent susceptible to one type of cancer upon being exposed to a particular environment. Other strains of mice are 100 percent resistant to this type of cancer under the same conditions. If these two strains were crossed with each other:
   a. What results would you expect among the F_1 progeny?
   b. Among the F_2 progeny?

4. Which of the following hypothetical concordances indicates (a) the greatest genetic component, (b) the least genetic component?

situation	monozygotic concordance	dizygotic concordance
A	60%	57%
B	7	0.5
C	83	83

5. The following sample data from three cities in England classify peptic ulcer patients according to blood group (A or O). The control data also organize non-ulcer patients by blood group. Do these data indicate any association between a disease and a human blood group? If so, which association is the stronger? Support your answers with appropriate calculations.

city	persons with peptic ulcer		control group without peptic ulcer	
	group O	group A	group O	group A
London	911	579	4578	4219
Manchester	361	246	4532	3775
Newcastle	396	219	6598	5261

Source: Woolf, B. 1955. *Annals of Human Genetics, London,* 19: 251–253.

REFERENCES

Aird, I. M., H. H. Bentall, and J. A. F. Roberts. 1953. A relationship between cancer of the stomach and the ABO blood groups. *British Medical Journal* i:799–801.

Dukes, C. E. 1954. Familial intestinal polyposis. *Annals of Eugenics* 117:1–27.

Emery, A. E. H. 1976. *Methodology in medical genetics.* New York: Churchill Livingstone Inc.

Mulvihill, J. J. 1977. Genetic repertory of human neoplasia. In *Genetics of human cancer,* ed. J. J. Mulvihill, R. W. Miller, and J. F. Fraumeni. New York: Raven Press.

Woolf, B. 1955. On estimating the relation between blood group and disease. *Annals of Human Genetics* (London) 19:251–253.

11

Consanguinity

Reproduction is the essential mechanism of hereditary transmission. Many mating systems exist among living organisms, and some of them exhibit nonrandomness. In humans, mating between relatives is of particular concern because of its medical impact and societal consequences.

Inbreeding is a broad term that refers to mating between individuals who are genetically more closely related than mates chosen at random. Naturally, the degree of relatedness—the intensity of inbreeding—can vary immensely. In human genetics, the adjective **consanguineous** describes matings between individuals who have a recent ancestor in common. Generally, consanguinity implies a common ancestor in the preceding two or three generations. **Incestuous** refers to matings between very closely related persons, such as father and daughter or brother and sister. Most human societies regard incest as taboo. Some animal species have even been observed to purposefully avoid mating between close relatives.

Many states have laws prohibiting marriages between individuals who are genetically related as first cousins or closer; however, other states do allow first-cousin marriages. Regardless of which levels of marriages are prohibited among relatives, states cannot always enforce their laws effectively. Various religious rules and customs are intertwined with the civil regulations in human societies. Many historians believe that the laws prohibiting consanguinity were initially set down for social reasons such as property considerations, and not necessarily for genetic reasons.

Parent-offspring matings and sib matings are the closest types of inbreeding possible in humans, and they are abhorred by most contemporary societies. On the other hand, second-cousin matings are generally viewed as acceptable by most societies. Various consanguineous mating patterns are illustrated by the pedigrees in Figure 11-1.

The frequency of first-cousin marriages in the United States and Europe is fairly low. Estimates range from 0.1 percent (1 out of 1000) to 0.5 percent (1 out of 200), although the rate varies considerably from one area to another. At one time, first-cousin marriages were looked upon with high favor in Japan, and the rate was more than 4 percent. This social acceptance is no longer the norm, and the rate has decreased considerably. Uncle-niece marriages are still favored in some regions of India, where their rates reach more than 10 percent. Overall, consanguineous marriages are decreasing because of the population's increased mobility, laws prohibiting such marriages, and changing attitudes. Rates are less than one percent in most soci-

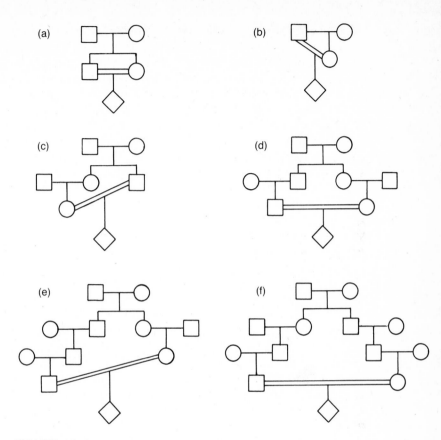

FIGURE 11–1
Examples of consanguineous matings: (a) brother-sister, (b) father-daughter, (c) uncle-niece, (d) first cousins, (e) first cousins once removed, and (f) second cousins.

eties. The reduction in consanguinity should decrease the genetic burden in the human population to some extent.

GENETIC CONSEQUENCES OF INBREEDING

The consequences of inbreeding emerge from the basic rules of Mendelian and population genetics. The principal genetic effect of inbreeding upon an individual or a population is an increase in **homozygosity** and a decrease in **heterozygosity**. The closer the relationship between parents, and the more alleles they share, the greater the tendency towards homozygosity. If like genotypes always mated (as when plants self-fertilize), the level of heterozygosity would decrease by half among the progeny of each successive generation, as shown in Table 11-1.

By increasing homozygosity, inbreeding fosters the expression of recessive alleles that are normally hidden in the heterozygous condition. Once the alleles are homozygous, the recessive phenotypes are expressed, and some of them are very detrimental. The number of deleterious alleles—recessive and dominant—per individual in a population makes up part of the population's **genetic load**. Researchers have estimated this load in terms of the mean number of **lethal equivalents** per individual. One lethal equivalent is a single recessive allele that, when homozygous, can cause death. Estimates range from one to 10, but most geneticists agree on three to four lethal equivalents per person. Inbreeding greatly increases the chance that such alleles will become homozygous. No matter how rare a particular recessive allele might be in the population, if a close relative carries it, there is a high probability that you carry it too. If the two of you mate, you face a high probability that your child will express the trait governed by the recessive allele. By similar reasoning, when a trait appears repeatedly in consanguineous matings, it is safe to assume that the trait is caused by a recessive allele following a Mendelian pattern of inheritance. Figure 11-2 illustrates the inheritance of such a trait. This relationship is especially obvious for extremely rare traits. However, the trait need not be recessive. Similar results can be observed for polygenic traits, even though their inheritance is much more complex.

TABLE 11–1

Increase in homozygosity resulting from mating between like genotypes, as in inbreeding

Genera-tions	Matings	Genotypes			Hetero-zygosity	Homo-zygosity
		AA	Aa	aa		
	Initial population	0%	100%	0%	100%	0%
I	100% Aa × Aa	25	50	25	50	50
II	50% Aa × Aa	12.5	25	12.5		
	25% AA × AA	25	0	0		
	25% aa × aa	0	0	25		
		37.5	25	37.5	25	75

244

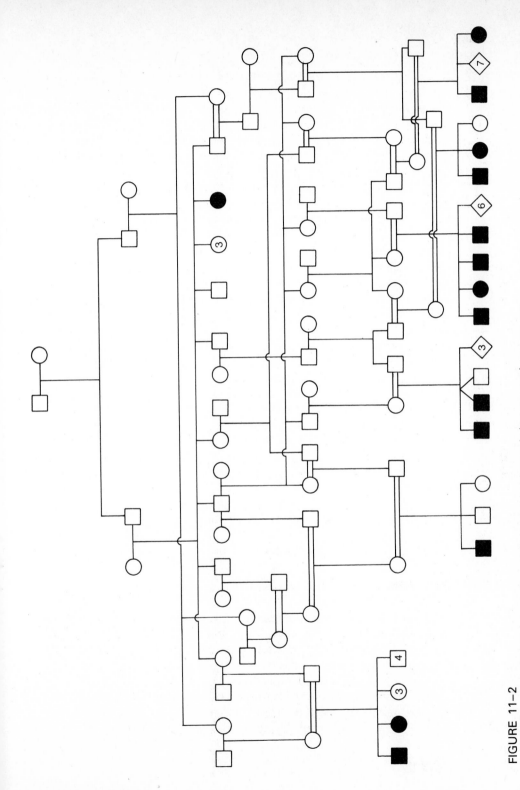

FIGURE 11–2
An example of a recessive hereditary trait (muscular dystrophy) expressed among consanguineous matings. The numbers within the symbols indicate the number of children with that phenotype. (Reprinted from *The American Journal of Human Genetics*, 10: 61–63, by David T. Hammond and Charles F. Jackson by permission of the University of Chicago Press. Copyright © 1958 by the University of Chicago Press.)

Plant and animal breeders use inbreeding to produce **inbred strains**. The mating of close relatives for many successive generations yields organisms that are homozygous at nearly all their loci. At that point, the chromosome segregation pattern at meiosis is always the same in parents, and the progeny are all identical to each other and to their parents. A number of plants and animals have been developed in this way for agricultural and other domestic purposes. Inbred strains have also been developed for laboratory research with mice, rats, rabbits, and other organisms.

Society is concerned about inbreeding because it increases the likelihood of genetic disease. Yet inbreeding by itself is intrinsically neither good nor bad. It simply increases homozygosity, and the effects of the recessive alleles it makes homozygous can be either beneficial or detrimental. They are more often the latter, but there is a chance that inbreeding will bring beneficial traits to the surface. Such success stories are common in domestic plants and animals. History also records examples of consanguinity without ill effects, as when the Egyptian pharaohs favored brother-sister matings. Yet the odds are not favorable, and the successes of plant and animal inbreeding call for careful culling of rejects. This part of the process is not acceptable to humans; we must care for all our progeny, live with them, and suffer with them if need be. Once our children are born, we cannot eliminate certain ones in the same fashion as plant and animal breeders.

The long term detrimental effects of inbreeding go beyond specific traits. Many organisms display a general lack of vigor after long, intensive inbreeding. This phenomenon is called **inbreeding depression**. We see its reverse when we cross two different inbred strains to produce a **hybrid** with much improved growth and vigor. Hybrid vigor, or **heterosis**, is of great importance in agriculture.

FIRST-COUSIN MATINGS AMONG HUMANS

First-cousins have two grandparents in common. Consequently, they share many alleles derived from these grandparents, as shown in Figure 11-3, and offspring from first-cousin matings can become homozygous for these alleles. If one member of a first-cousin marriage is definitely a carrier of a particular deleterious allele, the probability that the other first cousin carries the same allele, identical by descent, is one in eight (1/8). The probability that any child of this first-cousin mating is homozygous for the allele in question is then one in 32. (The probability that

246

FIGURE 11-3
A first-cousin consanguineous marriage can produce offspring homozygous for any particular allele in either of the great-grandparents.

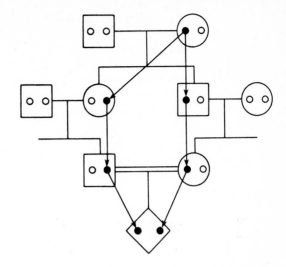

both cousins are heterozygotes is 1/8; the probability that two heterozygotes have a homozygous recessive child is 1/4; 1/32 is derived from 1/8 × 1/4.)

The pedigree in Figure 11-4 will help clarify the one in eight (1/8) probability that the one first cousin carries a particular allele carried by

FIGURE 11-4
The marriage involving III-2 and III-3 is between first cousins. For contrast, a marriage is also indicated between III-2 and III-1, who are nonrelatives.

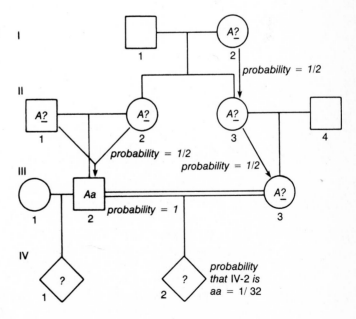

the other first cousin. It does not matter whether the allele is common or rare. Assume that III-2 is a known heterozygote, that is, a carrier for allele *a*. He received this allele from either his father (II-1) or his mother (II-2). The probability that his mother (II-2) is also heterozygous for *a* is then simply one in two (1/2). If II-2 is the heterozygous parent, the probability that her sister (II-3) is also heterozygous for *a* is also one in two (1/2), since either I-1 or I-2 must be heterozygous. Finally, if II-3 is heterozygous, the probability that III-3 is also heterozygous is one in two (1/2). All three conditions must be met for III-3 to be a carrier (assuming that she does not receive the allele from II-4), and the probability of each event is 1/2. Therefore, the overall probability is 1/2 x 1/2 x 1/2, or 1/8. Including the Mendelian probability of one in four (1/4) gives 1/8 x 1/4 or 1/32 as the probability for the trait to be homozygous in the offspring. That is:

III-2		III-3		Mendelian		IV-2
known carrier		first cousin		factor		probability of being homozygous
1	×	1/8	×	1/4	=	1/32

In a marriage between a known carrier (III-2) and a nonrelative (III-1), the probability assigned to III-1 depends upon the frequency of the allele in the population (q). If we assume Hardy-Weinberg equilibrium for this allele, the frequency of carriers in the population will be $2pq$. If we suppose that $2pq = 1/100$, then:

III-2		III-1		Mendelian		IV-1
known carrier		nonrelative		factor		probability of being homozygous
1	×	1/100	×	1/4	=	1/400

If III-2 is not a known carrier and he marries his first cousin, the calculation becomes:

III-2		III-3		Mendelian		IV-2
				factor		probability of being homozygous
1/100	×	1/8	×	1/4	=	1/3200

Finally, if neither III-2 nor his nonrelative mate is a known carrier:

III-2		III-1		Mendelian factor		IV-1 probability of being homozygous
1/100	×	1/100	×	1/4	=	1/40,000

The wide range in these probabilities should be enough to discourage most people from consanguineous marriages. The chances against such matings producing healthy children are too high.

MEASURING THE INTENSITY OF INBREEDING

Ultimately, the likelihood that a related couple will have a child with a genetic disorder comes down to whether the husband and wife are both carriers for the same deleterious allele. The probability depends on how closely related they are, because the risk involved in any consanguineous mating is proportional to the relatedness of the mates.

Determining such information requires some way to assign a number to relatedness, to measure consanguinity or the intensity of inbreeding. Once this number is developed, it can give both researchers and genetic counselors more ability to predict the outcomes of matings. Fortunately, there are several ways to obtain such numbers. One calculation is the **coefficient of inbreeding** (F), which simply states the probability that two related parents will have a child homozygous for a given allele, with the two copies of the allele identical by descent. This means that both copies of the allele must come from the same ancestor through two different routes in the pedigree.

We can use the calculations we have already performed to find the coefficient of inbreeding for first cousins. For any allele carried by one parent, the chance that the other parent also carries it is one in eight (1/8). Since only one of two alleles at any particular locus is transmitted to the progeny, the chance that the other parent will also transmit the same allele is $1/2 \times 1/8$, or 1/16. The value of F is thus 1/16 for first-cousin matings. It is the probability that the offspring will have two alleles identical by descent at any particular locus.

With brother-sister matings, the probability that the sister has an identical copy of any one of the brother's alleles is one in two (1/2), since a parent gives one of two alleles to each child. Because the sister can now transmit that allele to an offspring with the same probability of one in two (1/2), we have $1/2 \times 1/2$, or 1/4, for the inbreeding coefficient for **siblings**. This is a much more intense relationship than that

249

TABLE 11-2
Inbreeding coefficients for consanguineous matings

Mating Type	Inbreeding Coefficient
Siblings	1/4
Father-daughter	1/4
Mother-son	1/4
Uncle-niece	1/8
Aunt-nephew	1/8
First cousins	1/16
First cousins, once removed	1/32
Second cousins	1/64
Second cousins, once removed	1/128
Third cousins	1/256
Nonrelatives	0

calculated for first cousins. Table 11-2 lists the inbreeding coefficients for several other consanguineous matings. Note that the inbreeding coefficient drops to one in 256 for third cousins. Despite the last entry in the table, F never quite drops to zero, for we all share some ancestors in human history.

We should not confuse the inbreeding coefficient (F) with another measurement, the **coefficient of relationship** (r). This assessment of inbreeding refers to the proportion of alleles in any two individuals that we can expect to be identical by descent. The coefficient of relationship, r, is always twice the coefficient of inbreeding. When we calculated F, we first had to calculate r.

THE RISKS OF CONSANGUINITY

Although marriages closer than first cousins are not legal in North America, such matings do occur. For instance, it has recently come to public attention that father-daughter sexual abuse is much more common than most people imagine. Although there is no solid evidence that incest is more widespread in modern society than it was in the past, there is definitely more public recognition of the problem today. Perhaps this grows out of serious studies of child abuse of all types. In addition, the growth of the support provided by the women's movement may increase the likelihood that women will report incest within their families.

Defective offspring are too often the result of consanguineous matings. The data are limited, but they do support the theoretical consequences we have discussed. Children of related parents are more often mentally retarded, and there are enough data to show that related mates suffer more miscarriages, stillbirths, neonatal deaths, infant deaths, and juvenile deaths.

To be more specific, in 1971 E. Seemanova reported that in a study of women who had become pregnant by their brothers, fathers, or sons, 15 of 161 (9.3 percent) children were stillborn or died within the first year after birth, while 64 (39.8 percent) had serious physical or mental defects. In a control group fathered by nonrelatives, 5 of 95 (5.3 percent) children suffered the first fate and 4 (4.2 percent), the second. Generally, the risk to normal parents of having children with serious genetic defects is 3 to 4 percent. Considering the same types of defects, first-cousin matings have shown a risk of 6 to 8 percent. The percentages vary slightly from study to study, but the risks for first cousins are almost always twice as high as those for matings between nonrelatives. Some people do not consider the difference between first-cousin and nonrelative matings to be very significant, perhaps because they see the difference between 3 and 6 percent as only 3 percent; however, it is a twofold increase.

The difference is even more pronounced for the major genetic abnormalities and for closer consanguineous matings. Uncle-niece and aunt-nephew matings result in a three- to fourfold increase in birth defects, and an even higher risk (40 percent chance) is observed for parent-child and sib-sib matings. Eliminating consanguineous matings would certainly reduce the frequency of genetic defects and improve the survival rates of children.

CONSANGUINITY AND RECESSIVE DISORDERS

By studying the children of consanguineous marriages we can often determine whether a particular trait is hereditary. These observations are especially helpful in the cases of traits caused by rare recessive alleles. In such instances, the frequency of heterozygotes in the population is low, the frequency of marriages between unrelated heterozygotes is extremely low, and very few noninbred children with the trait are born. However, any heterozygous individual who marries a first cousin has one chance in 32 of having a child with that rare trait. Table 11-3 shows the probabilities of offspring being homozygous for recessive alleles with varying gene frequencies when the parents are unrelated and when they are first cousins.

251

TABLE 11-3

Probability of the offspring of unrelated parents and of first cousins being homozygous recessive for particular traits

Gene Frequency	Probability of Homozygosity in a Child of		Ratio
	Unrelated Parents ($F = 0$)	First-Cousin Parents ($F = 1/16$)	
0.1	0.01	0.016	1.6
0.01	0.0001	0.00072	7.2
0.005	0.000025	0.000335	13.4
0.001	0.000001	0.000063	63.0

Source: Crow, J. F., and Kimura, M. 1970. *An Introduction to Population Genetics Theory.* New York: Harper and Row.

The rarer a recessive allele is in the population, the greater is the proportion of homozygotes who owe their homozygosity to consanguineous parents. If the allele is common, most homozygotes will come from unrelated parents. In Table 11-3 we saw the relationship of gene frequency and homozygosity. Table 11-4 shows us that in a population in which 3 percent of the marriages are between first cousins, a trait with an overall frequency of 1 in 1,000,000 is due to first-cousin matings 65 percent of the time. Table 11-5 indicates how often children with some specific genetic disorders actually have parents who are first cousins.

Small populations, of course, are more subject to inbreeding, and some of the isolated religious communities scattered throughout North America offer excellent examples. By religious and social custom,

TABLE 11-4

The percentages of parents with a genetically abnormal child, who are expected to be first cousins

Frequencies of Genetic Abnormalities	First-cousin Marriages in the Population		
	0.5%	1.0%	3.0%
1/2500	1.5	3	9
1/10,000	3	6	16
1/100,000	9	16	37
1/1,000,000	24	38	65

TABLE 11–5

Percentage of first-cousin marriages among the parents of offspring with various recessive disorders

Genetic Disorder	First-cousin Marriages Among the Parents %
Phenylketonuria	5
Total color blindness	11
Tay-Sachs disease	15
Albinism	17
Congenital ichthyosis	24
Xeroderma pigmentosum	26
Alkaptonuria	33
Wilson's disease	47

their members seldom marry outside their group. Thus they have become genetic isolates by choice (other genetic isolates have existed because of geographic barriers). One study of Amish isolates found that 6.2 percent of the marriages were between first cousins or closer relatives, and 49 percent of them were between second cousins or closer relatives.

Some of these groups show a much higher than normal frequency of muscular dystrophy. One group shows a high frequency of a rare liver disease called **tyrosinemia**, which is virtually unknown elsewhere. Others display relatively high frequencies of other rare hereditary disorders. Today, some isolates are slowly dispersing into the general population. Consanguineous matings in modern societies are largely due to ignorance. Too often, the explanations of the hazards of inbreeding from medical and educational sources fall on deaf ears.

SUMMARY

Human inbreeding has both genetic and societal consequences. Consanguineous matings are those that occur between persons having a common ancestor in the preceding two or three generations. Incestuous matings are those between very close relatives. Despite state laws, religious rules, and a general societal disapproval, consanguinity cannot be considered rare. Some states in the United States allow first-cousin marriages, and some countries have laws and customs that allow various degrees of consanguinity.

253

Inbreeding tends to increase the amount of homozygosity in an individual or population. The closer the relationship of the mates, the greater the effect will be. The consequence of homozygosity is the expression of recessive alleles, many—but not all—of which are deleterious. Inbreeding greatly increases the chances of homozygosity for the very rare alleles that make up part of our genetic load. On the other hand, intense inbreeding under carefully controlled conditions has produced many fine and useful strains of plants and animals. Inbreeding can also cause a decrease in vigor, called inbreeding depression.

Most human consanguinity involves first-cousin matings. Basic to these matings is the probability of one in eight that both cousins carry the same allele at any particular locus. Therefore, a child whose parents are first cousins has a one in 32 chance of being homozygous for that same allele. The probabilities of producing children homozygous for a detrimental allele decrease drastically when the parents are nonrelatives.

The degree of inbreeding can be described numerically in several ways. The inbreeding coefficient (F) is the probability that an offspring will be homozygous at a given locus, with the two alleles identical by descent. The coefficient of relationship (r) is the proportion of alleles expected to be identical by descent in any two individuals. The F value is always twice the r value.

Generally, first-cousin marriages produce children with genetic defects at a frequency about twice that for marriages between nonrelatives. Uncle-niece and aunt-nephew matings result in twice as many children with genetic defects as first-cousin marriages. About 40 percent of the children of sib-sib and parent-child matings have genetic defects. All types of mortality are also higher in the offspring of consanguineous matings.

Religious and geographic isolates are often genetic isolates because of their high degree of inbreeding. These small populations show a large amount of genetic disease, often including extremely rare diseases. Consanguineous matings contribute a large proportion of all the children having rare genetic disorders, but lesser proportions of the more common genetic disorders.

DISCUSSION QUESTIONS

1. A friend comes to you for advice. His son and the neighbor girl wish to marry each other, but your friend is the father of the girl next door, unknown to everyone except the two biological parents involved. What advice would you give your friend?

2. The Brown brothers and the Green sisters are identical twins. Each of the Brown brothers marries one of the Green sisters. A boy results from one marriage and a girl from the other. Ultimately, these two first cousins wish to marry each other. The state of their residence allows first-cousin marriages, but should they be allowed to marry? Why or why not? What is the chance that they would have genetically defective children?

3. Why is consanguinity a good test for whether or not a trait is hereditary, especially for very rare traits?

4. Tay-Sachs disease is caused by an autosomal recessive allele. The percentage of cases resulting from consanguineous mating is 40 percent among non-Jews and 2 percent among Ashkenazic Jews. Why is the former figure so high and the latter so low? Explain fully.

5. In addition to inbreeding coefficients for individual matings, average inbreeding levels can be determined for an entire population. In most human populations, the average inbreeding coefficient is quite low. Why do you suppose this is so? Does the inbreeding coefficient ever reach zero?

6. Discuss the difference(s) between autosomal inheritance and X-linked inheritance with regard to inbreeding.

PROBLEMS

1. Diagram a pedigree showing the relationship for double first cousins and calculate the inbreeding coefficient.

2. Assume that a particular allele is deleterious when homozygous, and that the frequency of heterozygotes (carriers) in the population is one in 100. If a woman known to be heterozygous for this allele marries her first cousin, who has the normal phenotype:
   a. What is the probability that she will have a child with the trait?
   b. What is the probability that she will have a child with the trait if the frequency of carriers is one in 1000 rather than one in 100?
   c. What is the probability that she will have a child with the trait if the frequency of carriers is one in 10,000?

3. Joe and Mary, who are first cousins, are married to each other. Neither is a known carrier for the deleterious allele, *b*. However, Joe's brother Mike is homozygous for this allele.
   a. What is the probability that Joe is a carrier?
   b. What is the probability that Mary is also a carrier?
   c. What is the probability that their child will be homozygous for the *b* allele?

255

4. Demonstrate with appropriate pedigrees and calculations that the probability of producing a genetically defective child is the same for an uncle-niece mating as for a half-brother-half-sister mating.

5. A man marries his first cousin's daughter, and they have a son.
   a. Draw the necessary elements of the pedigree.
   b. Calculate the coefficient of inbreeding for the marriage.

6. A young man wishes to marry his second cousin, but he knows that he is heterozygous for a certain deleterious recessive allele.
   a. What is the probability that his second cousin has the same allele?
   b. If the second cousins do marry and have a child, what is the probability that their child would express the defective phenotype?
   c. What is the probability of producing a child with the defect if he marries a nonrelative?

7. A man who has married his first cousin has a brother with a recessive disorder. Both the first cousins have the normal phenotype, but what is the possibility that they are *both* carriers for the disorder?

8. Identical twin brothers marry identical twin sisters. The brothers are albinos. Could one couple have an albino child and the other a normally pigmented child? Illustrate your answer with gene symbols and simple pedigrees.

9. A boy and a girl are born into different families by AID (artificial insemination by a donor). In both cases, the donor is the same. Since the donor is anonymous, a chance exists that these two children could grow up and marry each other without realizing their relationship. If they did marry, what would the inbreeding coefficient be?

10. Consider a recessive genetic defect that occurs once in 10,000 live births in a population. How often would you expect to observe that defect among the offspring of first-cousin matings?

REFERENCES

Crow, J. F., and M. Kimura. 1970. *An introduction to population genetics theory.* New York: Harper & Row, Publishers Inc.

Hammond, D. T., and C. F. Jackson. 1958. Consanguinity in a midwestern United States isolate. *American Journal of Human Genetics* 10:61–63.

Jackson, C. E., W. E. Syman, E. L. Pruden, I. M. Kaehr, and J. D. Mann. 1968. Consanguinity and blood group distribution in an Amish isolate. *American Journal of Human Genetics* 20:522–527.

Laberge, C. 1969. Hereditary tyrosinemia in a French Canadian isolate. *American Journal of Human Genetics* 21:36–45.

Schull, W. J., and Neel, J. V. 1965. *The effect of inbreeding on Japanese children*. New York: Harper & Row, Publishers Inc.

Seemanova, E. 1971. A study of children of incestuous matings. *Human Heredity* 21:108–128.

12

Genetic Counseling and Genetic Screening

No geneticist would claim that we know everything there is to know about human genetics and its relationship to hereditary disorders. Certainly, we know the basic rules of heredity, how genes in general exert their effects through polypeptides, and how genes and chromosomes can be altered both biochemically and structurally. If we know gene frequencies in the population, we can predict the proportion of newborns with birth defects. Our predictions are more accurate if we are dealing with consanguineous matings or if we know that one or both parents carry a particular allele. The business of making genetic predictions and advising persons with questions about genetic problems is the purview of **genetic counselors**. These scientists are important members of the medical research community—their numbers, contributions to society, as well as their ethical dilemmas are steadily growing.

GENETIC COUNSELING

Genetic counselors are interpreters of medical and genetic information. Their job is to help parents, prospective parents, and family members to understand the biology and genetics of birth defects. They strive to explain risks and to clarify the available ways of preventing children with defects. As public awareness grows, so does the demand for genetic counseling services, and though there are several hundred genetic counseling centers in the United States today, more than 90 percent of the Americans who could benefit from professional genetic counseling are not receiving it. Many do not even know it is available. This is unfortunate, for genetic disorders account for 42 percent of all childhood deaths and 25 percent of all persons in hospital beds. Genetic counseling could enable people to eliminate many of these deaths and illnesses.

The majority of people seeking genetic counseling belong to the higher socioeconomic classes, although this situation is gradually changing. They are referred to genetic counselors by physicians, knowledgeable friends, and various media. As genetic counseling services become more well known, perhaps more physicians will refer all classes of patients and counseling will reach all who need it.

Who needs genetic counseling? Anyone might benefit from it, because any couple can have deleterious alleles that may result in an affected child. For most people, however, the risks are small. In certain groups the risks are high enough to benefit from counseling, as when one or both mates are known carriers of a deleterious allele or chromosomal defect, when the would-be mother is over 35 years old, when

260

the parents have already had one or more children with physical or mental defects, when there have been repeated fetal losses, when the parents' families have a record of genetic defects, or when the parents are closely related. Racially mixed marriages and family histories of mental illness or retardation often bring people to counselors, too. In each case, the counselor must explain the risks and options.

Genetic counselors often use biochemical tests to determine whether the parents are heterozygous for a deleterious allele, but often heterozygosity is not detectable. Also, some genetic diseases can have several genetic bases, a situation called **genetic heterogeneity**. In such cases, counselors base their advice on parental ages, family histories of disorders, spontaneous abortions and stillbirths, phenotypes of previous children, and other factors. They must explain why they cannot give precise probabilities for the recurrence of mental retardation and mental illness when the cause cannot be determined. Even when persons have the defective genotype, the severity or **expressivity** of the disease can vary greatly from case to case. Some people having the gene may not show the disease at all; that is, the trait lacks **penetrance**.

Genetic counselors provide parents with both biological and statistical information. They endeavor to correct misconceptions about heredity, explain just what the disorder is and how it arises, and give the risks in terms the parents can understand. Usually, they also instruct parents in how to avoid an affected child through prenatal diagnosis and selective abortion, artificial insemination, contraception, sterilization, and adoption, but most counselors try not to force a decision. Indeed, most counselors try to assure the well-being of both the future child and the family. The parents' decision depends partly on the counselor's effectiveness, but also on their knowledge of genetics and probability, level of education, and cultural background.

How effective is genetic counseling? Some researchers report that about half of those counseled achieve a good grasp of the information, one-fourth gain only part of the essentials, and one-fourth learn little, if anything. The success of the interaction depends on the counselor's skills and, more importantly, on the parents' individual willingness and ability to understand. Genetic counseling would have increased impact if more people knew more about genetics, biology, and medicine.

Genetic counselors strive to be as effective as possible. Only then can they help as they should—at the same time minimizing the inevitable legal entanglements. Physicians have already been sued for not telling patients that genetic counseling services were available. And

261

parents have even been sued by their children with genetic disorders for not preventing their birth.

Should we prevent the births of all children with defects? How can we objectively compare the quality of life with a defect to no life at all? Just what is the proper role of the genetic counselor? Are selective abortion, sterilization, and other reproductive manipulations moral or ethical? Should prospective parents, fully aware that they are both carriers for a deleterious allele, knowingly have children? When a woman is very likely to have spontaneous abortions or stillbirths, does she have a moral obligation to prevent these fetal deaths by preventing pregnancy? The ethical issues only become more complex with time and medical developments.

CHANCE AND RISK

One of the genetic counselor's most difficult tasks is making probability and risk understandable to those being counseled, whatever their educational level. This problem is challenging because recurrence risks are not usually easy to determine. The pattern of inheritance may not be fully known, or it may be very complex or ambiguous. Average risk tables are available for many disorders, as shown in Table 12-1, and they are used by genetic counselors; but each individual case is unique. The probabilities vary with parental age, the conditions of other family members, ethnicity, the history of spontaneous abortions, and many other factors. For instance, at birth the child of a parent with Huntington's disease has a 50 percent chance of having the disease someday. However, if the person reaches age 48 without expressing the symptoms of the disease, the odds of contracting it decline dramatically.

With some problems, of course, the genetics is straightforward and well known. The counselor's responsibility then is to make the information understandable to the clients. For example, many people have trouble grasping the exact meaning of a probability of "one in two" or "one in four." Some think that since any event will either happen or not, everything has one chance in two of happening. Consequently, these persons are not overly concerned about a one in two chance of having a defective child. Others think that a one in four recurrence risk means that since they have already had a defective child, they can now expect to have three normal ones. They do not understand that the probability of each successive event is independent of those that preceded it, that chance has no memory.

262

TABLE 12-1

Risk values for some common disorders

Disorder	Overall Incidence (%)	Incidence (%) for Normal Parents Having a Second Affected Child	Incidence (%) for an Affected Parent Having an Affected Child	Incidence (%) for an Affected Parent Having a Second Affected Child
Anencephaly	0.20	2	—	—
Cleft palate	0.04	2	—	—
Cleft lip with or without cleft palate	0.10	4	4	10
Clubfoot	0.10	3	3	10
Congenital heart disease (all types)	0.50	1–4	1–4	—
Diabetes mellitus (early onset)	0.20	8	8	10
Dislocation of hip	0.10	4	4	10
Epilepsy (idiopathic)	0.50	5	5	10
Hirschsprung's disease	0.02	—	—	—
Manic depressive psychoses	0.40	—	10–15	—
Mental retardation (idiopathic)	0.30	3–5	—	—
Profound childhood deafness	0.10	10	6	—
Pyloric stenosis	0.30	—	—	—
Schizophrenia	1–2	—	16	—
Scoliosis	0.22	7	5	—
Spina bifida	0.30	4	3	—

Source: Emery, A. E. H. 1975. *Elements of Medical Genetics.* Berkeley, CA: University of California Press.

Since people's perceptions of identical probabilities vary, they may understand risk quite well yet still make what strike others as unwise decisions. They may see one in four as too high a risk for an extremely deleterious disorder, but not for a less serious one. Some may be willing to accept a one in four risk, while others will reject a one in ten risk for a disease of comparable severity. People vary in disposition as well as comprehension; genetic counselors must recognize these differences and work with them.

PRENATAL DIAGNOSIS

Selective abortion, which we have already mentioned, is the medically induced abortion of a fetus that has been shown by some method of prenatal diagnosis to bear a defect. It is *not* the same as an elective abortion chosen solely for convenience. It is mostly motivated by the

263

desire to prevent parental heartbreak, medical expense, and the child's future suffering. Nevertheless, opposition to all kinds of abortions has limited the application of prenatal diagnosis and selective abortion.

Obviously, prenatal diagnosis need not result in abortion. Once informed that the fetus has a defect, parents can prepare to care for an abnormal child. Prenatal diagnosis can also alert physicians to be ready to treat newborns with specific defects. And when the result of the diagnosis is good news, it can greatly ease anxiety.

Amniocentesis

Several methods of prenatal diagnosis are now available. They involve observations of the fetus itself, karyotypes of fetal cells, or analyses of fetal chemical products in the amniotic fluid or in the mother's blood. Both fetal cells and fetal products can be obtained by sampling the **amniotic fluid** (the fluid that surrounds the developing fetus within the **amniotic sac**). The process of sampling this fluid is **amniocentesis**, and is illustrated in Figure 12-1.

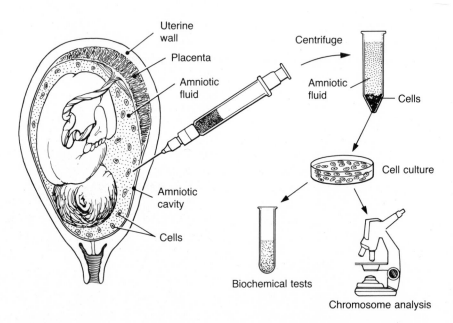

FIGURE 12–1
In amniocentesis, amniotic fluid and suspended fetal cells are withdrawn for analysis. (The cells in the amniotic fluid are greatly magnified in this illustration.)

264

Amniocentesis requires the withdrawal of 5 to 20 milliliters of amniotic fluid from the amniotic sac with a needle and syringe. The physician inserts the needle through the abdominal wall into the cavity of the amniotic sac, as shown in Figure 12-2. Most amniocenteses are performed during the sixteenth to eighteenth week of pregnancy. Few cells are found in the amniotic fluid before 14 weeks; from then until 16 to 18 weeks the number increases rapidly. Thereafter, the proportion of cells able to grow and proliferate decreases. The cells that are collected are those sloughed from the fetus' skin, gastrointestinal tract, respiratory tract, and other surfaces, and from the amnion. Thousands of these cells are found in every milliliter of amniotic fluid during the seventeenth or eighteenth week of pregnancy.

The feasibility of amniocentesis as a prenatal technique depends upon advances in several areas. Medical researchers first had to devise methods to extract the amniotic fluid safely, to culture the amniotic fluid cells successfully in a synthetic medium, and to test for genetic defects. The tests rely on chromosome analyses and biochemical determinations. Since the tests usually take two to three weeks, and they cannot begin before the sixteenth week of pregnancy, they leave little time for a decision to have an abortion.

It is now possible to test for any of the known chromosome abnormalities and more than 100 defects due to enzyme disorders. Some of the detectable biochemical and metabolic disorders are Fabry's disease, Tay-Sachs disease, certain types of Gaucher's disease,

FIGURE 12–2

Amniocentesis is the collection of amniotic fluid for use in diagnosing fetal disorders. A syringe is passed through the uterine wall into the amniotic cavity to extract the fluid. (Photograph courtesy of March of Dimes, Birth Defects Foundation.)

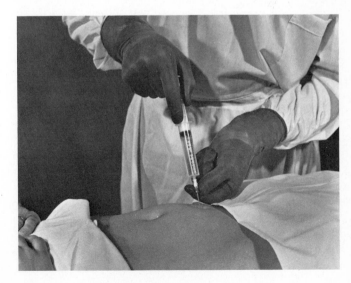

265

Hunter syndrome, galactosemia, cystinosis, sickle cell anemia, thalassemia, xeroderma pigmentosum, neural tube defects, and maple syrup urine disease. Many more disorders cannot be detected prenatally, but research in this area is very active and the list of diagnostic tests grows steadily.

Most amniocenteses and other prenatal diagnoses are performed on individuals who may have higher than normal risks for having genetically defective children. The frequency of Down syndrome in the offspring of older women is a constant concern, and many prospective mothers who are over 35 years old choose to undergo prenatal testing for this abnormality. Other high-risk parents are those who have had a number of previous spontaneous abortions, one or more children with defects, have genetic defects themselves or have other family members with defects, or are known carriers. Some prospective parents without any of these problems simply want to ease anxiety, or to learn the sex of their child. Amniocentesis for this last purpose might be considered unnecessary and frivolous, but sometimes it is important to know the child's sex, as when the prospective mother is a carrier for a detrimental X-linked recessive. In such cases, the mother might choose to abort sons but not daughters.

We should stress that if amniocentesis yields good news—which it does 90 to 95 percent of the time—that *in no way* guarantees the birth of a genetically healthy baby. The fetus is tested for only a few of all the possible genetic defects—one of the many other defects could occur in the child, just as in any offspring.

Amniocentesis is not problem-free. The risks are generally small, but any time tissue is invaded, there is a chance of infection. When the procedure was new, the fetus would occasionally be struck by the amniocentesis needle. That risk is almost nonexistent now, because ultrasonic devices are used to locate the fetus more precisely. Sometimes physicians have difficulty obtaining enough fluid or culturing the cells successfully; the procedure is usually repeated in such cases.

Very occasionally, a diagnosis is wrong. This can happen if the syringe draws in a few maternal cells, or if some of the fetal cells undergo genetic changes causing a mosaicism of cells in culture. A **false negative** occurs when the test indicates no defect, but the offspring does have the genetic defect; this usually causes great emotional difficulties for the parents, who feel misled. A **false positive** occurs when a fetus is normal, but the test wrongly reports that the offspring would be defective. If such a fetus is aborted, the discovery that it appears to be normal is not easy to accept. Both kinds of errors can be minimized by testing several different cultures from the same fetus. Overall, the reliability of amniocentesis is very high, between 99.4 and 99.7 percent.

Finally, there is a 1.5 to 3.5 percent risk of spontaneous abortion following the amniocentesis procedure. This is a great disappointment to parents of the 95 percent of fetuses that yield negative results for the genetic defects tested. In one study, 1040 pregnant women underwent amniocentesis; 36 (3.5 percent) fetuses were subsequently lost as a result of spontaneous abortions and stillbirths. A control group of 992 had 32 (3.2 percent) fetal losses during the same time period. However, 39 fetuses were selectively aborted following amniocentesis, while only one fetus from the control group was aborted in this manner. Since some of the 39 fetuses would probably have been lost spontaneously, the true post-amniocentesis spontaneous abortion and stillbirth rate should be somewhat higher than 3.5 percent. Nonetheless, at worst the risk is only slightly higher, and there are often very good reasons for taking that slight risk. Appropriate genetic counseling should always precede amniocentesis.

The use of amniocentesis varies greatly from state to state. It is more common in urban areas than in rural areas, and among better educated people. Although the cost varies considerably from area to area, the procedure is usually expensive enough that income and funding play a role in determining who will use it. Surprisingly, religious preference does not seem to influence which women will choose to undergo the procedure.

Other Methods of Prenatal Diagnosis

The fetus can also be diagnosed and monitored with techniques other than amniocentesis. **Radiology** uses X rays to detect gross bone and limb anomalies, but it is accompanied by the risk of radiation-caused mutation and malformation in the developing fetus. One also has to be concerned about causing leukemia in the newborn and mutation in the mother's ovaries.

Ultrasonography is a more widely used noninvasive prenatal diagnostic technique. It relies upon pulsed high-frequency sound waves of low intensity. As these sound waves echo from the body parts of the fetus, they are transformed into a visual display on a screen, as shown in Figure 12-3. The latest versions of the technique can give rather good detail of the heart, bladder, kidneys, liver, spine, limbs, and other body structures. In the hands of a skilled operator, ultrasonography can reveal microcephaly, cardiac malformations, cleft lip and palate, brachy-

267

FIGURE 12-3
Echogram of a fetus. (Used with permission of Girard Photography.)

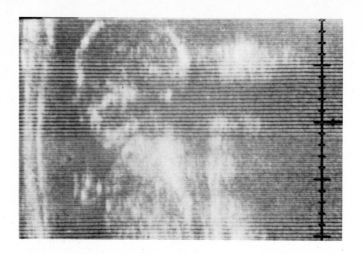

dactyly, and neural tube defects, as well as detecting fetal position prior to amniocentesis. So far, most studies have shown the technique to be relatively safe, but cautious persons would like to see more investigations before ruling out subtle and long-term problems.

Fetoscopy, a relatively new technique, is still considered experimental. A slender tube is inserted into the amniotic cavity, in much the same manner as the amniocentesis syringe. The tube is equipped with an optical system so that one can actually view the fetus and its surroundings within the amniotic sac. The physician can then see gross abnormalities and can also insert a syringe to draw small quantities of fetal blood, using the fetoscope to guide the needle to an appropriate vessel. Sometimes the fetoscope comes equipped with a sampling needle, as shown in Figure 12-4. A blood sample can be used for diagnostic tests that cannot be done on amniotic fluid or amniotic fluid cells. On

FIGURE 12-4
Fetal blood samples can be used for prenatal analyses that are not possible with amniotic fluid and cells. Fetoscopy makes it much easier to insert the sampling needle into a fetal blood vessel.

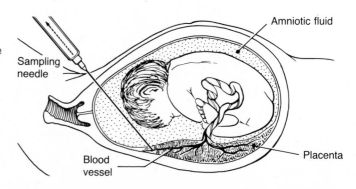

the negative side, the risk of a subsequent spontaneous abortion is high; for this reason, fetoscopy has so far been used only on a limited basis.

Neural tube defects in the fetus and the newborn can lead to paralysis, mental retardation, and death. They are typically accompanied by a rise in the level of **alpha-fetoprotein**, a substance detectable in the mother's blood serum during pregnancy. However, an elevated alpha-fetoprotein level in a pregnant mother's blood serum does not always mean she carries a fetus with a neural tube defect. Additional tests such as amniocentesis are sometimes necessary, and most of these will show the fetus to be normal.

Remarkably, one fetal cell in 500,000 maternal cells can be detected in maternal blood, on the basis of markers such as the Y chromosome in the cells of a male fetus. Methods of detecting and isolating such cells are currently being studied. If the cells can be sorted out accurately, and if they are capable of cell division, it might prove possible to use the mother's blood as a source of fetal cells for prenatal diagnoses. The potential is appealing because maternal blood is easily available for analysis.

Recently, a new technique for prenatal diagnosis has been developed called **chorionic villus biopsy** (or sampling). This procedure can be performed as early as the fifth week of pregnancy, but between the eighth and the tenth week is the best time. The method consists of obtaining chorionic villi by using a thin tube inserted into the uterus through the vagina. Chorionic villi are hairlike projections of tissue from the chorion, the outer epithelial wall of the membrane within which the embryo develops. The villi are composed of rapidly dividing fetal cells that can be cultured and analyzed for genetic disorders. Chorionic villus biopsy is believed to be safer than amniocentesis, although the question of safety is undergoing further study. The technique is still regarded as being experimental.

Another technique to diagnose genetic disorders using cells obtained through prenatal methods is rapidly emerging. The DNA obtained from amniotic fluid cells can be cut into variably sized pieces by special enzymes called **restriction enzymes**. Biochemical methods allow for sizing, identifying, and comparing these pieces of DNA with similarly treated DNA from persons with and without a particular disorder. Determining whether the fetal DNA matches the DNA from someone with a particular disorder or someone without it indicates the genotype of the fetus. Restriction enzymes are more fully discussed in Chapter 19. At this point, however, note that certain abnormalities can be detected by means of such comparisons. The detection of thalassemia has already

269

been demonstrated in this way. This technique is an important advance because it carries a risk much lower than that associated with attempting to obtain fetal blood by fetoscopy. Some progress has been made in the diagnosis of PKU and even in the detection of carriers for PKU. Certainly other genetic disorders will eventually be detected in this way.

MASS GENETIC SCREENING

Prenatal diagnosis is very useful when parents have reason to worry that their child might have a genetic defect. But it is also useful for parents who are not known carriers and whose families have no history of genetic disorder. In these cases, prenatal diagnosis offers a way of identifying unexpected genetic defects. It is then an example of **genetic screening**.

Retrospective genetic screening attempts to identify people who actually have a genetic disorder, in order to begin treatment in time to improve the condition or prevent symptoms. **Phenylketonuria (PKU)** is the prototype disorder for this approach. Screening for PKU in newborns, who do not yet show symptoms, began in 1960. This screening is also called **presymptomatic.**

Prospective or parental **genetic screening** checks would-be parents to identify heterozygotes for deleterious recessive alleles. It is feasible for only a few disorders, but it can indicate definitively whether the parents face a serious risk for having children with certain genetic disorders. If a risk exists, the parents can choose artificial insemination, prenatal diagnosis with selective abortion, or other options. A good example of this approach is the voluntary genetic screening of people of Ashkenazic Jewish descent that was begun in some areas in 1971. These people run a high risk of having children with **Tay-Sachs disease**, for one in 25 to one in 30 are carriers for the deleterious recessive allele. The screening programs, combined with educational programs and thorough prenatal diagnosis, have proved very successful, decreasing the incidence of Tay-Sachs disease by 65 to 85 percent in some populations between 1971 and 1981.

Heterozygotes for **sickle cell anemia**, who comprise one in eight to one in twelve black Americans, can be detected with an inexpensive blood test. In homozygotes, the disorder usually develops within six months of birth, and there is no effective treatment. Thus, a satisfactory method of preventing these births is highly desirable. The screening programs have not been very successful, probably owing to a lack of adequate educational efforts in the community. Several states

270

require a premarital sickle cell diagnostic test, but offer no appropriate genetic counseling. In addition, identified carriers have been discriminated against as evidenced by their loss of jobs, higher insurance rates, and being barred from service in the United States Air Force. Recently, the courts have put an end to this last abuse.

Generally speaking, it is often difficult to identify heterozygotes. The screening tests are usually based on finding a gene product, a metabolic intermediary, or a clinical phenotype. However, heterozygotes and normal homozygotes rarely differ significantly in the concentrations of gene products and intermediaries. When they do, their concentration ranges may still overlap, so that it is impossible to distinguish them reliably. For instance, heterozygotes for **Duchenne muscular dystrophy** and normal homozygotes overlap a great deal in their concentrations of the enzyme creatine phosphokinase. This observation is graphically illustrated in Figure 12-5.

Attempts to develop a screening program for **cystic fibrosis** have been particularly frustrating. This genetic disorder causes much chronic disability and death among Caucasian children. A severe pulmonary disease with many associated physical effects, it occurs in one in 1500 to one in 2500 Caucasian newborns. Thus, 4 to 5 percent of all Caucasians are carriers. There is no cure, although limited treatment is possible. Researchers have sought ways to detect carriers, but the specific biochemical defect remains unknown and current tests are incompletely reliable.

Reliable tests are available for other genetic disorders, and screening programs are becoming popular. Many states require screening of newborns and tests for many disorders. For instance, in 1981 newborns in New York state were tested for PKU, **galactosemia**,

FIGURE 12–5

A comparison of phosphokinase levels in blood serum from controls (probably homozygous normals) and obligate carriers (known heterozygotes) for Duchenne muscular dystrophy. (From Lubs, M.-L. 1979. In: Porter, I., and Hook, E. B. (eds.). *Service and Education in Medical Genetics*. Academic Press, New York. Used with the permission of the publisher and the author.)

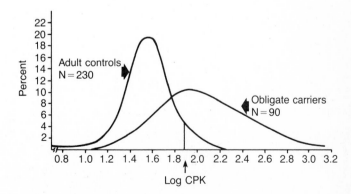

homocystinuria (mental retardation resulting from the lack of a certain liver enzyme), histidinemia (mental retardation and speech defects), sickle cell anemia, hypothyroidism (growth and mental retardation resulting from underproduction of the hormone thyroxin), maple syrup urine disease, and adenosine deaminase deficiency. The eight tests required but one blood sample, and the service was paid for by the government. The results prevented much suffering, for when detected early, PKU can be treated with a low phenylalanine diet, galactosemia with a low milk or milk-free diet, and hypothyroidism with thyroxin.

Everyone agrees that mass genetic screening, if we are going to establish it, requires sensitive, effective, and reliable tests. Some people add that there should be a treatment available for the disorder being tested. Foremost among the objectives of genetic screening should be the health benefits to individuals and families. Effective education and counseling should be essential components of screening programs. In addition, and to gain public support, the public must have a voice in setting up these programs. Lawmakers and scientists must work with the people they serve.

Mass Screening for Phenylketonuria

One of our first mass screening programs involved tests to detect phenylketonuria (PKU). It has been somewhat successful and we will discuss it more fully. PKU is an autosomal recessive disorder affecting one in 40,000 persons. PKU patients lack one particular liver enzyme, phenylalanine hydroxylase, and accumulate large amounts of the amino acid phenylalanine, which can cause abnormally slow physical development, severe, irreversible mental retardation, and other problems, such as convulsions. Fortunately, it is possible to take a small amount of blood from a newborn's finger or heel, analyze its phenylalanine level, and detect PKU before any symptoms appear. Newborns having PKU can be placed immediately on a low phenylalanine diet, which prevents the toxic buildup of the amino acid and minimizes the physical and mental effects. Limited breast feeding is possible, but only in conjunction with a special infant formula now being marketed. When the child is older, the diet is high in fruits and vegetables and is supplemented with amino acids (other than phenylalanine).

Today, more than 40 states require PKU testing of newborns, and the others have voluntary testing programs. As a result, 90 percent or more of all newborn Americans are screened for PKU. Since the test is relatively simple and inexpensive, most people agree that this genetic screening measure is of significant benefit, even though the disorder is rare. Nonetheless, some problems are still associated with PKU screening and treatment.

Technically, the PKU test detects high blood levels of phenylalanine, not PKU. The correlation of the two is strong, but other biochemical abnormalities can also cause high phenylalanine levels. So can a high-protein meal eaten by a normal infant just before the test, or a delay in the production of phenylalanine hydroxylase. Very occasionally, then, a child may be treated for the disorder before the false positive is discovered.

Usually, a child with PKU must stay on the special diet for five to seven years, and the diet must be carefully managed to optimize the phenylalanine level. Even then, however, the child will develop only *relatively* normal mental capacities. The eventual I.Q. will depend upon the timing of diagnosis and treatment, but it will almost always be slightly below the I.Q.'s of the parents and other non-PKU individuals in the family. The individual with PKU may also exhibit perceptual motor dysfunction, difficulties with math or reading, and even neurotic behavior. Some medical researchers advocate that the special diet be continued longer, for nine to ten years or even more. Clearly, we still have much to learn about the effective treatment of PKU.

When treated females with PKU grow up and bear children, their children show high incidences of mental retardation, congenital heart disease, microcephaly, and low birth weight. If the women return to the PKU diet before and during pregnancy, or if they stay on it continuously, these problems seem to diminish. This observation prompts scientists to wonder whether extending the dietary treatment to heterozygous PKU mothers before and during pregnancy might help reduce the incidence of physical abnormalities or severity of PKU.

Economics of Genetic Screening

The economics of prenatal diagnosis, selective abortion, genetic counseling, and mass genetic screening is irrelevant to many people, but it matters greatly to the federal and state agencies that must justify their programs. Unfortunately, we do not have much definitive information on the economics of genetic problems. Typically, researchers have developed cost-benefit analyses that take into account the incidence of the abnormality, the cost of screening, and the economic benefits of avoiding the abnormality. Such analyses basically compare the costs of lifetime care of persons with severe genetic disorders with the costs of mass screening, prenatal diagnosis, and selective abortion.

273

Down syndrome had cost families and government $1.7 billion per year before amniocentesis and selective abortion became widespread. This cost reflects $250,000 for lifetime support of each person with Down syndrome, assuming institutionalization and intensive care. Since it does not reflect inflation, this is a conservative figure. Nonetheless, compare these costs with the $600,000 per year required to conduct amniocentesis for 2450 pregnant women over 40 years of age and abort the 26 fetuses with Down syndrome that are identified. If we assume that the 26 individuals with Down syndrome would live an average of 20 years and that their care would cost $6,500,000 (at $250,000 each), the savings achieved by this program are $5,900,000.

Comparable calculations can be made for phenylketonuria. The lifetime care for all children with PKU in the United States has been estimated to be $19,250,000. Identifying all PKU newborns costs about $800,000, with another $50,000 required for the special diets to avert the full PKU syndrome. The difference is $18,450,000.

Medical care for the infant with Tay-Sachs disease costs a staggering $40,000 per year. Ironically, the more money that is spent, the better the care, and the longer the infant lives, thus driving the total costs ever higher. If all Jewish persons under 30 years of age who intend to marry were screened for the Tay-Sachs allele, the program would cost $5,730,281. An initial screening of only one member of each married couple would be a genetically useful measure, and the cost would drop to $3,122,695. Over a 30-year period, the testing could avert 990 cases of Tay-Sachs disease that would otherwise cost $34,650,000. The tremendous savings are obvious, even without considering inflation.

Other cost-benefit analyses could be presented, but the point should already be obvious. Generally, such calculations show at least a tenfold savings for families and taxpayers, and they do not take into account such nonquantifiable factors as anxiety, suffering, marital conflict, separation, and divorce.

Cost-effective genetic screening programs have certain identifiable characteristics. In the case of phenylketonuria, a very low cost is associated with the test itself, but a very high cost can be expected for the care of a child with PKU who is not properly treated. Regardless of the fact that almost all newborns are checked, the cost-effectiveness can be realized. The Tay-Sachs and Down syndrome screening programs are cost-effective for a different reason. The people who participate in these programs tend to belong to high-risk groups. Therefore, enough cases are detected to make the programs cost-effective. Table 12-2 shows the tremendous difference between the costs of detecting Down

274

TABLE 12-2
Maternal age and specific costs of detecting a fetus with neural tube defects or Down syndrome

Maternal Age at Delivery	Probability That an Affected Fetus has Down Syndrome (fetuses per 1000)	Probability That an Affected Fetus has a Neural Tube Defect (fetuses per 1000)	Total Probability (Down Syndrome Plus Neural Tube Defect) (fetuses per 1000)	Number of Pregnancies Theoretically Necessary to Be Monitored to Identify One Affected Fetus	Cost of One Detection
30	0.70	1.55	2.25	444	$202,800
31	0.89	1.55	2.44	410	190,600
32	1.15	1.55	2.70	370	176,100
33	1.47	1.55	3.02	331	160,100
34	1.88	1.55	3.43	292	144,600
35	2.42	1.55	3.97	252	128,200
36	3.10	1.55	4.65	215	110,600
37	2.98	1.55	5.53	181	95,200
38	5.10	1.55	6.65	150	80,100
39	6.54	1.55	8.09	124	67,200
40	8.39	1.55	9.94	101	55,100
41	10.76	1.55	12.31	81	44,800
42	13.81	1.55	15.36	65	36,200
43	17.71	1.55	19.26	52	29,000
44	22.71	1.55	24.26	41	23,000
45	29.14	1.55	30.69	33	18,600

Source: Sadovnick, A. D., and Baird, P. A. 1981. A cost-benefit analysis of prenatal detection of Down syndrome and neural tube defects in older mothers. *American Journal of Medical Genetics,* 10:367–368.

syndrome fetuses in women ranging in age from 30 to 45 years old. The cost per detection can be entirely too high for very rare diseases, but we should note in this table that even a maximum cost of $202,800 per detection compares favorably with the $250,000 cost of lifetime care.

Opposition to Genetic Screening

Mass screening programs are sometimes controversial. Opponents point to safety hazards and false positive and false negative diagnostic errors. Most tests are fairly safe, and rechecking makes the diagnostic errors rare. When they do happen, however, the errors can have severe consequences.

Some people are concerned about stigmatization, for being called a "carrier" can provoke anxiety and fright. Identified carriers could suffer psychological restrictions with regard to marriage and reproduction. Probably worse is the attitude that others may show toward carriers, as if they were different or tainted. Those involved in genetic screening have actually been charged with changing the frequencies in certain gene pools. For example, sickle cell anemia affects mostly blacks and Tay-Sachs disease affects mostly Jews, and the screening programs for these disorders do have the effect of reducing reproduction.

Genetic screening is being used in a new way in industry, where some companies have developed tests to detect which workers are especially susceptible to certain chemicals. They have called this a protective measure, since they transfer sensitive workers to other jobs. Critics, however, charge that such practices can only catalyze more stigmatization and discrimination. Certain people will be labeled as unfit for certain jobs. The most adamant critics assert that the entire workplace should be made safe for all workers, regardless of cost.

Many critics object to the use of genetic screening to justify selective abortion. They base their objections both upon morality and the potential impact of selective abortion on other comparably handicapped individuals who were not aborted. They also worry that selective abortion might become too widespread and too arbitrary, with the result that many fetuses with very minor disorders would be eliminated.

Spina bifida is a neural tube defect in which bone, skin, and muscle fail to form properly over the spinal cord. High maternal blood levels of alpha-fetoprotein during pregnancy indicate the possible presence of this defect. Usually one in 500 tests will give a positive result, but many of these are false positives and additional tests are needed to confirm spina bifida. Opponents of genetic screening believe that many persons will choose to abort based upon the first positive test without waiting for further tests, and many of their fetuses will not have spina bifida. The test also detects fetuses with slight neural tube defects which could, according to some, be surgically mended.

Some opponents of genetic screening object on economic grounds. Cost-effectiveness is virtually impossible to achieve, they point out, with very rare diseases. It may also be impossible to achieve if screening programs expand to cover genes involved in heart disease and other common conditions. They add that even with cost-effective programs, many of the aborted defective fetuses are later replaced by children without genetic disorders, who then impose their own costs on society; on the other hand, people without severe genetic disorders presumably contribute more to society than they take from it.

Some people oppose mass genetic screening because the accompanying regulations and restrictions invade their privacy and infringe on their parental rights and individual liberties. They feel that all genetic screening programs should be voluntary, not compulsory, and that it is not sound policy for the government to force such programs on people by law. The programs take too long—many generations—to reduce the frequency of deleterious alleles significantly, and their admitted benefits are not worth the loss of freedom they entail. Still other opponents ask such questions as: "What is a severe defect?" "Who should be classified as unfit?" "Does the fetus have rights?" "Should a defective fetus be given the same legal protection as the handicapped?" and "Who knows what is or is not genetically best?" Many others, however, counter with assertions that there are crucial differences between rights and needs. They argue that having children *is not just the business of the parents*, but the business of all of society, and that in some situations one cannot have complete freedom to procreate. Certainly the societal issues that surround genetic screening must be openly addressed.

EUGENICS

In some ways, **eugenics** is related to genetic screening, and it easily arouses thoughts of racism, sexism, and class superiority. The literature is full of opinions on this subject, even though few people are sure just what eugenics and its counterparts, **euthenics** and **euphenics**, really mean. In addition, historical atrocities have helped give the subject a bad name. For example, between 1934 and 1945 the Nazis of Germany exterminated more than six million Jews, East Europeans, gypsies, and persons with defects, and sterilized 350,000 to 2,000,000 other people considered to be defective, all as the result of a 1933 Act for Averting Descendants Afflicted with Hereditary Disease. Against that bitter background, it is no wonder that eugenics often evokes strong negative feelings.

The goal of euthenics is to improve the physical and mental qualities of the human population by controlling environmental factors. We have already touched on some aspects of euphenics, which deals with efforts to correct and modify abnormal phenotypes; it does not affect genes, and it leaves people fully able to transmit their defective genes to their descendants.

The aim of eugenics is to improve the future physical and mental qualities of human beings by controlling and manipulating those factors involved with heredity. Eugenics technically encompasses any and all attempts to manage the human gene pool. It resembles the selective

277

breeding so successful with plants and animals. **Negative eugenics** refers to the removal or reduction of defective genes from the gene pool, that is, the elimination of undesirable traits. **Positive eugenics** refers to measures that increase desirable traits in the population. Both can be put to inhumane social and political uses.

Germany is not the only nation to have had a eugenics movement. By 1917, sixteen states in the United States had compulsory sterilization for the insane and feeble minded, rapists, habitual criminals, paupers, and others believed to be hereditarily unfit. More than 30,000 sterilizations had been carried out by 1935. Administrators segregated the inmates of penal and mental institutions to ensure that they could not reproduce. However, it was never certain that the traits for which they were institutionalized or imprisoned were hereditary. Certainly, many of them have a large environmental component.

The eugenics movement also affected immigration policies. The Immigration Act of 1924 set ethnic quotas that strongly favored northern Europeans over other groups. The Act is believed to have resulted from an almost worthless Army mental examination used during World War I. Although the test failed to take into account that scores could vary as much with ethnic background or inability to understand English as with intelligence, eugenicists used the results to further their own ideas about how American society should be molded. The entire movement became extremely political. Most biologists, especially geneticists, remained aloof from the affair, but a few biologists and psychologists supported the Immigration Act of 1924. Some historians of science have been quite critical of those who remained silent even though they knew that many of the Act's measures were unnecessarily harsh and unwarranted.

The American Eugenics Society was officially incorporated in 1926. However, by 1940 a mounting disapproval of some eugenics activities in this country and of the racial madness in Germany had largely obliterated organized eugenics in the United States. Even so, some eugenic practices have continued with little real regard for the common good of all concerned. Some states continue to push sterilization on the poor, those on welfare programs, blacks, and other minorities, sometimes making sterilization much easier to obtain than prenatal care. The state programs supposedly required informed consent before anyone could be sterilized, but critics have frequently charged that on many occasions the consents were obtained from young people under duress. Also, sterilization was often suggested by the physician, not requested by the patient. One naturally wonders about the degree of coercion in

278

such cases. Steps have been taken to counteract the abuses, but some question their effectiveness.

Sterilization is a method of effecting negative eugenics. It is also a potent method of birth control, and this is the main reason why, by 1976, some 65,000,000 couples had been sterilized worldwide, often voluntarily. Other negative eugenic methods include prenatal diagnosis followed by selective abortion of any genetically defective fetuses, and even the encouragement given to carriers of genetic disorders to have their children by adoption or AID. Many such programs have been successful, and they are often hailed as genetically sound.

Positive eugenics is less readily accepted than negative eugenics. One example of this approach might be sperm (or ova) banks that use only prestigious men, such as Nobel Prize winners, as donors and only carefully selected women as recipients. One such exclusive program, recently begun in California, was criticized by scientists and the general public, who called the participants "conceited elitists."

The late H. J. Muller, an eminent geneticist and himself a Nobel Laureate, advocated the preservation of sperm from outstanding men for later use as long ago as 1935. Interestingly, his first list of superior men included Lenin and Marx; by 1959 (after spending time in the Soviet Union), he had replaced them with Einstein and Lincoln. Muller's list of desirable traits included such subjective attributes as joy of life, humility, empathy, emotional self-control, fortitude, and other vague characteristics. Others might list other traits, but the questions about what is good and what is bad, what is the ideal human genotype or phenotype, what role the environment plays, and how broadly defined traits are inherited, remain crucial. We can rely on probability to a large extent, but we cannot be sure how the genes from two parents will interact with each other. Plant and animal breeders have known this for a long time, and the former remind us of the cross between a radish and a cabbage, which produced a worthless plant with the top of a radish and the root of a cabbage instead of the desired radish-rooted and cabbage-topped hybrid.

The same problems of judgment exist with negative eugenics. Should we cull the schizophrenics who have given us much of our superior art and great ideas? Overzealous eugenics might well reduce valuable variability in the population. Too, many of the more desirable traits (if we can decide upon them) may have heavily environmental components. Even a genetically sound eugenics program will change the human gene pool very slowly. The problems and the uncertainties are endless, which has brought many people to believe that there should be

279

absolutely no germinal choice at all. In other words, chance alone should influence the outcome of human reproduction.

In contemporary society, technology is sometimes used for political reasons and as a disguise for socioeconomic prejudices and social manipulation. Geneticists by themselves, of course, do not have the right nor a broad enough perspective to establish the goals to guide society. Perhaps a resonable system of goals will evolve as society's institutions gradually learn to deal with the new ethical questions that grow out of technological advances.

SUMMARY

Parents can now intervene prenatally in the genetic characteristics of their children. Genetic counselors interpret genetic information and explain various reproductive alternatives to concerned parents. Increasing the number of counselors would be feasible since many people who need counseling are not yet receiving it. Counselors face a wide variety of biological and psychological problems, but explaining recurrence risks for genetic disorders is probably the most common. They must also evaluate genetic heterogeneity, differences in expressivity and penetrance, and multigenic traits. It is their responsibility to clarify the role of probability in each client's genetic situation. Depending on the situation, they can offer various approaches to avert the problem: prenatal diagnosis, artificial insemination, selective abortion, contraception, sterilization, and adoption.

Amniocentesis and other methods of prenatal diagnosis make genetic counseling more definitive. These techniques permit the analysis of fetal chromosomes and cellular biochemistry using cells found in the amniotic fluid. Based on these tests, medical researchers can detect most known chromosome abnormalities and more than 100 metabolic diseases. Other prenatal diagnostic techniques include radiology, ultrasonography, fetoscopy, chorionic villus biopsy, and the analysis of maternal blood. However, there remain some drawbacks to prenatal diagnosis, including safety, subsequent spontaneous abortion, occasional wrong diagnoses, and other errors. All of these hazards are relatively rare.

Some genetic screening is compulsory in most states. Retrospective genetic screening identifies persons who have a disease before the symptoms appear; prospective genetic screening identifies carriers of deleterious alleles. Screening for Tay-Sachs heterozygotes among Jewish people has been very successful. Screening for sickle cell heterozygotes among blacks has been less successful. Screening for phenylketonuria has worked fairly well, but further research is needed to define the optimum treatment of the disease. A good test for cystic fibrosis screening does not exist; heterozygotes cannot be distinguished from homozygous normal individuals.

Screening programs are usually justified partly in terms of costeffectiveness. The most successful programs result in monetary savings of about tenfold. Scientists derive this factor by comparing the costs for operating the screening program with the costs for lifetime care of the persons who would be born with the defect if the program were not in place. The largest savings are associated with those programs with inexpensive tests, a clearly identified high-risk group to screen, or both. Other factors justifying screening programs include relief from anxiety, suffering, marital conflict, and other emotional problems.

Opponents of genetic screening point to the hazards of diagnostic errors, safety problems, misinformation, and stigmatization of and discrimination against "carrier" individuals. Some opponents object to abortion on principle, others to the invasion of privacy and infringement of liberty that accompanies compulsory screening programs. On the other hand, some people feel that procreation is everyone's business, not just the concern of individual parents.

Eugenics attempts to improve the future physical and mental qualities of humans. The discipline has been controversial because of people's horror of Nazi practices and even resistance to the eugenics movement in the United States earlier in this century. This latter movement supported sterilizations, immigration laws, segregation, and other related measures. Genetic screening is a mild form of eugenics.

Technology is often used for social manipulation and other political purposes. The fear of this happening with genetic screening and other eugenic measures is an important factor in how many people react to these innovations.

DISCUSSION QUESTIONS

1. Do parents have the right to decide what is genetically best for their offspring? Defend your position on this matter.

2. Do people have a responsibility to disclose genetic information to relatives who may be at high risk for carrying a deleterious allele? Discuss.

3. Do parents have a duty to avoid bearing children with serious genetic defects? Why or why not?

281

4. Should reliable and cost-effective genetic screening be mandatory or voluntary? Defend your position.

5. What does it mean to proclaim that having children is not just the business of the parents?

6. We do not have a clinically available prenatal test for Huntington's disease. If such a test were available, and assuming that prospective parents were not opposed to selective abortion, why might it still be difficult to decide to abort a fetus with Huntington's disease? Think through the same predicament for cystic fibrosis.

7. Should a physician be held liable for malpractice for failing to give genetic counseling to a patient? Discuss on the basis of several hypothetical situations.

8. Many scientists and nonscientists do not believe that scientists can say exactly when someone becomes a human being. What do you suppose is the substance of their argument?

9. Identify and discuss some of the possible long-range effects of intensive genetic screening practices.

10. Explain why prenatal diagnosis and selective abortion of autosomal recessive diseases that cause death in infancy cannot change the gene pool. Also explain why selective abortion can have little effect upon nonlethal autosomal recessive alleles in the gene pool.

11. There are tests that can indicate whether two person contemplating marriage are carriers for certain very deleterious alleles. If both persons are identified as carriers, which of the following measures, if any, do you favor? Discuss.
    a. The couple should be given the facts in a very neutral manner.
    b. The couple should be discouraged from having children.
    c. The couple should be penalized for having children by a fairly modest measure, such as not allowing the children to be used as tax deductions.
    d. The couple should be forbidden by law to marry each other or to have children if married.
    e. Sterilization of the couple should be mandatory.

12. If you learned that your next child had a one in ten chance of a severe handicap, would you change your plans to have the child? A one in ten chance of a mild handicap? Would the number of children you already have affect your decision?

13. How would you explain and/or demonstrate to a 19-year-old son of a patient with Huntington's disease that his own son has approximately a 25 percent chance of having the disease someday? Assume that the 19-year-old is a high school graduate of average intelligence.

282

14. Should prenatal diagnosis be denied to a couple who refuse to consider an abortion? What situations, if any, would make you qualify your answer in either direction?

15. Which do you think is the greater—the right of a child born with a serious genetic disorder to procreate upon reaching sexual maturity, or the right of the public to be protected against increases in the number of genetically disabled persons and the associated consequences. Support your conclusion.

16. A man with a 50 percent chance of having Huntington's disease is married to a normal woman, and they want to have children. What alternatives are open to them?

17. A malformed child grows to adulthood and sues his parents because, although they were counseled that they had a 50 percent chance of having a malformed child, they decided to take the chance and have the child. He is now suing them for not accepting the advice not to have children and claiming substantial physical and mental suffering in his lifetime. As a jury member in this case, would you award damages to the plaintiff? Why or why not?

18. In an actual case, a pregnant woman was exposed to German measles, but her physician did not tell her that the disease gave her child a 25 percent chance of being malformed. After having a malformed child, the parents sued the physician for negligence. The court rejected the suit on the grounds that the child would have chosen to live. Do you agree with that decision? Why or why not?

19. Assume that a series of simple, fast, and inexpensive tests can determine for everyone whether he or she is heterozygous for each of 150 different genetic diseases. In detail, explain what advantage such information would have for you. For society as a whole? Can you think of any disadvantages?

PROBLEMS

1. A normal couple has a child with an autosomal recessive disorder.
   a. What is the chance that their next child will have the same disorder?
   b. If they have three normal children in succession, what is the chance that their next child will have the disorder?

2. What genetic situations could give the following risks of having an abnormal child?
   a. 100 percent
   b. 75 percent

c. 50 percent
d. 25 percent.

3. Acting as a genetic counselor, try to determine the modes of inheritance in the following pedigrees:

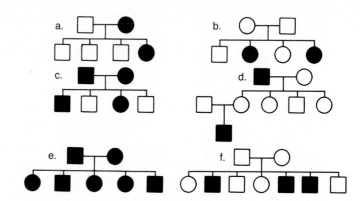

4. Assume that a newborn has a genetic disease caused by an autosomal dominant allele that shows 100 percent penetrance (it is always expressed). Both parents are normal for the trait. What is the risk that the parents' other children will have the disease?

5. A young man wishes to marry his mother's half sister, who is close to his own age. He confers with a genetic counselor to learn the risks that such a marriage will produce children with genetic defects. What is the degree of inbreeding? What risks are involved?

6. In the pedigree of an autosomal recessive disease below, which persons are most likely to be carriers? Which are least likely to be carriers?

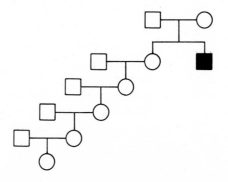

7. The following pedigrees are of an X-linked recessive trait. In each case, assess the probability that the II-2 person is a carrier.

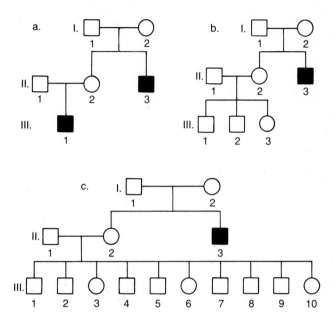

8. Experimental tests for cystic fibrosis are now being offered on a limited basis to the spouses of normal siblings of children with cystic fibrosis. Show the rationale for this procedure using diagrams of genotypes and crosses. Use *Cf* for the normal allele and *cf* for the cystic fibrosis allele.

9. Between 1971 and 1976, about 150,000 Jewish adults voluntarily took part in Tay-Sachs screening programs. Over 6000 heterozygotes for the Tay-Sachs gene were detected, and 125 couples were identified in which both persons were heterozygous for that gene. Use the Hardy-Weinberg principle (see Chapter 5) to determine whether these data are fairly close to expectation, assuming random mating.

10. Medical and social practices are bound to affect the frequencies of genetic diseases and their responsible alleles. For example, consider an autosomal recessive trait and its allele frequency in conjunction with the possible medical and social practices listed below. In each case, say whether the frequencies of the disease and its allele would generally increase or decrease.

	practice	disease frequency	allele frequency
a.	improved treatment of the disease		
b.	selective abortion without replacing with another child		
c.	limitation of family size in carrier marriages		

REFERENCES

Emery, A. E. H. 1975. *Elements of medical genetics*. Berkeley: University of California Press.

Etzioni, A. 1974. Genetic fix. *New Scientist* 61:139.

Gould, S. J. 1982. A nation of morons. *New Scientist* 94:349–352.

Guthrie, R. 1961. Blood screening for phenylketonuria. *Journal of the American Medical Association* 178:863.

Guthrie, R., and A. Susi. 1963. A simple phenylalanine method for detecting phenylketonuria in large populations of newborn infants. *Pediatrics* 32:338–343.

Harmsen, H. 1955. The German sterilization act of 1933: "Gesetz zur Verhu- tung erbkranken Nachwuchses." *Eugenics Review* 46:227–232.

Hoehn, H., E. M. Bryant, L. E. Karp, and G. M. Martin. 1975. Cultivated cells from diagnostic amniocentesis in second trimester pregnancies. II. Cy- togenetic parameters as functions of clonal type and preparative tech- nique. *Clinical Genetics* 7:29–36.

Kelly, S., J. Burns, and L. Desjardins. 1974. Incidence of galactosemia at birth in New York state. *American Journal of Epidemiology* 99:8–13.

Leonard, C. O., G. A. Chase, and B. Childs. 1972. Genetic counseling: A con- sumer's view. *The New England Journal of Medicine* 287:433–439.

Lippmann-Hand, A., and F. C. Fraser. 1979. Genetic counseling: Provision and reception of information. *American Journal of Medical Genetics* 3:113– 127.

Lubs, M.-L. 1979. Carrier screening in hemophilia and Duchenne muscular dys- trophy: Economical and psychological consequences. In *Service and education in medical genetics*, ed. I. Porter and E. B. Hook. New York: Academic Press Inc.

Mahoney, M. J., and J. C. Hobbins. 1977. Fetoscopy and fetal biopsy. In *Ge- netic counseling*, ed. H. A. Lubs and F. de la Cruz. New York: Raven Press.

286

REFERENCES

Marks, J. H. 1979. Masters level training programs for genetic counselors: An eight year report. In *Service and education in medical genetics*, ed. I. Porter and E. B. Hook. New York: Academic Press Inc.

Milunsky, A. 1975. Risk of amniocentesis for prenatal diagnosis. *The New England Journal of Medicine* 293:992–993.

NICHD National Registry for Amniocentesis Study Group. 1976. Midtrimester amniocentesis for prenatal diagnosis: Safety and accuracy. *Journal of the American Medical Association* 236:1471–1476.

Rose, R. J., D. W. Fulker, J. Z. Miller, C. E. Grim, and J. C. Christian. 1980. Heritability of systolic blood pressure. Analysis of variance in MZ twin parents and their children. *Acta Geneticae Medicae et Gemellologiae* 29:143–149.

Sadovnick, A. D., and P. A. Baird. 1981. A cost-benefit analysis of prenatal detection of Down syndrome and neural tube defects in older mothers. *American Journal of Medical Genetics* 10:367–378.

Seidenfeld, M. J., and R. M. Antley. 1981. Genetic counseling: A comparison of counselee's genetic knowledge before and after (part III). *American Journal of Medical Genetics* 10:107–112.

13
The Judicial Uses
of Genetics

Now we are ready to apply genetics to broader questions. We will discuss environmental issues, evolution, behavior, and genetic engineering, all of which interact with political issues to some extent. The issues we have raised so far—concerning genetic counseling and screening, the technologies of reproduction, consanguinity, and so on—seem to have a more personal focus. But they also involve judges and lawmakers as well as politicians.

Genetics plays an important part in many social controversies, in education, in medicine, in the courts, and in legislatures. We will see this again and again in the later chapters. On the other hand, genetics can serve the courts as a tool, for a knowledge of genetic principles can be an immense aid in settling questions of disputed parentage or relatedness; genetics thus becomes **medicolegal science**. As **forensic genetics**, it can help settle criminal cases.

MEDICOLEGAL SCIENCE

Disputes over biological relationships often come to court. Occasionally, relatedness must be established to settle a will, or it may be an issue in a kidnapping case. Although the event is very uncommon, when two mothers are given the wrong babies in the maternity ward, the legal system must have some way of establishing which baby belongs to which mother.

The same kind of question arises more often in divorce and child support cases, which can be decided according to whether the putative parent—usually the father—is actually the genetic parent of the child. This question can usually be answered by carefully comparing parent and child for genetic characteristics, such as blood types. The characteristics must be clearly Mendelian (simply inherited, with a well-documented mode of inheritance), fully expressed at birth, and unchanging throughout life.

Even though such comparisons can sometimes result in totally unambiguous answers, a few states have been slow to accept genetic data in cases of disputed parentage. Initially, the results of blood group testing were used to *exclude* individuals as possible parents. Sentiment from both outside and inside the judicial system, however, was in favor of test information that would identify the parent with very high probability. Genetic data are now being used in just this way in many states, though there is a big difference between *proving* a man is the undisputed father of a child and *disproving* the parenthood of a putative father. Blood tests can be used to definitively rule out a particular candidate as the father, but establishing someone as the father can only be done in terms of probability, never with absolute certainty.

The available immunological and biochemical tests are becoming more accurate and conclusive, and their use will probably increase greatly in the future. Many people dislike having the government support children whose fathers have not been identified. More cases might be settled promptly, without long court battles, if paternity tests were more definitive.

Blood Groups as Tests of Paternity

Many blood groups have characteristics that make them suitable as paternity tests. Their genes are inherited in Mendelian fashion, and the tests produce clearcut results. In addition, the gene products or **genetic markers** remain constant throughout the life of the individual.

In 1928, K. Landsteiner and Philip Levine were the first to suspect that human blood groups could be useful in forensic medicine and in cases of disputed paternity. By 1932, blood groups were being used for paternity testing in Great Britain. Soon thereafter, the practice spread to other countries, although American courts were slow to accept blood groups as valid evidence. One of the first American human geneticists to become involved in testing parentage was Laurence H. Snyder. While at Ohio State University, he provided testing and interpreted the results in ten court cases. The first occurred in 1934.

Basic biology and genetics dictate the usefulness of blood groups in paternity testing. For instance, the ABO blood group system is but one of many different blood group systems. It is a **multiple allelic** system, controlled by one gene with three alleles, *A*, *B*, and *O*; these are alleles of the **isoagglutinin locus**, I, and are often written as I^A, I^B, and I^O. The system is based upon **antigens** and **antibodies**. Antibodies are complex protein molecules that can combine with antigens. Usually, antigens are molecules foreign to the body. These non-self molecules are capable of stimulating the production of antibodies by the body's immune system. In this case, however, antigens and antibodies are ever-present in the body, but in combinations that do not react with each other. Blood type A persons have the A antigen on their red blood cells; they also have **anti-B antibodies** in their blood **serum** (the liquid part of the blood). Type B persons have B antigens on their red blood cells and **anti-A antibodies** in their serum. If type A and type B bloods are mixed, as in a poorly planned transfusion, the anti-A antibodies bind to the A antigens and the anti-B antibodies bind to the B antigens, and the mixture will **agglutinate**, forming clumps of red blood cells. The

291

TABLE 13–1

Antigen-antibody relationships of the ABO system

Phenotype (blood type)	Genotype[a]	Cellular Antigens	Serum Antibodies
A	$I^A I^A$ or $I^A I^O$	A	anti-B
B	$I^B I^B$ or $I^B I^O$	B	anti-A
AB	$I^A I^B$	AB	neither
O	$I^O I^O$	neither	anti-A and anti-B

Note: [a]The symbols I^A, I^B, and I^O are often used to designate the alleles in the ABO system.

FIGURE 13–1

Results of typical blood tests with anti-A and anti-B sera that yield blood type information. The anti-A serum is obtained from B blood and the anti-B serum is obtained from A blood.

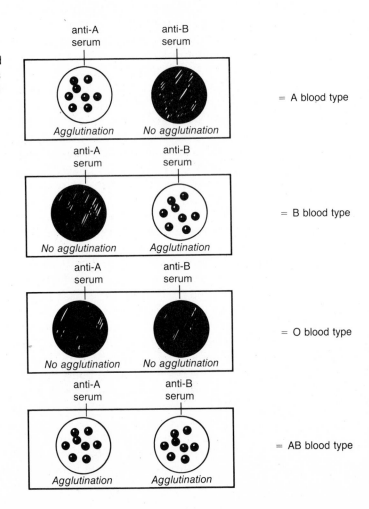

292

anti-A antibodies do not cross-react with the B antigens in type B blood, and the anti-B antibodies do not cross-react with the A antigens in type A blood.

Individuals with type AB blood have both A and B antigens on their red blood cells and neither antibody in their serum. Individuals with type O blood carry neither A nor B antigens, but their blood serum has both anti-A and anti-B antibodies. Table 13-1 shows these antigen-antibody relationships. Figure 13-1 shows how mixing a blood sample with anti-A and anti-B serums reveals the blood type of the sample. Figure 13-2 shows the basic concept of agglutination.

Blood group A has now been separated into at least two subgroups, A_1 and A_2. Both of these *A* alleles are dominant to *O*, and A_1 is dominant to A_2. The *B* allele is also dominant to *O*. Blood groups A and B are **codominant**, since when the alleles are together in an AB genotype, both antigens, A and B, can be detected on the red blood cell. These dominance and codominance relationships determine the offspring any given mating can and cannot produce, as shown in Table 13-2.

Now that we know the basic heredity of the ABO blood groups, we can consider a specific example. A type O woman has a child by her type A husband, and the newborn is found to have type B blood. Is some other man the father? Or was there a rare mutation of the I locus? A third possibility has been the answer on rare occasions. People synthesize their A and B antigens from a third antigen, the **H substance**, which is made, in turn, from a precursor, with each step requiring an appropriate enzyme. This process is diagrammed in Figure 13-3. Most people have the homozygous dominant genotype (*HH*) for generating H substance, so that their blood type depends on whether they convert

FIGURE 13–2
A diagrammatic simplification of the agglutination mechanism that could occur with a mixing of blood types A and B. Agglutination of the A red blood cells could also occur in a similar manner.

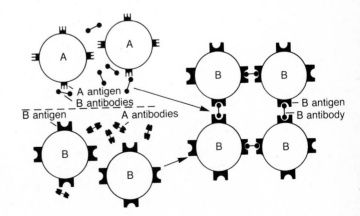

TABLE 13–2
Possible and impossible ABO phenotypes of offspring resulting from all possible matings with respect to ABO blood groups

Phenotypes of Matings	Possible Offspring	Impossible Offspring
A × A	A, O	B, AB
A × B	A, B, AB, O	none
A × AB	A, B, AB	O
A × O	A, O	B, AB
B × B	B, O	A, AB
B × AB	B, A, AB	O
B × O	B, O	A, AB
AB × AB	A, B, AB	O
AB × O	A, B	AB, O
O × O	O	A, B, AB

the H substance to the A or B antigens. Some, however, are homozygous for the rare recessive allele h, and they cannot synthesize H substance. Consequently, even if they do have the enzymes for the next step, turning H substance into A or B antigens, they cannot do so. Therefore, they show the O blood type. If such a person mates with a person who has the HH genotype, as did the woman in our example, the resulting child can have an apparently anomalous blood type. Our type O woman had the metabolic blocks 1 and 2 shown in Figure 13-3, and the genetics of the cross with her husband is:

mother (blood type O) $hh(B)O$ × father (blood type A) $HHAO$

$HhBO$
child
(blood type B)

The mother's (B) represents the B allele she cannot express. Technically, the father could have had an Hh genotype, but since the recessive allele is rare, we assume that his genotype is HH. Persons lacking the H antigen are said to have the **Bombay blood type**, so called because it was first discovered in that city.

Another blood group commonly used for medicolegal purposes is the **MN system**. M and N are codominant alleles of the L (Landstei-

294

FIGURE 13–3

The pathway for the production of ABH antigens, and the phenotypic results of various metabolic blocks in this pathway.

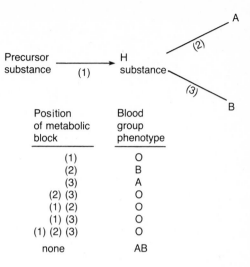

Position of metabolic block	Blood group phenotype
(1)	O
(2)	B
(3)	A
(2) (3)	O
(1) (2)	O
(1) (3)	O
(1) (2) (3)	O
none	AB

ner) locus, and they can comprise three genotypes, *MM*, *MN*, and *NN*, and three corresponding phenotypes, the M, MN, and N blood groups. If, for now, we ignore some complicating factors in the system, the basic genetics is very simple. Table 13-3 lists possible and impossible offspring from all types of matings involving the MN blood group. Another pair of alleles, *Se* and *se*, that control the presence of an Se antigen, is very tightly linked with the L locus. The *Se* or *se* allele is usually transmitted with the *M* or *N* allele in the combinations MSe, Mse, NSe, and Nse. The combinations increase the likelihood of determining parentage because a greater number of phenotypes is possible. Still other alleles, although rare, have been found at the L locus itself.

TABLE 13–3

Possible and impossible MN phenotypes of offspring resulting from all possible matings with respect to MN blood groups

Phenotypes of Matings	Possible Offspring	Impossible Offspring
M × M	M	MN, N
M × MN	M, MN	N
M × N	M N	M, N
MN × MN	M, N, MN	none
MN × N	MN, N	M
N × N	N	M, MN

The **Rh blood group** is also complex, but still a medicolegally useful system. At first, geneticists recognized only two phenotypes, **Rh positive** (Rh+) and **Rh negative** (Rh−). Actually, this is a complex locus with many subgroups discernible with appropriate testing, but still, everyone can be classified initially as Rh+ or Rh−, with Rh+ dominant to Rh−. Again, the many possible phenotypes add to the usefulness of the system in parentage disputes.

Still other blood groups are known, and some of them can be used in assessing parentage, although problems are associated with their use. These include the **Kell** blood group, the **Kidd** system, and the **Duffy** system.

Most courts now accept certain genetic information for the testing of parentage. Blood groups, along with biochemical tests, enable one to calculate the probability of excluding non-fathers. Technically, this calculation indicates the probability of excluding a wrongfully accused male if such a male were randomly selected from the general population. The calculations are based upon the gene frequencies in the general population. The more tests performed, then, the greater will be the probability of exclusion. The possibility of mutation is usually ignored because of its rare occurrence (approximately once in 100,000 or 1,000,000 times). With each trait tested, the probability of exclusion increases. If the male in question is *not* excluded, the probability of his being the father increases. The probability of exclusion can be as high as 95 percent or even more than 99 percent in some series of tests. Generally, a probability of 90 to 95 percent is regarded as likely, 95 to 99 percent as very likely, 99.1 to 99.75 percent as extremely likely, and 99.8 to 99.9 percent as "practically proved." In 1980, a New Jersey court was the first to assign paternity when a series of tests that resulted in a 99.8 percent probability of excluding a wrongly accused man did not exclude the man. However, assignment of paternity can never be *absolutely* certain. Probability always leaves room for exceptions, such as rare coincidence, mistakes, and the small chance of mutation.

Other Useful Genetic Markers

Probabilities of exclusion can reach the "practically proved" level of 99.8 to 99.9 percent only when traits other than blood groups are tested. One complex group of such traits comprises the antigens of white blood cells, including the **human-leukocyte-associated (HLA) antigen** system. At least 60 different white blood cell markers are known, and many more are believed to exist. The HLA antigens are controlled by a series of multiple allelic genes at several different loci. Because each combination of HLA alleles is practically unique, a match between a putative parent and a child is taken as strong evidence for parentage.

HLA testing is definitely more powerful than blood group testing, and the method is rapidly revolutionizing the establishment of parentage. The blood group tests can quickly and convincingly exclude certain persons as the parent, and such tests are advocated for that purpose. However, to *establish* paternity positively, HLA testing is necessary. Already, HLA testing has shown in one case that dizygotic twins had two different fathers.

Other tests can also be used to settle disputed parentage. In some cases, parent and child may share peculiar patterns of chromosome banding. Biochemical tests are available, and they too can be used, often in conjunction with some of the tests we have already mentioned. Because their standard modes of operating have not caught up with technical advances, courts do not avail themselves of all possible genetic tests.

Any genetic test to be used for medicolegal purposes must first gain a high level of acceptance in the scientific community. Only then will the judicial system accept the test results as trustworthy information. Courts must be satisfied that the test is reliable and that it is performed by competent persons. Most states now admit evidence based on HLA testing in disputed parentage cases. Wrong assignments of paternity, although rare, remain possible; however, many within the judicial system feel that the overall value of the paternal assignments outweighs the harm done by an occasional error.

FORENSIC GENETICS

The application of genetic knowledge and methodology to solving crimes is called forensic genetics. Most often, this work centers on the identification of a certain person. Therefore, testing relies upon the genetic diversity of individuals, just as it does in disputed parentage cases.

The materials on which the forensic geneticist works are, of course, clues. Violent crimes can leave blood stains, although the quantities of blood are often so small that analysis is difficult. Rape victims can have bits of their attacker's skin under their fingernails, and semen samples may be available. In some cases, the geneticist can identify a criminal's sex by finding fluorescent Y chromosomes or sex chromatin in white blood cells. Given an adequate blood sample, he or she can report ABO, MNSese, and Rh blood types. With both blood and semen samples, if they are large enough, he or she can report HLA antigen analyses. Even hemoglobin and various enzymes have been analyzed in some cases. This field certainly has a great deal of potential. It has been slow to develop because only large investigative laboratories are equipped for forensic genetics, and because only a few reliable methods have so far been devised.

SUMMARY

Many genetically related problems are reaching the courts today, and there undoubtedly will be more in the near future. Genetics can be a valuable aid to the judicial system in resolving parentage disputes and identifying individuals in criminal cases. Genetics probably plays its largest role in parentage cases, where inherited traits in a child and a putative parent (usually the father) are matched. The characters must be simply inherited, and the tests must be reliable and conclusive. At one time, such tests only exonerated falsely accused persons, but today they are allowing positive assignments of paternity.

Some blood groups are useful in parentage tests. The multiple-allelic ABO system is the most commonly used. Blood can be typed easily according to the presence or absence of agglutination when antibodies are mixed with the blood. Blood types A and B are dominant to O, but codominant with each other. In addition, an H antigen is involved; when it is absent, unexpected ABO types can result. Other blood types often used in this work are the MN, Rh, Sese, and Kell systems, and occasionally the Kidd and Duffy systems.

Still other medicolegally useful genetic markers are found in the HLA (human-leukocyte-associated) antigen system. This system consists of many alleles at each of several loci, and it is the basis for a powerful test. It can permit the exclusion of wrongly accused persons with probabilities as high as 99.9 percent. With such accurate, reliable results, courts in many states are now *assigning* paternity rather than simply *excluding* the innocent. HLA antigen paternity testing is gaining widespread acceptance and use.

Forensic genetics is the application of genetics to the solution of crimes. Its major role, again, is identifying individuals. It uses genetic markers, usually tested in small samples of blood or semen. The problems in forensic genetics are still large, and additional reliable methods need to be developed.

298

DISCUSSION QUESTIONS

1. A hemophilic woman had a hemophilic mother and a normal father (hemophilia is a recessive X-linked genetic disorder). When the woman with hemophilia attempted to collect her "father's" huge estate after his death, the next relative in line contested the inheritance, claiming that she could not possibly be the deceased man's daughter.
   a. What is the basis for saying the woman is not the man's daughter?
   b. What is the best genetic argument the woman can use on her own behalf? Is there any way she can prove her argument? (Review the Lyon hypothesis.)

2. Assume that a mulatto woman gave birth to dizygotic twins, of which one was white and the other was mulatto. Some people would immediately conclude that the twins had two different fathers, which does occasionally happen. How could the event be attributable to one man? (Review multiple genes.)

3. A fairly well-documented case is that of a woman who developed from a rare fusion of two fertilized eggs into one individual early in embryonic development. Discuss how she could have children whose blood types would indicate that the children were not her own.

4. Explain fully why the failure to be excluded with a probability of 99.9 percent makes a person a very good suspect for a paternity accusation.

5. As disputed parentage now stands, many states use HLA antigen and other test data to assign paternity despite a small chance for a wrong assignment (one in 1000, or less). The attitude is that the occasional mistake is justified by the benefits resulting from all the correct assignments. What are these benefits? Do you agree that the mistakes are justified? Why or why not?

6. Explain how probabilities of exclusion must depend on allele frequencies in the population. Offer a good hypothetical example.

7. Do you think we will ever have genetic tests that would give 100 percent accurate assignments or exclusions of paternity? If so, give a general description of what such tests must be like.

299

PROBLEMS

1. In a paternity case, the child has type O blood, the mother has type A, and the putative father has type B.
   a. What inference can you draw from this situation?
   b. What inference could you draw if the putative father were type AB?

2. A type B mother has a type O child.
   a. What is the child's genotype?
   b. What is the mother's genotype?
   c. What are the father's possible genotypes?
   d. What genotypes could the father not have?

3. How many different ABO system alleles can one normal individual have? How many can a population have?

4. Given fresh samples of all four ABO blood types, we can mix them in 12 distinct combinations. In the spaces below, mark the combinations that will result in agglutination.

	A	B	AB	O
A				
B				
AB				
O				

5. The *MN* and *Sese* alleles are tightly linked, so that alleles of the two pairs are transmitted together as couplets. Considering these two pairs of alleles,
   a. How many different genotypes are possible?
   b. How many different phenotypes are possible?
   c. How many different matings are genotypically possible?

6. Considering the ABO and MN blood groups,
   a. How many different phenotypic combinations exist?
   b. List these phenotypes.
   c. How many different phenotypes exist if you also consider the Rh system?

7. One night in a very small hospital, mass confusion causes a complete mix-up in the assignments of four newborns to their parents. The babies' blood types are:

300

a. O,
b. A,
c. B, and
d. AB.

The blood types of the four pairs of parents are:

a. O and O,
b. AB and O,
c. A and B, and
d. B and B.

Assign each baby to its own parents.

8. A woman of blood types O and M bears dizygotic twins, one of types B and M and one of types A and MN. She claims to have had relations with two men at about the same time. One has blood types B and M; the other has types A and MN. Who is responsible for fathering these twins? Demonstrate your conclusion with diagrams of the appropriate genetic crosses.

9. Listed below are six of the first paternity cases in the United States that used the ABO and MN blood groups for exclusion purposes. In which cases did the data exclude the alleged father?

case	mother	child	alleged father
1	A,N	O,N	O,N
2	A,MN	O,M	O,M
3	A,MN	A,MN	O,M
4	B,MN	B,M	O,M
5	A,MN	O,MN	A,N
6	A,MN	B,MN	O,M

10. The type O mother of a type B child is seeking child support from a type A man.
    a. The man's lawyer presents the blood type data in court. Show how these data could help the man win the case.
    b. The woman's lawyer counters by suggesting a mutation. How can the man's lawyer respond to this contention?
    c. The woman's lawyer counters again by saying the man could still be the father because of the Hh system. What would the genotypic situation have to be to support a sound argument?

REFERENCES

American Medical Association, Committee on Transfusion and Transplantation and American Bar Association, Section on Family Law, Committee on Standards for Judicial Use of Scientific Evidence in the Ascertainment of Paternity. 1976. Joint AMA-ABA guidelines: Present status of serological testing in problems of disputed parentage. *Family Law Quarterly* X(3):247–285.

Hymen, H. S., and L. H. Snyder. 1936. The use of the blood tests for disputed paternity in the courts of Ohio. *The Ohio State University Law Journal* 2:203–219.

Landsteiner, K., and P. Levine. 1928. On the inheritance of agglutinogens of human blood demonstrable by immune agglutinins. *Journal of Experimental Medicine* 48:731–749.

Terasaki, P. I., D. Gjertson, D. Bernoco, S. Perdue, M. R. Mickey, and J. Bond. 1978. Twins with two different fathers identified by HLA. *The New England Journal of Medicine* 299:590–592.

14

Genes and Environmental Chemicals

We are surrounded by thousands of different chemicals. More than 7000 synthetic chemicals are currently in use, and approximately 1000 new ones are added to the list each year. They are used in agriculture, medicine, and industry, as food additives, drugs, pharmaceuticals, and pesticides. Some are **toxic** to certain plants and animals, including humans. Others, the **carcinogens**, cause cancer. **Mutagens** cause **mutations**, heritable changes in DNA. **Clastogens** cause chromosome breakage. **Teratogens** interfere with embryonic development and cause birth defects. Some chemicals have less drastic effects, such as those causing low sperm counts and infertility in men.

Chemicals can pose serious genetic and other hazards to the population. The extent of the hazards is clear from the accumulated statistics. Approximately 250,000 children are born in the United States each year with birth defects, many resulting from parental exposure to chemicals. One-fourth of the population will eventually develop cancer, and 80 to 90 percent of those cancers may be due to agents in the environment. The health problem is severe and widespread, but we cannot treat it in depth in this book. We can only review the basic concepts relating to these problems.

Some chemicals in the environment are classified as mutagens; they cause mutations, which affect the population's **genetic load**. Genetic load is the number of deleterious mutations and chromosome aberrations in the population's gene pool. More technically, it is the average loss of fitness of a population attributable to selection against deleterious alleles. Any increase in the mutation rate might go unnoticed in the short term, but it would eventually increase the genetic load. Since most mutations are recessive, the effects, shown diagrammatically in Figure 14-1, would not appear for several generations. In addition, **somatic mutations** (in body cells, not germ-line cells) can lead to many types of cancer; their effects may not appear for 10 to 30 years or more. The mutagenic effects of substances in the environment threaten the genetic health of both present and future generations.

Some scientists call this problem environmental mutagenesis or genetic pollution. Only recently has much effort gone into assessing the total significance of environmental mutagenesis in terms of human health. In essence, the problem is as serious as it is because large numbers of people are exposed to so many different chemicals, and thorough testing often takes years with extremely large costs. Of the few chemicals whose effects have been tested, some hold serious risks. In such cases, the risks and the benefits must be carefully

304

FIGURE 14-1

Each grid links the frequency of gametes with a specific mutated allele to the frequency of homozygotes for the mutation in the population. Each twofold increase in the mutation frequency in the population results in a fourfold increase in homozygotes. Consequently, increases in genetic load can be disastrous.

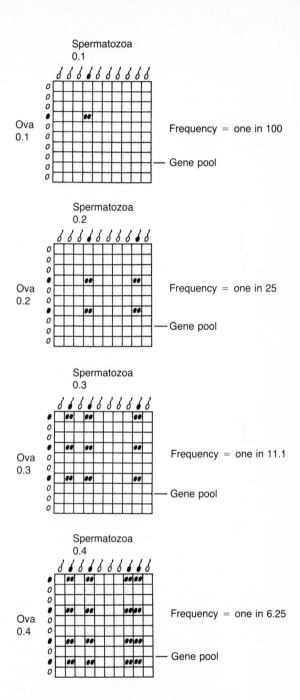

weighed before we can allow the chemical to be used—the ultimate decision can be difficult. Some people feel that we should ban all chemicals shown to have any genetic effects, regardless of their benefits.

MUTAGENESIS, CARCINOGENESIS, AND TERATOGENESIS

A mutation is an abrupt, heritable change in the genetic material of an organism. **Point mutations** are very small changes—such as a single base change at some point in the DNA. **Chromosomal aberrations** are gross changes of chromosomal number or structure. Both types of mutations can be caused by various agents in the environment. And because, in general, organisms are so well adapted to their environments, most mutations are harmful to them, at least in their present environments. The 1 to 2 percent that are beneficial contribute to the genetic variability of all species, including humans. (Nonlethal mutations that do no actual harm may also become valuable if the environment changes.)

Some mutations are promptly repaired by the cell's machinery. Those that are not repaired can have various effects. Mutations in gametes can have inconsequential, slight, or pronounced phenotypic effects; the worst are lethal. Many scientists believe that somatic mutations can cause cells to lose control of their growth and become cancerous. They may also be involved in triggering benign tumors. Many researchers are convinced that **mutagenesis** and **carcinogenesis** have a great deal in common.

Various kinds of changes affect the genetic material, including breaks, rearrangements, and other gross chromosomal abnormalities. Many agents, called clastogens, have been shown to be capable of breaking chromosomes. Various types of radiation (see Chapter 15) and many chemicals are clastogens. Exposure to them is so common that in karyotyping studies of white blood cells from normal individuals, scientists find that 2 to 3 percent of the cells contain one or more chromosomal abnormalities, with the percentage increasing with the individual's age. If chromosomal damage occurs in white blood cells, it is also likely to occur in the germ-line cells that produce gametes.

Teratogenesis is the production of congenital structural malformations. The developing embryo is extremely sensitive to radiation, chemicals, and dietary deficiencies, especially during the critical first three months of its development when its tissues and organs are differentiating and developing. Teratogens may exert their effects by causing

306

point mutations and chromosomal aberrations, or by interfering with the intercellular signals that control development. Either way, the result is some degree of deformation, usually followed by much suffering, and even premature death.

In 1958, an example of teratogenesis appeared in the frog population of a small lake in the southern United States. A large proportion of these frogs had extra legs and were quite clumsy. The most likely cause of their malformation was thought to be the insecticides that drained into the lake from neighboring cotton fields, but no one has been able to prove the connection.

From time to time, human beings are afflicted with similarly severe malformations. **Thalidomide** is probably one of the best known human teratogens. About 1960, it was marketed as a tranquilizer and anti-nauseant, especially for use by pregnant women. In tests, it had shown no ill effects on mice or rats; however, many mothers who had taken the drug, especially during the second month of pregnancy, gave birth to severely deformed children. Often, the infants were born with only flipper-like stumps for arms, or with no arms at all. A typical thalidomide baby appears in Figure 14-2.

FIGURE 14–2
An infant with the congenital malformation caused by the mother taking thalidomide during the second month of pregnancy. (World Wide Photo.)

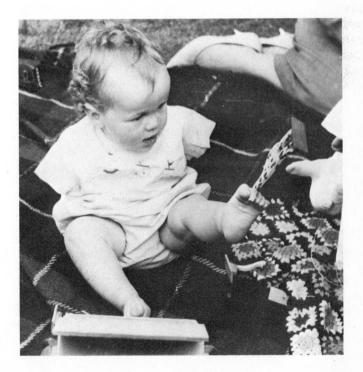

Some substances have already been indicted as mutagens, clastogens, carcinogens, or teratogens. Researchers strongly suspect that many others function as one or more of these genetic hazards. For others, the data are contradictory. We will briefly discuss each group below, but we cannot provide an exhaustive treatment here. In fact, most chemicals have not yet been tested.

Foods and Food Additives

Many studies have cast suspicion on certain foodstuffs, food additives, and preservatives as mutagens and/or carcinogens, or even teratogens. Researchers disagree on the best methods for testing substances and on how best to interpret the statistics essential to evaluating the small differences between tests and controls. For these reasons, test results are not always unambiguous. One of the most vigorous methodological debates is over the relative usefulness of *in vitro* vs. *in vivo* tests. *In vitro* experiments are conducted with populations of cells growing on or in a synthetic medium in glass vessels. *In vivo* data come from tests on intact organisms. The two approaches are different biologically, but they can both be useful under particular circumstances.

There is little doubt that some substances are genetically hazardous, and hazards of others are strongly indicated. **Nitrites**, for instance, are used to cure various meats, in which they react chemically with **amines**, natural meat constituents, to produce **nitrosamines**. Nitrosamines are believed to be carcinogens. Other carcinogens appear in well-cooked meats (especially in charcoal-broiled meats), and recently a special committee of the National Academy of Sciences declared that meats and fats, in general, may increase the risk of some cancers. Certain factions of the meat industry, of course, vigorously contest this assertion.

A particular red food dye and various other food additives have also been implicated in carcinogenesis. So has **caffeine**, a stimulant found in substantial amounts in coffee, tea, colas, chocolate, and certain over-the-counter and prescription drugs. Caffeine seems to exert its effect by interfering with cellular repair mechanisms of many species, perhaps including humans, so that mutations and chromosomal breaks are left free to affect the organism. Pregnant women are now cautioned to use caffeine sparingly because the molecule, once in the bloodstream, can cross the placenta and enter the embryo, and because caffeine has caused birth defects when given to pregnant rats. Unfortunately, this assessment of the dangers of caffeine is complicated by the fact that humans and rats metabolize caffeine differently. High coffee intake has also been linked to cancer of the pancreas. However, many studies of caffeine consumption have shown no adverse effects. Uncertainty remains, but we should perhaps be wary of such practices as using caffeine to improve the motility of human sperm for artificial insemination.

Saccharin, an artificial sweetener, has been the subject of an intense controversy. Various reports both indicate and refute that it is a possible mutagen, carcinogen, or clastogen. Some reports suggest that the usually slight effects observed are due to impurities in the saccharin, but the United States Food and Drug Administration (FDA) has nevertheless proposed to ban the sale of saccharin. Congress, however, decided to block the proposed ban.

The initial decision to ban saccharin depended chiefly on the results of a Canadian experiment in which 100 rats were given doses of saccharin amounting to about 5 to 7 percent of their total diet. This is the rough equivalent, for humans, of drinking 800 cans of diet-type soft drinks per day. Three of the rats developed bladder tumors, but 14 percent of the F_1 offspring of the treated rats developed bladder tumors. In light of the data, the FDA proposed its ban and provoked an immediate furor. The general public criticized the unpopular decision, claiming it was a result of the arbitrary nature of big government and asserting that people should be free to eat what they want. Many of the 10 million diabetics in the United States rely on saccharin, and critics pointed out that the ban might have grave effects on some of them. The effects of ordinary refined sugar were deemed to be just as unhealthy—consider dental caries, obesity, and bowel problems—even for non-diabetics. Nonetheless, the FDA felt obliged to ban saccharin in accordance with the Delaney Clause, passed in 1958, which prohibits chemicals that have caused cancer in animals or humans from being used in American-produced foods. Without congressional intervention, the only recourse of the critics is to have the Delaney Clause repealed. The saccharin controversy is a good example of a complex social situation without easy answers.

Drugs

People consume large amounts of many different drugs, many of them for medical reasons. Some of them can cause cancer and/or genetic damage. Thalidomide is only one example. Women are warned not to become pregnant immediately after they stop taking contraceptive pills; the female sex hormones in the pills remain in the body for a time and can damage the fetus. **Diethylstilbesterol**, or **DES**, was once used to prevent miscarriages. The daughters of women who took the drug show a high incidence of vaginal cancer. Penicillin, aspirin, and cyclamates (another banned artificial sweetener) are clastogenic *in vitro*. Even excessive amounts of vitamins A and D can cause genetic damage.

The genetic effects of many drugs are largely unknown, especially in the long term, and people may have little choice about whether to use them. "Recreational" drugs are another matter, for they can easily be avoided, and perhaps they should be. **Ethanol** (grain or beverage alcohol) has serious effects on the physical and mental health of drinkers; it is believed to be mutagenic, carcinogenic, clastogenic, and teratogenic. Heavy drinkers show a significantly higher incidence of cancers of the tongue, larynx, esophagus, and other oral structures. Alcoholic beverages are mutagenic when tested on certain bacteria, and alcohol is a demonstrated clastogen *in vitro* and in rats.

Fetal alcohol syndrome, a teratogenic effect in newborns, is due to alcohol consumption by the mother during pregnancy. It affects one or two newborns per 1000 with symptoms of varying severity. The symptoms include central nervous system problems, growth deficiencies, characteristic facial features, and other minor malformations. The child usually has lower than average mental abilities (mentally retarded in many cases), and sometimes exhibits **microcephaly** (small head). The severity of the syndrome depends on the amount of alcohol ingested, the timing of the intake during pregnancy, and the genetic make-up of the fetus. According to recent estimates, an average of six substantial alcoholic drinks per day taken by a pregnant woman constitutes a major risk for the syndrome.

An estimated 60 percent of all 18- to 25-year-olds used **marijuana** or its concentrated form, **hashish**, to some extent during 1978. The marijuana components that enter the body are quite fat soluble and therefore tend to concentrate in the gonads and brain, where they can remain for long periods. It takes at least a week to remove half of these chemicals. The many reported physiological effects of marijuana include listlessness, reduced sexual function, decreased immunity, and impaired brain function.

Some researchers have claimed that marijuana does genetic damage, but proof has been difficult in humans, *in vivo*. Marijuana has been shown to cause chromosome damage and other cellular effects in mice, rats, and monkeys, and it is highly mutagenic in certain bacteria. In humans, however, the data are inconclusive, largely because of technical problems such as small sample size. More investigations are needed. Only then will we know for sure whether marijuana should be decontrolled in our society. So far, most data indicate that it would not be wise to legalize marijuana.

LSD, lysergic acid diethylamide, may also cause genetic damage. *In vitro* and *in vivo* studies have shown that it can cause chromosome damage, although some *in vitro* studies do not support this conclusion.

Other Substances

Soot was blamed for cancer of the scrotum as early as 1775, when a high incidence of that cancer was noticed among chimney sweeps. Today, we know that components of smoke, tar, and soot are all definite carcinogens. Many mutagens have been identified in cigarette smoke, and the increase in the risk of lung cancer for smokers is well known. Lung cancer has increased a startling 20-fold over the last 40 years. The increase in the number of women smoking is accompanied by a parallel increase in lung cancer among them. The graph in Figure 14-3 presents details of this relationship. Several studies even indicate that a hazard exists for non-smokers when they become "passive smokers" in the presence of smokers, but several other studies dispute this.

Some pesticides, especially insecticides, appear to be mutagenic. Several studies have found no mutagenic effects for a few herbicides tested with microbial test systems, but others have indicted the weed killer **trichlorophenoxyacetic acid** (2,4,5-T and 2,4-T) as mutagenic and teratogenic, probably owing to contamination by **dioxin**. **DDT** (dichlorodiphenyl trichloroethane) is a fairly effective insecticide, but possible human health hazards and toxicity to nontarget species forced it off the American market. Its carcinogenicity and other long-range effects on humans have not been definitively settled, however.

Industrial chemicals can also cause cancer or genetic damage. For example, rubber industry workers are exposed to several hundred chemicals, and researchers have found that they have a significant number of cells that show the effects of exposure to mutagens. The cells of these workers show many more **sister chromatid exchanges** (discussed later in this chapter) than those of controls, a phenomenon geneticists have correlated with a substance's ability to cause mutation.

Asbestos-using industries have received a great deal of attention because asbestos is a potent carcinogen. Asbestos, a naturally occurring substance, has been used in brake linings, insulation, and construction materials for years. Tiny airborne asbestos particles are

311

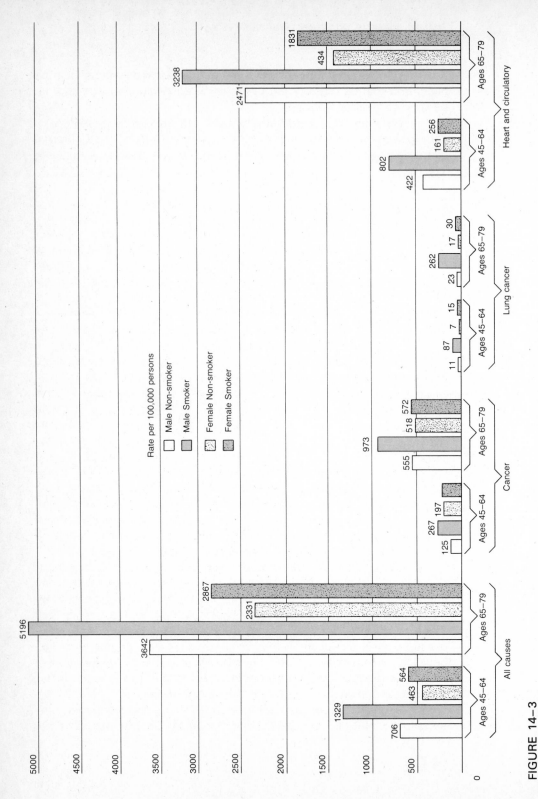

FIGURE 14–3

Death rates for smokers and non-smokers. (Adapted from Hammond, E. Cuyler. 1966. In Haenszel, W., ed. *Epidemiological Approaches to the Study of Cancer and Other Chronic Diseases*. National Cancer Institute, Monograph 19. U.S. Dept. of Health, Education, and Welfare.)

believed to collect in the lungs, where years later they cause cancers of the lungs and the linings of the chest cavity. Asbestos workers have a tenfold higher risk of developing such cancers than non-asbestos workers. Those who also smoke have an 80-fold higher risk than non-asbestos workers. The most dramatic difference in risks—90-fold—exists between asbestos workers who smoke and non-asbestos workers who do not smoke. The case is so well documented that many ex-workers with cancer are suing their past employers. One large manufacturer of asbestos products had so many suits waiting in the courts that it decided to file for bankruptcy, saying that the company could not pay so many claimants.

In some parts of the country, asbestos particles enter the air naturally, especially where local rock containing asbestos has been crushed and used for roads, driveways, patios, and playgrounds. In such areas, the concentration of asbestos particles in the air can be 1000 times higher than normal. However, no one has yet linked living in such areas to a high incidence of asbestos-caused cancer.

PCBs (polychlorinated biphenyls) are found mostly in electrical capacitors and transformers. When released into the environment, they may cause cancer and reproductive dysfunction. **Formaldehyde** is a chemical that was initially considered a carcinogen and later reprieved. **Vinyl chloride** was banned as being a carcinogen in 1974. Other chemicals shown to be mutagenic and/or carcinogenic are ethyl methane sulfonate, alkylating agents, nitrous acid, flame retardants, benzene, some hair dyes, urethane, spray adhesive, arsenic compounds, cadmium compounds, nickel, and ethylnitrosourea, among others. Ethylnitrosourea has been called the most potent mutagenic chemical ever tested on mice.

Many industrial chemicals find their ways into hazardous waste dumps. Thousands or millions of tons are buried in forgotten locations all around this country. When these wastes seep into ground water, as they did when the dump beneath the Love Canal housing area in Niagara Falls, New York, sprang a leak, they can become serious health hazards. Researchers have found it difficult to prove that genetic damage has occurred in Love Canal residents as a result of their exposure to the waste disposal site. But it seems likely that the damage exists and will eventually find expression as cancer or recessive birth defects. Other "Love Canals" also exist in the United States, and many more will undoubtedly be discovered. Hazardous waste is one of society's serious problems.

Not all chemicals capable of doing genetic damage come from human inventions and factories. Nature produces some dangerous sub-

313

stances without our help. Certain fungi can produce **aflatoxin B$_1$**, a powerful liver carcinogen. There are also carcinogens in some plant extracts; they include the **flavonoids** of the bracken fern.

Human beings, as we have seen, are exposed to a tremendous variety of chemicals, many of them genetically hazardous. It is not alarmist, but only sensible, to develop effective methods to determine which chemicals are dangerous.

IDENTIFICATION OF GENETICALLY DAMAGING SUBSTANCES

The United States uses an estimated 75,000,000 mammals in laboratories each year as part of an expensive effort to detect mutagens and carcinogens. Much of society recognizes the serious need to identify these hazards, along with clastogens and teratogens. Questions remain, however, concerning how many chemicals should be tested, how promptly, and at what expense.

A good testing system must have certain characteristics. If the objective is to test many substances, it should be simple, efficient, and inexpensive. It should allow testing of a chemical with large numbers of organisms in a short time. It should also be sensitive, accurate, and reproducible.

Good controls are also essential, so that toxic effects can be confidently attributed to the substance tested. For example, rats are sensitive test organisms, developing tumors even in response to very mild mutagens and carcinogens. However, they also develop tumors spontaneously, and it takes carefully designed experiments and controls to show that if rats exposed to a chemical develop a few more tumors than unexposed rats, those tumors are due to the chemical. Researchers would like a test system that would give more clearcut results. They would also like a test system that would reliably detect all types of genetic damage.

Types of Test Systems

The wide variety of test systems now in use can be separated into three groups, according to the organisms used.

1. Some tests use fungi or microbes such as bacteria. *Neurospora* (pink bread mold) is one of the most frequently used experimental organisms. The immense advantages of mi-

314

    crobes include small size, large numbers available, and short generation times.

2.   Other tests use higher plants and animals. Some of the animals that have been used in chemical testing schemes are *Drosophila melanogaster* (fruit flies), frogs, mice, rats, hamsters, cats, dogs, and other mammals, including primates such as monkeys. Plant tests have used the root tips of onions and broad beans, among others.

3.   Still other tests use mammalian (including human) cells grown in tissue culture. They have become very popular, but they are limited by the unnatural conditions of the system. *In vitro* systems do not use the organs of an intact body and cannot simulate the natural organ and cell interactions. There is no way to concentrate the substance in the blood, excrete it through a urogenital system, or detoxify it by action of the liver. Still, many of the test systems show promise and warrant further development.

One rather interesting system relies on sister chromatid exchange and involves staining sister chromatids differently to tell them apart. Sister chromatids result when chromosomes replicate, producing two identical chromatids that remain attached at their centromeres (kinetochores). Occasionally, the sister chromatids exchange parts, and this occurs more frequently upon exposure to mutagens or carcinogens. The differential staining of the sister chromatids reveals the exchanges, whose frequency is at least a rough measure of mutagenicity or carcinogenicity. Figure 14-4 shows sister chromatid exhanges observed in human blood lymphocytes. Some scientists consider this test powerful and valuable.

Unfortunately, none of our present testing schemes is without problems. The main difficulty is simply the time involved in many of the tests. Expense is also a problem, because testing one chemical has taken up to five years and can cost as much as $1,000,000. Other problems relate to the testing itself, for scientists are not always sure that the chemical being tested is reaching the right organ in the test organisms. Humans and test organisms may get their exposures in different ways, unknown interactions may take place, and organisms seem to differ significantly in their responses to mutagens, even within the same species. Most tests require large doses of the chemical over short times, while humans are generally exposed to small doses over long periods. Many scientists believe that such high doses may overwhelm

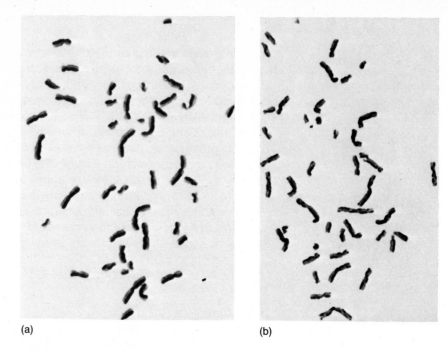

(a)

(b)

FIGURE 14-4

Sister chromatid exchanges. (a) 96-hour culture of human peripheral blood lymphocytes. (b) 96-hour culture of human peripheral blood lymphocytes in medium supplemented with mitomycin-C. (Courtesy Richard J. Green, Centers for Disease Control, Atlanta, Ga.)

cellular repair mechanisms whereas smaller doses would not, so that the chemical may only be hazardous in "test" doses. Still another problem is how to interpret the benign tumors that sometimes occur when animals respond to test chemicals. And another is the long latency period of many cancers—by the time a chemical is identified as a carcinogen, many persons may have already been exposed to it and "seeded" with cancers that will appear decades later. Our goal must be to identify hazardous chemicals before the public is exposed to them.

Perhaps the most basic problem of all is how to extrapolate results from test organisms to humans. The effects are not always the same on different test organisms or on test organisms and humans, because different organisms can differ greatly in metabolism and other physiological functions. Both false positives and false negatives have oc-

316

curred. Thalidomide serves as a classic example of the latter—the drug was not teratogenic for mice and rats, but it was for humans.

At this time, the ultimate test for too many chemicals is their day-to-day application to humans *in vivo*. The responses of the human population must be monitored carefully, even though it will be difficult to study large numbers of people. However, it does not seem ethical to develop large bodies of knowledge about the damaging effects of chemicals based on people's long-term experiences with the substances. Furthermore, even these kinds of observations will not tell us for sure which chemicals are to blame, and in what doses. We need a test system that is fast, efficient, reliable, and predictive for the human population. So far, the best candidate may be the Ames test, a microbial test showing great promise.

The Ames Test

Bacteria are attractive organisms for testing mutagens, because of their small size, large numbers, short generation time, low expense, and good sensitivity. B. N. Ames and his colleagues have developed a sophisticated test system, using the bacterium *Salmonella typhimurium*. The test is called the **bacterial-liver extract test**, the *Salmonella*-**mammalian liver test**, or simply the **Ames test**.

The design of the Ames test system is based on a number of carefully considered factors. The *Salmonella* test organism has several useful mutations. One blocks the bacterium's ability to synthesize a **lipopolysaccharide** substance that normally coats its surface. Lacking this lipopolysaccharide, the bacterium can more readily absorb test chemicals. Another mutation blocks the bacterium's DNA repair mechanism, thereby allowing genetic damage to be detected. A third mutation prevents the organism from making its own **histidine**, an essential amino acid. The "*his –*" organism therefore cannot grow on a medium lacking histidine. If it is placed on a histidine-free **selective medium**, only *his +* (histidine synthesizing) mutants will survive. Since any mutagen will produce a few *his +* mutants (among many other mutants), the use of the *his –* bacterium offers a very sensitive test for mutagenicity. The number of *his +* mutants, easily counted as growing colonies on the selective medium, is the measure of mutagenicity.

The chemical being tested and an extract from rat liver are mixed with *Salmonella* and spread on a plate of selective medium, as shown in Figure 14-5. The rat-liver extract (the **microsomal fraction**) contains many of the smaller structures and soluble enzymes of the liver cells. Its role is to process the test chemical much as it would be metab-

317

FIGURE 14–5

The principal components of the Ames *Salmonella*-mammalian liver test for chemical mutagens.

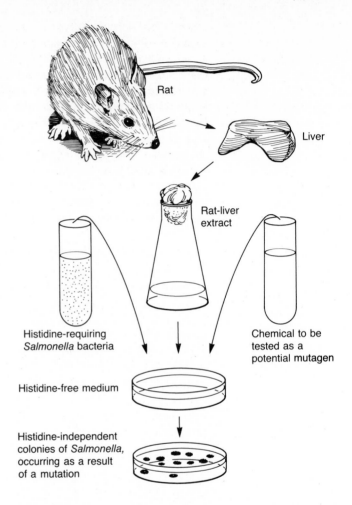

Rat

Liver

Rat-liver extract

Histidine-requiring *Salmonella* bacteria

Chemical to be tested as a potential mutagen

Histidine-free medium

Histidine-independent colonies of *Salmonella*, occurring as a result of a mutation

olized in a mammalian body, for while many chemicals are not themselves carcinogenic or mutagenic, their metabolites are. The extract is thus a critical part of the testing system.

The Ames test has several advantages. It is very sensitive and relatively inexpensive, even though data are collected on millions of organisms. Because of the small size of the bacteria, space is not a limiting factor. Furthermore, results can be obtained within two or three days, and the test has correlated very well with other types of testing.

The Ames test has already been used on more than 5000 chemicals. Many substances, especially those strongly suspected of being carcinogens, have caused a significant increase in the mutation rate. Of those chemicals, many required the rat-liver extract to activate their mutagenicity. Marijuana, many alcoholic beverages, and tobacco condensates, for example, have yielded positive mutagenic results. The test has even shown that tobacco smokers, but not most non-smokers, have mutagenic urine. Strangely enough, it has shown that the water extracts of the feces of vegetarians are less mutagenic than those of meat-eaters.

The Ames test seems to be a reasonable test, but like all tests, it has its limitations. It does not work well with all classes of chemicals; metals, for example, give very poor results. And, because bacteria are certainly not closely related to humans, it may not be predictive. A few chemicals that show a great effect on bacteria do not have much effect on mice, which are much more closely related to humans.

Nevertheless, the Ames test holds promise as a screening technique. It will become especially valuable if a strong correlation exists between carcinogens and mutagens, both of which damage DNA. The test shows that most tested carcinogens—90 percent—are also mutagens, at least with respect to *Salmonella*, while few noncarcinogens—12 of 108 (11 percent) tested—are mutagens. We already have good reason to believe that most carcinogens are mutagens. Next we need to demonstrate that mutagens are carcinogens. Indeed, many scientists are already convinced of that relationship.

THE ROLE OF SOCIETY

Gene mutations and chromosomal changes are the raw materials of adaptation, survival, and evolution. However, excessive genetic change could be disastrous to the human population. Mutation can be considered a precious genetic resource, but we should not dramatically increase the mutation rate by polluting our environment.

What is society doing to protect the environment? Congress has mandated the Clean Air Act, the Clean Water Act, the Toxic Substances Control Act, the Consumer Product Safety Act, and the Food, Drug, and Cosmetic Act, among others. The Environmental Protection Agency, the Food and Drug Administration, and the Occupational Health and Safety Administration are segments of government directly charged with enforcing these acts by regulating carcinogens and pollutants. Unfortunately, protecting the environment and public health is very expensive.

The objectives of our laws, regulations, and agencies make sense to most of us, because we want to reduce our exposure to carcinogens, as well as other hazards. Unfortunately, our knowledge is imperfect. Hazardous wastes leak from dumps into ground water partly because we do not know how to handle the wastes properly. We are not sure how to monitor possible hazards or assess the risks. Data may be available, but we do not always know how best to use them. More research is necessary, concentrating in particular on monitoring, disposal mechanisms, and testing.

Genetics will undoubtedly play an important part in ongoing efforts to protect the environment. Carefully designed studies will permit us to understand how people differ in their susceptibilities and to develop our knowledge of how carcinogens and mutagens affect the genetic material biochemically.

Society can do a great deal to protect its genetic resources. If many cancers are due to chemicals in the environment, we can reap significant health benefits by reducing our exposure to these substances. However, even as researchers document the hazards of environmental chemicals, individuals often fail to respond with cooperation. Current estimates indicate that 30 percent of all cancer deaths are due to smoking, and some scientists believe that abolishing cigarette smoking would nearly eliminate lung cancer, a common killer. Still, a high proportion of individuals in our society continues to smoke. An estimated 70 percent of all cancer deaths are due to smoking, alcohol, and diet. One preliminary study indicated a possible two- or threefold increase in pancreatic cancer among heavy coffee drinkers. Later studies, however, countered this conclusion. Science identifies the environmental hazards and assesses the probabilities of damage occurring. Society is responsible for the ultimate decisions, and like it or not, its members may have to make some sacrifices to reduce carcinogenesis and mutagenesis.

SUMMARY

Many toxic and genetically damaging chemicals permeate our environment and affect humans as carcinogens, mutagens, clastogens, and teratogens. Because most mutations are harmful, the accumulated mutations within the human population increase genetic load. Somatic mutations may cause cancer. Changes in the genetic material caused by chemicals may be small point mutations or chromosome rearrangements. Normal cells, however, rely on repair mechanisms that can sometimes restore the changed gene or chromosome to its normal form.

Teratogens such as thalidomide, a tranquilizer anti-nauseant, cause congenital structural malformations in the developing embryo or fetus. Foods and food additives suspected of being mutagens, carcinogens, or clastogens include nitrosamines, saccharin, and caffeine. Caffeine is believed to interfere with the cell's repair mechanisms. The Delaney Clause requires the removal of carcinogenic food additives from the open market, but it cannot guard against the many natural chemicals that are highly carcinogenic. Many drugs and other chemicals are also strongly indicated to be genetic hazards. Among them are ethanol, marijuana, synthetic sex hormones, possibly LSD, soot, smoke, tobacco, some pesticides, asbestos, PCBs, and vinyl chloride. Many industrial chemicals enter the environment from leaking waste dumps.

Genetically damaging substances can be identified in several ways. Good methods should be relatively simple, efficient, inexpensive, fast, sensitive, and repeatable. The current tests use microbes, higher plants and animals, and tissue cultures of mammalian cells. Mutations, cancer, and abnormal chromosomes are the effects most often monitored. Sister chromatid exchange also indicates mutagenicity. None of the tests is completely free of problems. The bacterial-mammalian liver extract (Ames) test has many advantages and good potential for reliable results. It capitalizes on several mutations in a strain of *Salmonella* bacteria and uses rat-liver extract to activate test chemicals so that their presumed mutagenic forms enter the bacteria. The many scientists who equate carcinogens with mutagens are the most active advocates of the test.

Regardless of the scientists' success in identifying genetically hazardous chemicals in the environment, members of society do not always make use of this information. For example, people typically ignore the evidence that tobacco, alcohol, and other substances are hazardous.

DISCUSSION QUESTIONS

1. Why is it difficult to recognize mutational events in humans?

2. Why should we be very careful about applying *in vitro* data gathered on chemical hazards to *in vivo* situations?

3. Do you think we will ever be able to prove that a substance is completely safe? Why or why not?

321

4. The cost of testing a new chemical can be as much as $1,000,000. Do you think this cost is worth it? Does your answer depend on specific circumstances? If so, what are they?

5. Why is it not a simple matter to test each new chemical compound or drug for deleterious effects before putting it on the market?

6. Is there a safe dose for a mutagen? A carcinogen? Discuss both questions.

7. When certain substances are administered to an organism together with caffeine, genetic damage is often greater than the sum of the genetic damage caused by the two substances separately. Give a plausible reason for this synergism.

8. The incidence of cancer in Seventh Day Adventists, who eat low-fat diets, is approximately 70 percent of the risk reported for other Americans. Give several possible reasons for this observation.

9. Some people feel that we regulate too many foods, drugs, and chemicals, while others believe we regulate too few. What is your position on this matter?

10. What is the relationship, if any, between a mutagen and a carcinogen?

11. Some carcinogens have a much greater mutagenic effect on bacteria than on mammals such as mice. Can you give any reasons for this difference?

12. How should we deal with jobs that expose workers to hazardous substances? Should we simply point out the hazards, offer extra pay to exposed workers, or allow workers to make their own choices about where to work?

13. Is it possible to simultaneously achieve high industrial production and employment while satisfying the requirements of the Clean Air Act? How stringent should the regulations be?

14. Safe levels of exposure of many chemicals are much lower for fetuses than for adults. Should working conditions be made completely safe for a developing fetus? If not, should all women with reproductive potential be barred from jobs that entail exposure to hazardous substances?

15. What form of hazardous waste storage and/or disposal do you favor? Why?

16. What do you suppose the following statement means? "We don't need tighter pollution regulations. We need tougher rats!"

17. Would you prefer a mutation in a somatic cell or a mutation in one of your gametes? Qualify your answer if you wish, but explain your choice fully.

18. Do you think the incidence of cancer is higher in the human population today than in earlier times, or are we simply better at diagnosis and more aware of cancer? In general, do you think the environment of the past held fewer chemical hazards? Qualify your answers if you wish, but explain your reasoning.

19. Might natural selection play any kind of role in human carcinogenesis? Fully explain your ideas on the matter.

REFERENCES

Ames, B. N. 1979. Identifying environmental chemicals causing mutations and cancer. *Science* 204:587–593.

Ames, B. N., W. E. Durston, E. Yamasaki, and F. D. Lee. 1973. Carcinogens are mutagens: A simple test system combining liver homogenates for activation and bacteria for detection. *Proceedings of the National Academy of Sciences* 70:2281–2285.

Ames, B. N., F. D. Lee, and W. E. Durston. 1973. An improved bacterial test system for the detection and classification of mutagens and carcinogens. *Proceedings of the National Academy of Sciences* 70:782–786.

Ames, B. N., J. McCann, and E. Yamasaki. 1975. Methods for detecting carcinogens and mutagens with the *Salmonella*/mammalian-microsome mutagenicity test. *Mutation Research* 31:347–364.

Bloom, A. D. 1980. Issues in human mutagenesis and teratogenesis. In *Genetics and the law II*, ed. A. Milunsky and G. J. Annas. New York: Plenum Press.

Bruce, R. W., and J. A. Heddle. 1979. The mutagenic activity of 61 agents as determined by the micronucleus, *Salmonella*, and sperm abnormality assays. *Canadian Journal of Genetics and Cytology* 21:319–334.

Carson, R. 1962. *Silent spring*. Boston: Houghton Mifflin Co.

Hammond, E. C. 1966. Smoking in relation to the death rates of one million men and women. In *Epidemiological approaches to the study of cancer and other chronic diseases*, ed. W. Haenszel. National Cancer Institute Monograph 19. Washington, DC: U.S. Department of Health, Education, and Welfare, U.S. Public Health Service, National Cancer Institute.

Husfield, R. T., and L. Kuczma. 1978. *Karyotyping human chromosomes using microtechniques and the frequencies of chromosome aberrations*. B. A. thesis, St. Mary's College, Winona, MN.

International Commission for Protection Against Environmental Mutagens and Carcinogens. 1979. Cigarette smoking: Does it carry a genetic risk? *Mutation Research* 65:71–81.

Kuhnlein, U., D. Bergstrom, and H. Kuhnlein. 1981. Mutagens in feces from vegetarians and non-vegetarians. *Mutation Research* 85:1–12.

323

Lenz, W., and K. Knapp. 1962. Thalidomide embryopathy. *Archives of Environmental Health* 5:100–105.

MacMahon, B., S. Yen, D. Trichopoulos, K. Warren, and G. Nardi. 1981. Coffee and cancer of the pancreas. *The New England Journal of Medicine* 304:631–633.

McCann, J., and B. N. Ames. 1976. Detection of carcinogens as mutagens in the *Salmonella*/microsome test: Assay of 300 chemicals: Discussion. *Proceedings of the National Academy of Sciences* 73:950–954.

McCann, J., E. Choli, E. Yamasaki, and B. N. Ames. 1975. Detection of carcinogens as mutagens in the *Salmonella*/microsome test: Assay of 300 chemicals. *Proceedings of the National Academy of Sciences* 72:5135–5139.

Muneer, R. S. 1978. Effects of LSD on human chromosomes. *Mutation Research* 51:403–410.

Perry, P., and H. J. Evans. 1975. Cytological detection of mutagen-carcinogen exposure by sister chromatid exchange. *Nature* 258:121–125.

Ronen, A., and M. Marcus. 1978. Caffeine and artificial insemination. *Mutation Research* 53:343–344.

Rowan, A. N. 1981. The test tube alternative. *The Sciences* 21:16–19, 34.

Russell, W. L., E. M. Kelly, P. R. Hunsicker, J. W. Bangham, S. C. Maddix, and E. L. Phipps. 1979. Specific-locus test shows ethylnitrosourea to be the most potent mutagen in the mouse. *Proceedings of the National Academy of Sciences* 76:5818–5819.

Sorsa, M., J. Maki-Paakkanen, and H. Vainio. 1982. Identification of mutagen exposures in the rubber industry by the sister chromatid exchange method. *Cytogenetics and Cell Genetics* 33:68–73.

Volpe, E. P. 1981. *Understanding evolution*. Dubuque, IA: Wm. C. Brown Group.

Yamasaki, E., and B. N. Ames. 1977. Concentration of mutagens from urine by absorption with the nonpolar resin XAD-2: Cigarette smokers have mutagenic urine. *Proceedings of the National Academy of Sciences* 74:3555–3559.

15

Radiation,
Genes, and
Chromosomes

S ay "radiation"—and people may think of the Three Mile Island nuclear power plant incident of 1979. Or they may think of Hiroshima and Nagasaki and the immense stockpiles of nuclear weapons that could destroy all life on earth. On the positive side, they may think of medical uses of radiation to detect broken bones, treat cancer, and participate in dozens of other diagnoses and treatments. Radiation is both a curse and a boon, a bane and a blessing. In this chapter, we will focus on its hazards to humans.

FUNDAMENTAL NATURE OF RADIATION

The two main types of radiation are both forms of energy. **Nonionizing** radiation is visible light, ultraviolet light, infrared, radio, and other long-wavelength forms of **electromagnetic** radiation, consisting of bundles of energy (without mass) called **photons**. They do not penetrate well into solid matter; in particular, they cannot transfer sufficient energy to an atom such that the atom will lose an **electron**, or **ionize**.

The short-wavelength, high-energy electromagnetic radiations that can ionize atoms are **gamma rays** and **X rays**. Other **ionizing radiations** are composed of particles, which have both mass and energy. Such **particulate radiation** is generally made up of subatomic particles, electrons, **protons**, and **neutrons, liberated from atoms; beta radiation**, for example, consists of streams of electrons. It can also be composed of atomic nuclei, as in **alpha radiation** (helium nuclei, consisting of two protons and two neutrons) and **cosmic radiation** (larger nuclei).

When ionizing radiation strikes an atom, it can knock an electron loose, producing a charged ion. The electron carries a negative charge, and as the freed electron collides with other atoms it can knock more electrons loose. The original radiation strike yields a chain of electrons and ions that can take part in many chemical reactions. This chain is called a **linear energy transfer (LET)**, and it does the damage that is characteristic of ionizing radiation. We will pay special attention to ionizing radiation, its sources, and its biological importance in this chapter.

Biological Effects of Radiation

The effects of radiation can be due to either **acute** or **chronic** exposures. Acute exposure means the subject experiences high levels of radiation over a short period of time. Chronic exposure is low-level radiation exposure over a long period of time (sometimes years). Acute

326

exposures can bring about changes in blood count, loss of appetite, nausea, vomiting, diarrhea, loss of hair, burns, and even death. **Acute radiation syndrome** can strike people almost immediately after exposure; such reactions have been seen after severe radiation accidents and the World War II atomic bombings. The effects of chronic exposures are less severe, but they can cause cancers, birth defects, mutations, and chromosomal aberrations.

As early as 1897, skin ulcers and burns were noted as results of high radiation exposures. Clarence Dally, who helped Thomas Edison develop the fluoroscope, often placed his hand or body in X-ray beams. Eventually, both his arms had to be amputated, and he finally died of cancer. In 1927, geneticist H. J. Muller reported that X rays can cause genes to mutate. X rays became an important tool for geneticists, but Muller recognized the dangers of X-ray treatments and issued appropriate warnings. He pointed out both the direct and the long-term effects of radiation.

When radiation passes through a cell, depending on what cellular structures it hits, it may do no damage at all, or the cell can repair the damage that occurs, or the damage can go unrepaired. This last result can have dire consequences. If radiation causes mutations in gametes or in the germ cells that give rise to gametes, these mutations can be inherited by offspring. An example of the end result of this sequence of events appears in Figure 15-1. When radiation affects somatic cells, the cells may survive, but may eventually multiply to form a clone of malignant cells that spread rapidly in the body. On the other hand, the cells may be sufficiently damaged to die immediately or to die when they try to undergo cell division. In some cases, the effects of radiation on somatic cells may be latent; that is, cancer or cell death may occur long after the actual exposure. Mature **lymphocytes** (white blood cells) are quite sensitive to radiation. So are the thyroid gland, lungs, gastric glands, and the female breast. Radiation can also cause **cataracts** (small opacities in the lens of the eye). Many scientists believe that radiation accelerates aging and shortens the life span.

The energy particles of ionizing radiation enter the cell at random. Therefore, all parts of the cell have an equal chance of encountering radiation and being damaged by direct hits or by the linear energy transfer resulting from the ionization of nearby molecules. Large doses of radiation can destroy any cellular structure. However, the major site of radiation damage leading to mutation, cancer, or cell death is the nucleus of the cell. Since DNA is the primary constituent of the nucleus, it is the radiation target that demands our greatest concern.

327

FIGURE 15–1
Abnormal wing structure in the fruit fly, caused by a radiation-induced mutation. (Used with the permission of the Brookhaven National Laboratory.)

THE TARGET THEORY

The most satisfactory theory that explains how radiation affects the cell emphasizes that ionization within a small area of the cell can directly damage the cell. The particular area of the cell in question is the target, and the overall concept is the **target theory** (or **hit theory**). Clearly, the most vulnerable targets are the most essential cell structures. Most target theorists view DNA molecules as the targets. However, as we have seen, a direct hit on the DNA may not be necessary, because a hit in the vicinity of the DNA can cause ionizations that damage the DNA.

Understandably, the greater the quantity (or **dose**) of radiation passing through a cell or population of cells, the greater the chance for a hit upon one or more targets. With very low doses, there is little probability of hitting the target. With higher doses, there is a higher proba-

328

bility of hitting the target. One might logically expect a linear relationship between the dosage and the number of hits. However, at very high doses, some hits will be "invisible," since they will be second or third hits on targets that are already damaged. The dose-response curve, which plots dosage vs. the fraction of targets hit, will level off. This relationship is shown in Figure 15-2.

Let us assume that one hit is enough to have a certain effect, such as the complete inactivation (or death) of one cell. This is a **single-hit hypothesis**. From this information it should be possible to estimate the size of the radiation dose that is just large enough to inactivate all the cells in a population. However, when this dose is administered, statistics show that it will inactivate only 63 percent of the cells. The other 37 percent escape unscathed because some of the inactivated 63 percent are hit two or more times. We cannot expect each cell to be hit exactly once. Those cells that are not inactivated are called the **surviving fraction** of the population of cells.

We can base our experiments and analyses on different assumptions. We might, for example, use a **multi-hit hypothesis** that states that more than one hit is required to inactivate a particular cell. The mathematical analysis of this hypothesis is much more complex than that required for the single-hit hypothesis.

For the time being, let us assume that only one hit is required to damage a cell significantly. Also assume that whether or not the hit occurs is simply a matter of chance. Then the chances of hitting the target with four equal doses of radiation at four different times are equal to the chance of a hit occurring when four times the dose is applied at one time. These possibilities are presented schematically in Figure 15-3.

FIGURE 15–2
A radiation dose-response curve. As the dosage increases, the proportion of targets hit also increases in an almost linear manner at the lower levels. However, when larger doses are used, some targets are hit two or more times, resulting in a deviation from a linear relationship.

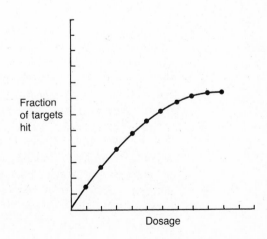

Fraction of targets hit

Dosage

329

FIGURE 15-3

Ignoring any complicating factors, the probability of a hit is the same for a low radiation dose spread over a long period of time (a) and for a high dose administered during a short period of time (b).

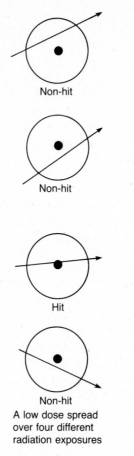

Non-hit

Non-hit

Hit

Non-hit
A low dose spread over four different radiation exposures

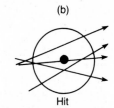

(b)

Hit
A dose four times higher than in (a), administered over a very short period of time

This simplification ignores any other factors that might be operating, but the point is that a target can be hit any time it is exposed to radiation. The greater the exposure, the greater the chance of a hit, regardless of the timing of the exposure. Many factors can complicate this situation. For example, the repair mechanisms of the cell may work more effectively at lower doses over longer periods of time than at higher doses over shorter periods.

The target theory, if correct, has important ramifications, especially if the target only has to be hit by one packet of radiation energy or one ionizing particle. Each hit then has a high probability of causing serious consequences such as mutation, and there may be no such thing as a safe, low dose of radiation. The issue of low-level radiation is controversial, as we will see later in the chapter.

SOURCES OF RADIATION

In order to compare the amounts of radiation humans receive from different sources we need a unit of measurement. There are several common units for measuring radiation. The **rad** is the amount of radiation that deposits 100 **ergs** of energy per gram of tissue. However, different types of radiation can cause different amounts of damage with the same dose in rads. The **rem**, or **rad equivalent man**, takes into account these differences in the type of radiation. A rem is a rad multiplied by a correction factor, the **relative biological effectiveness (RBE)** or **quality factor (QF)**. The **millirem** (mrem) is simply one-thousandth (1/1000) of a rem.

There are three main sources of radiation: background, medical, and other artificial sources. The average person's whole-body dose is about 178 to 200 mrems per year, with a range of 40 to 300 mrems. The dosage depends on where a person lives—altitude, rock types, and amount of snow cover—occupation, medical needs, and many other factors. **Background radiation**, which is virtually unavoidable, provides the greatest dosage (98 mrems per year), with medical sources not much less, averaging 73 mrems per year. Other artificial sources of radiation amount to approximately 7 mrems per year. The various sources contributing to the average yearly dose of whole-body radiation are compiled in Table 15-1.

Natural background radiation is both external (from radioactive materials outside the body) and internal (from radioactive materials in-

TABLE 15–1
Average whole-body radiation dose per year from all sources

Source	Dose (mrem per year)
Background: External	
Cosmic	28
Terrestrial	32
Fallout	3
Background: Internal	
Natural radioactive substances in the body (e.g., potassium-40)	34
Fallout radioactive substances in the body (e.g., strontium-90)	1
Total average background	98
Medical sources	73
Other artificial sources	7
	Total 178

side the body). One major external source is **cosmic rays**, which come from outer space. The rays can penetrate directly to the earth's atmosphere and collide with gas atoms, causing secondary radiation. Since the amount of cosmic radiation increases with altitude, people in Denver, Colorado, receive more than twice as much cosmic radiation per year as do people in Atlanta, Georgia. Annual values for the whole-body radiation doses associated with a number of cities appear in Table 15-2. High-flying airline pilots receive about 1200 mrems of cosmic radiation per year. **Terrestrial radiation** comes from the radioactive decay of elements in the soil, rocks, and minerals of the earth. A related—though small—source of radiation in this category is **fallout** that lingers after nuclear weapons tests.

Much of the radiation that makes up each person's total annual exposure comes from sources associated with industry and modern culture. The amounts are small, but they add up, and people vary tremendously in the doses they receive. The radium dials on wrist watches, radioactive materials in consumer products, color TV sets, microwave ovens, airport inspection units, certain jewelry (cloisonné-type), and smoke detectors all generate some radiation. The fluoroscopes once found in shoe stores for observing feet inside shoes emitted a high dosage, approximately 200 mrems per viewing. Sleeping with another person instead of alone increases the annual dose by about 0.1 mrem.

Ionizing radiation is not the only hazard. Nonionizing sunlight, especially its ultraviolet component, is mutagenic and is a potent natural cause of skin cancer. Fluorescent lighting has also been tentatively

TABLE 15–2
The effect of altitude on annual whole-body radiation dose from cosmic rays

City	Elevation (ft)	Dose (mrems per year)
Denver	5280	67
Albuquerque	4958	60
Salt Lake City	4260	54
Oklahoma City	1207	34
Phoenix	1090	33
Atlanta	1050	33
Kansas City	750	32
Chicago	579	31
San Francisco	65	28.5
Baltimore	20	28

linked to skin cancer, at least by one preliminary study, though how such a low-energy emitter could have this effect is puzzling.

Among the man-made sources of radiation, nuclear power is controversial because certain factions of the public see it as a high radiation risk. In reality, however, nuclear power is responsible for only 0.003 mrems of whole-body radiation per person per year. Much of the perceived risk seems to stem from the fear of possible catastrophic accidents at nuclear power plants.

As we will see, medical diagnoses and treatments are by far the greatest source of man-made radiation. Persons in certain occupations also receive considerable amounts of work-related radiation in addition to the average annual dose that comes from other sources. And, of course, nuclear weapons stand in a class by themselves as a frightening source of whole-body radiation.

Medical Radiation

When X rays were discovered in 1895, they were almost immediately applied to humans, and as a result, their harmful effects became apparent. Today, radiation—mostly X rays—is commonly used in medicine for diagnostic and treatment purposes. In fact, diagnostic X rays comprise the largest man-made source of radiation exposure in the United States. The average doses of radiation associated with common diagnostic X-ray procedures are shown in Table 15-3. All types of medical radiation account for 90 percent of the exposure of humans to man-made radiation. Each year, 700 million X rays are taken and 70 million nuclear medicine procedures are performed in this country. Many of the older X-ray instruments that are used are inaccurate and in poor overall condition. Consequently, they deliver more radiation than necessary.

As patients, or consumers of medical services, we must keep in mind that medical radiation can cause the same health problems as any other form of radiation. Penetrating the body and its cells, it can cause gene mutations, chromosome breaks, cancer in somatic cells, and other cellular malfunctions. The effects may be delayed for long periods of time. Children who receive high X-ray doses have been shown to have an increased frequency of chromosomal aberrations. Other studies have shown medical personnel to have a 1.7 times greater risk of leukemia than the general population. In the past, radiologists have had a dramatic ninefold greater risk of leukemia than the general population.

Very large doses of radiation can, of course, be lethal. Based upon severe radiation accidents, an acute dose of 300 to 450 rads of whole-body radiation is the **lethal dose-50 (LD$_{50}$)** for humans. That is,

TABLE 15-3

Average doses for typical diagnostic X-ray examinations in millirads

	Estimated Whole-body Equivalent Dose per Exam	Estimated Annual Excess Deaths per Million Exams[a]
Upper gastrointestinal	400–800	30–100
Lower gastrointestinal	300–700	25–80
Cholecystography	200–600	20–70
Intravenous pyelogram (IVP)	100–500	15–50
Thoracic spine	100–500	15–45
Lumbar spine (LS)	100–500	15–45
Abdomen or KUB	100–200	5–25
Mammography	100–200	5–20
Lumbo-pelvic	100–200	5–20
Hip or upper femur	50–150	5–20
Cervical spine	25–75	2–8
Skull	25–75	2–7
Shoulder	25–75	2–6
Chest (photofluoroscopic)	20–60	2–6
Dental (whole mouth)	10–30	2–6
Chest (radiographic)	5–10	<1
Dental (bite wing)	<5	<1
Extremities	<5	<1

Source: Laws, P. W. 1974. *Medical and Dental X-rays. A Consumer's Guide to Avoiding Unnecessary Radiation Exposure*. Washington, DC: Public Citizen Health Research Group. Pages 56–57.

Note: [a]National Academy of Sciences, National Research Council. 1972. *The Effects on Population of Exposure to Low Levels of Ionizing Radiation* (BEIR Report). Washington, DC: NAS/NRC. Page 91.

300 to 450 rads is a dose fatal to half of the exposed persons within a specified period of time. One rad is approximately equal to one rem for some of the common types of radiation, such as gamma rays and X rays. The basis of the LD_{50} concept is the wide biological variation in radiation susceptibility and resistance among individuals. As you might expect, the LD_{50} varies considerably from species to species.

Studies of mice and other animals have shown an increased sensitivity to radiation in developing embryos and fetuses. Radiation can cause death and various abnormalities, especially in the early stages of development. Many researchers in this area feel justified in extrapolating the animal data to humans and concluding that pregnant women should avoid diagnostic X rays whenever possible.

People often wonder how the same radiation that can cause cancer can be used for the therapeutic purpose of combating the disease. One important factor is that cancerous cells undergo rapid cell division, and cells in division are generally more easily killed by radiation. Apparently, the nuclear condition during mitotic division favors the formation of abnormalities and deficiencies. Many of these effects, in turn, are probably due to chromosome damage that causes the deaths of the cells attempting to undergo division. This is also why bone marrow is more susceptible to radiation, for bone marrow cells divide rapidly to generate new blood cells. These observations make it clear why pregnant women should avoid radiation. Another factor affecting the use of radiation as a cancer treatment is that a therapist can sometimes rotate a patient's body to ensure that the cancerous part of the body is always struck by the radiation, but the healthy parts are exposed to a much lesser degree. In some cases, a new technique is used in which a radiation-emitting object is embedded directly into a tumor.

Medical radiation has proved invaluable and essential in both diagnosis and therapy. The benefits usually outweigh the hazards. Still, many signs of misuse exist. A few physicians may order X rays less for the sake of the patient's health than to pay for their expensive X-ray equipment, or more probably as a defense against potential malpractice suits. Some feel that whether or not to use X rays—for whatever reason—is their prerogative, and the questioning consumer sometimes meets with a thoughtless reaction.

A number of measures can reduce the exposure of the general population to X-ray radiation. Only necessary X rays should be taken, not routine, useless ones. Patients should be able to take existing X rays with them when they switch doctors. When X rays are to be taken, patients should wear lead aprons for protection, if the situation permits. This precaution would help protect against machine scatter, which results in an estimated 14 percent of a chest X ray reaching a patient's gonads. Better X-ray machines and better trained X-ray personnel can also reduce unnecessary exposure. One survey showed that over a third of all dental X-ray machines emit unacceptably high levels of radiation. Almost half of the chest X-ray units surveyed exhibited the same problem.

Patients can help, too. They should not move during an X ray, for if the X ray does not turn out well, it will be repeated. Women who know they are pregnant should definitely inform their physicians whenever X rays are contemplated. People should be better informed and more assertive. Together with responsible and responsive physicians, they could probably reduce the total X-ray exposure of the general population by an impressive 50 percent.

335

Hazardous Occupations

People work to live, but many occupations carry high risks of disease, injury, and death. An estimated 100,000 American workers die each year as a result of all types of occupational illnesses. Some of these illnesses are especially prevalent in industries that expose their workers to radiation. One ten-year study found an increase in chromosome damage among workers exposed to above-average doses of radiation on the job. Other specific consequences have been documented for certain occupations.

An estimated 20 percent of all uranium miners will die from lung cancer. Uranium deposits emit large doses of radiation, and miners have been subjected to this hazard in the underground tunnels of the mines. The cost of compensation to their families will probably exceed $20,000,000. Other types of miners are also exposed to more radiation than the average worker.

In the 1920s, women factory workers painted radium onto watch dials so the dials would glow in the dark. Often, the workers used their tongues to give their brushes a fine point and ease their work. For reasons that are clear to today's scientists, these workers later had numerous cancers and a higher than normal mortality rate due to cancer.

Naval shipyard workers have been studied rather thoroughly. Dock workers are exposed to varying degrees of radiation, with those refueling nuclear submarine reactors probably subjected to the most radiation. The cancer rate among dock workers is twice the national average. The death rate from leukemia is 4.5 to 5.5 times greater than in the general population.

During the early days of nuclear bomb testing, many participating workers probably received higher than normal doses of radiation. Thus far, eight of the participants in one 1957 test have developed various cancers; nine have developed leukemia. A random group of the same size would ordinarily have approximately 3.5 people with each affliction. This is still a highly controversial situation; nonetheless, a recent court decision in Utah went against the United States government.

A number of occupations are classified as requiring nuclear work. These occupations include workers at nuclear plants, but 49 percent of nuclear workers are health workers such as radiologists. Many other occupations are accompanied by some level of hazard due to radiation. Even airline pilots are subjected to above-average levels of radiation because of their greater exposure to cosmic rays resulting from many hours of flight time.

The United States government has established 5 rems (5000 mrems) as the legal limit of radiation exposure per year for nuclear workers and 0.5 rems (500 mrems) as the limit for non-nuclear workers. These *maximum allowable levels* seem excessive to many people, who are calling for reductions. Yet reducing the maximum allowable levels by a factor of ten might cost industry as much as $500 million per year, while saving only about ten lives. Some industries hire temporary employees for high-radiation jobs such as repair work. The objective of these industries is to spread the radiation dose among as many people as possible.

Nuclear Bombs

In August 1945, single atomic bombs were dropped on Hiroshima and Nagasaki, Japan. Within two kilometers of each bomb's "ground zero," only ashes remained. Approximately 140,000 of the 300,000 people living in Hiroshima died immediately or within a few weeks. Three days later, another 70,000, out of 270,000, died immediately in Nagasaki. Both cities fell into chaos. Many other people, not killed instantly, received high doses of radiation, as did the relief teams and the first people who entered the area seeking friends and relatives. Such persons became the secondary victims.

Extensive studies were undertaken to determine the effects of atomic bomb radiation on humans. Before these efforts could yield any results, their sponsors had to overcome a number of problems. First, the United States and Japan had to cooperate during the difficult years immediately after World War II. Researchers had to identify the persons affected, incorporate enough persons in the studies to make the data significant, find good control groups of people, and determine the doses of radiation the affected persons received. The studies were designed to continue over very long periods of time since many effects might be latent, requiring that subsequent generations be observed to detect mutations (usually recessive, and hence unexpressed in F_1 heterozygotes) and chromosome aberrations. Researchers were technically limited because sensitive methods of studying chromosomes and genes were not available then; the excellent techniques we use today were developed mostly in the 1960s and 1970s. Nonetheless, the Atomic Bomb Casualty Commission (ABCC) was formed and numerous investigations were begun, many of which continue today.

The immediate effects upon the inhabitants close to ground zero were devastating, as shown in Figure 15-4. Those who did not die immediately from the bomb blasts suffered burns and damage to the gastrointestinal tract, bone marrow, and other organs. Common aftereffects included hemorrhages, nausea, vomiting, diarrhea, and fever. Still other victims lost their hair, became anemic, and contracted infec-

337

FIGURE 15–4
A 45-year-old woman who was 1.6 kilometers from ground zero when the atomic bomb was dropped on Nagasaki, Japan, on 9 August 1945. She died from severe burns on 15 October 1945. (Photograph by Masao Shiotsuki. From HIROSHIMA AND NAGASAKI: *The Physical, Medical, and Social Effects of the Atomic Bombings* by the Committee for the Compilation of Materials on Damage Caused by the Atomic Bombs in Hiroshima and Nagasaki. Copyright © 1981 by Hiroshima City and Nagasaki City. Reprinted by permission of Basic Books, Inc., Publishers.)

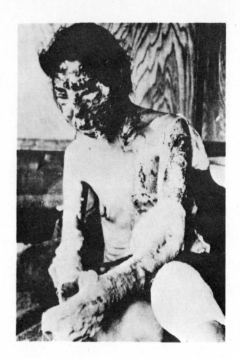

tions. Many of these secondary effects also proved fatal. Additional long-term effects are still appearing.

Most studies of children born to parents who were exposed to the bombings—the F_1 generation—have not demonstrated great differences from control groups. Their rates of mutation, congenital malformation, stillbirth, birth weights, and their mortality at early ages are comparable. However, although the differences from the control group were too slight to be statistically significant, the children of exposed parents did have higher frequencies of most of those effects. Also, because most mutations are recessive, additional studies may be needed to determine their frequency. Many Japanese exposed to the bombs' radiation have purposely not married or had children because they fear genetic damage. Some men have reported that their fiancées left them after their exposure to the bombings became known.

The survivors of the atomic bomb attacks showed definite increases in chromosome aberrations such as breaks, rings, and complex chromosome rearrangements. Table 15-4 summarizes some of these data and also shows how dramatically the incidence of chromosome aberrations rises with age for both survivors and controls. (Many other

338

TABLE 15-4

The frequencies of persons having complex chromosome aberrations among bomb survivors and control groups

Age Group at the Time of the Bomb	Exposed			Control	
	Dose (rad)	Number Examined	% Affected[a]	Number Examined	% Affected[a]
≤ 30 years (both sexes)	200+	94	34	94	1
Over 30 years (both sexes)	200+	74	61	80	16
In utero (maternal dose)	100+	38	39	48	4
Not yet conceived (maternal dose)	150+	103	0	—	—
Not yet conceived (dose received by at least one parent)	100+	25	0	—	—

Source: Miller, R. W. 1969. Delayed radiation effects in atomic-bomb survivors. *Science,* 166:569–574.

Note: [a]One or more complex chromosome abberations.

studies not related to radiation exposure have shown the same relationship of age to frequency of chromosome aberrations.) Table 15-4 also shows that the frequency of affected survivors is much greater than the frequency of affected controls at any age. Since these studies were done many years after the exposures, we can conclude that the cytogenetic effects of radiation persist for a long time.

Embryos still in the womb when exposed to the atomic radiation were also cytogenetically affected, with 39 percent of these persons having one or more complex chromosome aberrations, vs. only 4 percent in the control group. This is not the case, however, for children who were not yet conceived by parents exposed to the radiation. Although the chromosome aberrations persist in the parents, they do not seem to be readily transmitted to the offspring.

Other effects appeared frequently among newborns who were *in utero* at the time of the bombings. A significant number had microcephaly (exceedingly small head), mental retardation, or both. These effects were especially pronounced among those who had been conceived fewer than 15 weeks before the bombings. Such abnormalities are also known to result from the X-ray radiation of embryos.

One of the first effects observed among the bomb survivors was the high rate of leukemia. Later, the survivors showed significantly

339

higher incidences of thyroid, breast, lung, and salivary gland cancers. And undoubtedly, many persons died before they could become part of the study. Many of these latter deaths may have been due to the effects of radiation; it would not be surprising to find that radiation damage makes individuals more susceptible to various other diseases and injuries. In other words, they would escape cancer by dying of something else first. Figure 15-5 summarizes the incidence of cancer associated with various sources, including the atomic bomb.

Many other physical and medical problems are still prevalent among the survivors. These people seem to have more eye problems such as opacities, lesions, and cataracts. Anemia, blood disorders, and other diseases occur in greater frequencies. The mortality rate is also 15 percent higher, and the psychological effects on the survivors have been understandably immense.

The results of these studies have been used to set radiation standards for industry and to calculate risks. Recently the risks have been revised upward based upon additional data gleaned from the atomic bomb studies. Because of what we have learned about the effects of atomic bombs, citizens of many countries are calling for the abolition of nuclear weapons. A nuclear war, say physicians and many others, would be the "final epidemic."

Severe Radiation Accidents

The bomb data, animal experiments, and the results of excessive exposure of medical patients to therapeutic radiation have illustrated the effects of radiation on humans. In addition, some severe radiation accidents have occurred. One was the 1954 Bikini atomic bomb test, which caused very heavy fallout over the Marshall Islands. There have also been radiation accidents at scientific laboratories and various nuclear research plants in the United States and other countries. Such events are unwanted experiments, but they too have given us information about acute radiation exposure.

Since 1945, there have been only a few dozen severe radiation accidents. Although the exact exposures have not often been known, researchers have made fairly sound estimates. In a few cases, workers have received from 10,000 to 12,000 rems in a single dose of radiation. Because doses of 500 rems or more are generally fatal (the LD_{50} for humans is estimated to be 300 to 450 rems), acute radiation exposure has caused more than a few deaths. Some exposed individuals have lived for 30 to 60 days after an accident, but the most severely exposed victims have died within two to four days.

340

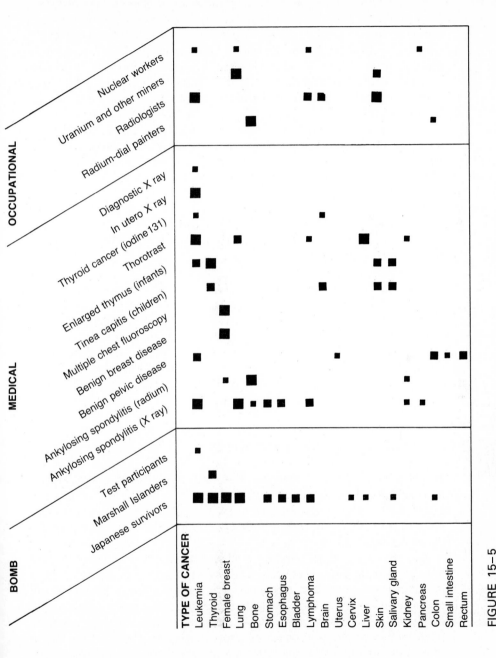

FIGURE 15-5
The incidence of cancer associated with radiation from three main sources: the atomic bombs, medical diagnoses and therapies, and occupational exposures. Strong associations are shown by the largest squares, fairly sound associations by medium-sized squares, and suggested but unconfirmed associations by the smallest squares. (From The biological effects of low-level ionizing radiation by A. C. Upton. Copyright © 1982 by *Scientific American*, Inc. All rights reserved.)

A fairly good description of **radiation sickness** has emerged from available data. As expected, the severity of the radiation sickness depends largely upon the dose. The symptoms include headaches, dizziness, nausea, vomiting, diarrhea, hemorrhaging, fever, and other physiological effects. Three main organ systems seem to be the most important in radiation sickness: (1) the central nervous system, (2) the gastrointestinal system, and (3) the **hematopoietic** (blood-forming) system in the bone marrow.

The central nervous system is affected by doses of several thousand rems or more. Few accidents have been reported in this category. Within minutes, the victim goes into shock and semiconsciousness. The lymphocytes disappear in about six hours. Death is inevitable, usually within hours, or within a few days at most.

An exposure of 500 to 2000 rems is enough to damage the gastrointestinal system. This type of radiation sickness is marked by loss of appetite, vomiting, diarrhea, and lethargic, depressed behavior. Death usually occurs within a week or two.

Hematopoietic problems generally occur with doses of several hundred to five hundred rems. Again, some vomiting, nausea, and diarrhea can occur, but often the victim may seem to recover for a short time. Two or three weeks later, however, symptoms such as chilling, fatigue, and oral ulcers appear. A decrease in the white blood cells makes the victim very susceptible to infection. This radiation dose is not always fatal, but if death does occur, it is usually between three and six weeks after exposure. Those who survive may experience other physical and genetic effects later.

THE LOW-LEVEL RADIATION CONTROVERSY

The effects of high and moderate doses of radiation are now well known and accepted by the scientific community. Whether or not effects can occur at very low doses, however, is highly controversial. In fact, statisticians with excellent qualifications can study the same data and disagree with each other.

The main issue is the existence of thresholds. If no safe level of radiation exists, even near-zero doses have effects. In that case, there is no threshold. Instead, there is a completely linear relationship between the effects of radiation and the dose. Advocates of this model graph the data at the higher doses and extrapolate the line over the lower doses to the zero dose, as shown in Figure 15-6A. If there is a threshold, effects occur only *after* the radiation reaches a certain level.

342

FIGURE 15-6
The relationships between the effects
of radiation and variable doses of radiation. In a completely linear relationship, the line is extrapolated to zero.
In the threshold model, no effects
occur below a certain threshold dose
of radiation.

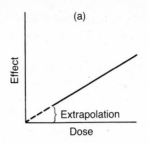

Linear, nonthreshold concept

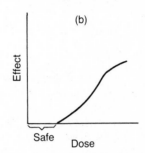

Nonlinear, threshold concept

The threshold model yields a very different graph of the relationship between effect and dose, as shown in Figure 15-6B. The linear nonthreshold concept means that there is no safe level of radiation exposure. The nonlinear threshold concept assumes that there is a certain
safe level of radiation exposure.

One might wonder why the controversy even exists. Do we not
have enough data and experience to resolve the issue? The problem is
one of statistics. With very low radiation exposures, we can expect an
extremely low frequency of effects whether there is a threshold or not,
simply because of chance. To distinguish which model is correct, we
need to study very large samples. Otherwise, too many questions about
coincidence and chance will arise.

For example, consider a hypothetical case, with an actual risk
of one effect of some kind in 100. Now, if 100 individuals are observed,
it is not unlikely that you would observe no affected individuals. But this
greatly underrates the real risk. Nor is it difficult to observe two affected
individuals in a population of 100. But this greatly overrates the real risk.
The crux of the problem is the effect of chance on the small sample
size. An accurate risk estimate must require 1000 or more times as

343

many people, but we already know that studies involving large sample sizes are not easy to perform with humans.

Do low levels of radiation have effects? Various studies support both views. Studies on large populations of *E. coli* (bacteria), *Drosophila* (fruit flies), and other organisms indicate that no level of radiation has a zero risk. Estimates based on studies of mice, and extrapolations of mouse data to humans, suggest that one rad (approximately 1000 mrems) of exposure to parents will result in 22 to 30 affected offspring per 1,000,000 live births. Based on studies of both animals and humans, calculations have established a risk of one to two malignant cancers per 10,000 exposed individuals per rad of radiation. If the target theory is correct and only one hit is necessary, one can easily rationalize the validity of a no-safe-limit concept.

The conviction that no safe limit exists for humans is based mostly on extrapolation, not on definitive data on humans. Opposing it are reports that many people exposed to high levels of background radiation (due to geographic conditions) do not have higher incidences of harmful effects. Background radiation can vary from 130 mrems to 2800 mrems for geographic and other reasons. Areas with high levels of background radiation have been found in Brazil, India, China, and the United States. The lack of effects may be explained if cellular repair mechanisms have the ability to keep up with the damage under these relatively low-level radiation conditions.

If risks exist at low levels of radiation, they are comparatively slight. Some people blame the news media for sensationalizing these risks. They accuse the media of concentrating on reports from scientists who think the risks are high, thereby stirring up more public fear than is warranted. In an effort to educate the public and reduce their fears, scientists have pointed out that the background exposure in the Three Mile Island area was increased by only 1.5 mrems per person per year. They have further calculated that this increase in radiation exposure will cause only one additional cancer case among the people living within 50 miles of the Three Mile Island site during the next 20 years.

The scientific consensus seems to indicate that a threshold for radiation effects does not exist. The possibility of effects from low-level radiation is of great concern to specialists in occupational safety. Some researchers think the maximum allowable levels for occupational exposures are presently too high and the risk benefit concept is overused. Yet if we accept the no threshold hypothesis, we should not describe the level decided upon as a safe limit, but rather as a maximum permissible or a maximum acceptable level. What should this level be? And

344

how do bioethical principles influence the decisions? Such issues may not be settled for some time.

SUMMARY

Radiation is a form of energy. Some types of radiation can transfer their energy to living tissues by causing ionization, which in turn causes cellular damage and burns, skin ulcers, cell death, and other severe physiological effects. Low levels of radiation can cause gene mutations, chromosome breaks, malignancies, and other cellular damage.

The target theory is based on the concept that direct hits on certain parts of the cell cause damage and elicit a response by the cell. Most scientists agree that DNA (the genetic material in the nucleus) is the target of greatest importance. According to this theory, there is a linear relationship between genetic effects and lower dosages. At higher doses, when some targets are hit more than once, the relationship is nonlinear. One implication of the one-hit target theory is that there may not be a safe level of radiation exposure.

Humans are normally exposed to radiation from the background, medical sources, and other artificial sources such as color TV sets, radium watch dials, and smoke detectors. Background radiation from cosmic rays, rocks and soil, and minerals within our bodies gives us our greatest exposure; medical radiation ranks a close second. At this time, fallout and nuclear power plants expose us to very little radiation. Radiation as a result of medical diagnoses and therapy can be considerable, and exposures should be limited only to those that are necessary. Embryos and fetuses are particularly sensitive to radiation. Occupations shown to have radiation hazards include uranium mining and naval shipyard work, among others.

Scientific investigations of the effects of the two atomic bombs dropped on Japanese cities during World War II continue today. Researchers find no increased mutation rate among survivors' children, but higher than average rates of chromosome aberrations, leukemia, cancer, other kinds of ill health, and mortality among the survivors themselves. Severe radiation accidents have also yielded data on the human body's response to large amounts of radiation. Acute radiation sickness is marked by headaches, diarrhea, vomiting, dizziness, hemorrhaging, fever, loss of white blood cells, infection, ulcerations, and loss of appetite, among other responses. Death, of course, generally occurs within days or weeks following very high doses.

345

A controversial issue is whether very low levels of radiation can cause damage. One concept is that no safe level of radiation exists. The opposing concept is that a threshold does exist below which radiation exposure poses no risk. The resolution of this problem will require cumbersome studies of immensely large populations.

DISCUSSION *QUESTIONS*

1. To what sources of background radiation are you exposed? Do you think you receive more or less radiation exposure than the average American?

2. Name several ways a somatic mutation could cause physical death in an organism.

3. Occasionally, lawsuits arise because someone worked in a laboratory that used extensive radiation and subsequently developed cancer or gave birth to a malformed child. In whose favor do you think such lawsuits should be settled?

4. Enumerate some of the data that suggest a link between radiation exposure and cancer.

5. Cancer cells, bone marrow, and embryos are all highly sensitive to radiation. What common factor or factors might make them so radiosensitive? Explain.

6. Do you think the mutational effects of radiation threaten the survival of the human species? Explain.

7. The text stated that reducing radiation exposure of employees of some industries tenfold would cost $500 million per year and save about ten lives. Assuming these figures are correct, how do you feel about the situation? Should we improve standards for workers by lowering the maximum allowable radiation exposure? At what price?

8. Discuss whether an increased mutation rate, presumably due to radiation exposure, could be beneficial.

9. How would you explain the observation that rates of leukemia and certain other cancers are very high among survivors of the World War II atomic bombings, but that the number of mutations among their children seems no higher than in control groups?

10. How would you explain the observation that the frequency of chromosome aberrations is significantly higher than average in the bomb survivors, but not in their children?

11. Since 1945, scientists have thoroughly studied the Japanese people exposed to the atomic bombs at Hiroshima and Nagasaki. Do you think that
we can finally conclude these studies, or should we continue them? On a
scientific basis, explain why or why not.

12. One sometimes hears that if a worldwide nuclear war were to occur, the
insects would inherit the earth. What do you suppose is the biological basis for this assertion?

PROBLEMS

1. Which of the following graphs indicates that a safe level of radiation does
not exist? Which graph indicates that a safe level of radiation does exist?
Label the safe level on the latter graph.

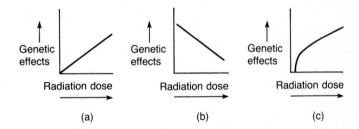

(a)　　　　　　　　(b)　　　　　　　　(c)

2. How would you interpret the following graph?

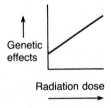

3. A 4000-rad X-ray dose produces X-linked lethal recessive mutations in 12
percent of treated fruit fly gametes. If the frequency of mutations and the
dosage in rads followed a linear relationship, what would be the expected
mutation frequency for 1000 rads? For 6000 rads?

4. Three different populations of cells were irradiated as follows: (1) 50 rems
each week for 6 consecutive weeks; (2) 15 rems each week for 20 consecutive weeks; (3) 300 rems administered as one dose.
   a. Which of these cell populations received the greatest amount of radiation as a result of the treatment?

347

b. Which cell population do you think will show the greatest genetic damage? Why?

5. Interpret the following graph.

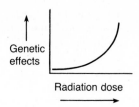

6. The hypothetical data below give the percentages of populations of a certain organism that die within 30 days after various radiation doses. What is the approximate LD_{50}?

dose	% dead within 30 days
150	0
250	5
275	15
300	35
325	55
350	85
400	100

REFERENCES

Adams, R., and S. Cullen, ed. 1981. *The final epidemic: Physicians and scientists on nuclear war*. Chicago: Educational Foundation for Nuclear Science, Inc.

Awa, A. A., A. D. Bloom, M. C. Yoshida, S. Neriishi, and P. G. Archer. 1968. Cytogenetic study of the offspring of atom bomb survivors. *Nature* 218:367–368.

Bloom, A. D., S. Neriishi, and P. G. Archer. 1968. Cytogenetics of the *in utero* exposed of Hiroshima and Nagasaki. *Lancet* 2:10–12.

Bloom, A. D., S. Neriishi, A. A. Awa, T. Honda, and P. G. Archer. 1967. Chromosome aberrations in leucocytes of older survivors of the atomic bombings of Hiroshima and Nagasaki. *Lancet* 2:802–805.

Bloom, A. D., S. Neriishi, N. Kamada, T. Iseki, and R. J. Keehn. 1966. Cytogenetic investigation of survivors of the atomic bombings of Hiroshima and Nagasaki. *Lancet* 2:672–674.

Caldicott, H. M. 1982. The final epidemic. *The Sciences* 22:16–21.

Cohen, B. L. 1979. What is the misunderstanding all about? *The Bulletin of the Atomic Scientists* 35:53–56.

REFERENCES

Committee for the Compilation of Materials on Damage Caused by the Atomic Bombs in Hiroshima and Nagasaki. 1981. *Hiroshima and Nagasaki: The physical, medical, and social effects of the atomic bombs*. Translated from the Japanese by E. Ishikawa and D. L. Swain. New York: Basic Books Inc., Publishers.

Evans, H. J., K. E. Buckton, G. E. Hamilton, and A. Carothers. 1979. Radiation-induced chromosome aberrations in nuclear-dockyard workers. *Nature* 277:531–534.

Laws, P. W. 1974. *Medical and dental X-rays: A consumer's guide to avoiding unnecessary radiation exposure*. Washington, DC: Public Citizen Health Research Group.

Miller, R. W. 1969. Delayed radiation effects in atomic-bomb survivors. *Science* 166:569–574.

Muller, H. J. 1927. Artificial transmutation of the gene. *Science* 66:84–87.

National Academy of Sciences, National Research Council. 1972. *The effects on populations of exposure to low levels of radiation* (BEIR Report). Washington, DC: National Academy of Sciences.

Rugh, R. 1962. Low levels of X-irradiation and the early mammalian embryo. *American Journal of Roentgenology* 87:559–566.

Russell, L. B. 1950. X-ray induced developmental abnormalities in the mouse and their use in the analysis of embryological pattern. I. External and gross visceral changes. *Journal of Experimental Zoology* 114:545–602.

Russell, L. B., and W. L. Russell. 1952. Radiation hazards to the embryo and fetus. *Radiology* 58:369–377.

Russell, L. B., and W. L. Russell. 1954. An analysis of the changing radiation response of the developing mouse embryo. *Journal of Cellular and Comparative Physiology* 43(Suppl.):103.

Upton, A. C. 1982. The biological effects of low-level ionizing radiation. *Scientific American* 246:41–49.

16
Heredity and Aging

"First we ripen, then we rot," according to Leonard Hayflick, a leading researcher in the field of aging. Like the rest of us, he asks why we must inevitably age and die. Why can we not live on and on, as long as we consistently eat well, breathe well, eliminate wastes well, exercise appropriately, and escape disease and accident?

Aging is the gradual decline of most of the body's functions. It begins not long after sexual maturity in animals, and in humans the signs are plain to see by age 30 or so. For older people and for society in general, this is an issue of great concern, because it has an impact on the medical and the social costs of caring for the aged. Yet only recently have the problems of aging attracted the attention of more than a few researchers and funding agencies.

Many scientists believe that large, long-term, multidisciplinary approaches to the study of aging will be required before any significant progress will be achieved. The problems inherent in studying aging include:

1. inbreeding is not permissible in humans;
2. environmental diversity among humans is immense;
3. the human life span is very long;
4. the interaction between aging and disease is difficult to delineate; and
5. investigators are limited by ethical considerations in their research on humans.

What factors affect aging? Heredity is certainly one, though its importance is controversial. There are many other hypotheses, and much disagreement among the scientists considering them. One current idea is that the cells of organisms are genetically programmed to die after a certain span of time. This hypothesis, which has an important hereditary component, will occupy our attention through most of this chapter.

HYPOTHESES OF AGING

Of the many hypotheses regarding mechanisms of aging, none is easy to test. Aging is a complex process. Researchers have proposed many mechanisms for aging, but they continue to debate which are the primary or secondary factors involved.

352

According to the **somatic mutation theory**, the main mechanism of aging is the gradual accumulation of gene mutations in an organism's somatic cells. These cells are constantly exposed to environmental mutagens such as chemicals and low-level radiation. Somatic mutations are inevitable, and they could well alter the flow of information in the cell, thereby affecting cellular functions. According to this hypothesis, the most important mutations would be the ones occurring in the cell's many regulatory genes. The hits that affect aging are probably dominant mutations. (Recessive mutations are not normally expressed in somatic cells because homozygosity rarely occurs.) In addition, mutations causing aging might not be reparable; otherwise, the cellular repair mechanisms would prevent their effects. Finally, the mutation rate acting in somatic cells would have to be much higher than the rate we typically observe in germ-line cells. Mutations do not normally occur at high rates, so some factor would have to exist to make the rate of somatic mutation high enough to bring about aging.

There is evidence both for and against the somatic mutation theory. Low-level radiation, for example, has been shown to cause accelerated aging in animals such as mice. On the other hand, some of the most radiation-resistant animals, the insects, for example, have very short life spans. Conflicting observations like these have caused the somatic mutation theory to remain questionable.

Closely related to the somatic mutation theory is the idea that the DNA repair mechanisms of cells become less efficient as we age. We have already noted that DNA is the primary target in the cell for environmental insults, and there is little doubt that certain chemicals and physical agents can damage DNA. A gradually increasing inability to repair damaged DNA would be noticeable as an accumulation of genetic mutations in somatic cells, erroneous messenger RNA molecules, nonfunctional enzymes, and general cellular degradation. This hypothesis has also been questioned.

In 1963, L. E. Orgel proposed his novel **error catastrophe hypothesis**. He suggested that aging might be due to errors in protein synthesis. Mutations are not necessary. In other words, errors might occasionally occur during the mechanics of protein synthesis even though the genetic material is correct. However, only a small proportion of any given population of protein molecules would contain errors. The errors supposedly accumulate over time, and each error creates other errors. Proper cellular functions depend not only on having the correct genetic material, but also on having the correct enzymes and other cellular components. This hypothesis is very attractive to some scientists, but it is opposed by others.

353

Some researchers prefer to blame aging on **free radicals**, highly reactive substances formed when atoms and molecules within a cell lose electrons. Free radicals seem to accumulate in the nucleus and cytoplasm of an organism's cells during life. Highly energetic, they chemically attack and change other molecules in the cells, perhaps resulting in cellular malfunctions.

Other body systems have also been directly or indirectly implicated in aging. There is evidence that the immune, endocrine, and central nervous systems deteriorate with age, so that the body's defenses and homeostatic controls become less efficient. These systems are complex and difficult to interpret, and a considerable research effort is currently underway in these areas.

Another hypothesis states that the body's cells contain an inherent aging program. This interesting idea is based upon a substantial amount of experimental data. The hypothesis does not rely on randomness and accidents, but on a fairly precise, genetically determined program for cellular aging, different for each species.

LIFE SPAN VS. LIFE EXPECTANCY

To appreciate the idea of programmed aging, consider the difference between **maximum life span** and **life expectancy**. The former is the biological limit of survival for the members of a species. The life span is the potential to live as long as the oldest surviving member of a species. The life expectancy, on the other hand, is the *average number of years* that the members of a population are *expected to live* from the time of birth.

The maximum life span for humans is estimated to be about 113 years. Humans have the longest life span of all mammals. Tables 16-1 and 16-2 provide data on life spans of a variety of mammals. The life expectancy in the United States in 1980 was 70.0 for males and 77.7 for females. The difference in life expectancies for the two sexes is not a uniquely human characteristic. Females have also been shown to live longer in the housefly and in several strains of the white rat.

Such life-span observations are among the first clues to the possible existence of programmed aging. Another clue, and a crucial one, is the observation that the maximum human life span has evidently changed by no more than a few years over the past 100,000 years. Cleanliness, safety measures, vaccines, and medical breakthroughs have not appreciably affected maximum life span. They have, of course, increased life expectancy. In 1900, the average human life ex-

354

TABLE 16–1

Maximum life span for some
mammalian species

Common Name	Observed Maximum Life Span in Years
Pigmy shrew	1.5
Field mouse	3.5
Opossum	7
Mongolian horse	46
Camel	30
Cow	30
Giraffe	34
Elephant (India)	70
Mountain lion	19
Domestic dog	20

Source: Cutler, R. G. 1978. The origin, evolution, and causes of aging. In
Behnke, J. A., Finch, C. E., and Moment, G. B. (eds.). *The Biology of Aging.*
New York: Plenum; and Cutler, R. G. 1979. Evolution of longevity in ungulates
and carnivores. *Gerontology,* 25:69–86.

pectancy was only 47.3 years, compared with 73.8 in 1980. The improvement is largely due to decreases in infant and childhood mortality.
Today, more people reach older ages, but once they reach age 75,
their life-expectancy prospects are changed very little. The increase in
life expectancy once a person reached the age of 75 was only 2.2 years
between 1900 and 1969.

The rectangular curve shown in Figure 16-1 demonstrates this
concept of life expectancy versus life span. Most people in this country
survive until age 75. Only a few years later, nearly everyone who

TABLE 16–2

Maximum life span for some primates

Species Scientific Name	Common Name	Maximum Life Span in Years
Homo sapiens	Man	113
Pan troglogytes	Chimpanzee	55
Cebus capucinus	Capucin	40
Galago senegalensis	Galago	25
Saimiri sciurens	Squirrel monkey	21
Tarsius syrichta	Tarsier	12
Perodicticus potto	Potto	12
Urogale everetti	Philippine tree shrew	8

Source: Cutler, R. G. 1976. Evolution of longevity in primates. *Journal of Human Evolution,* 5:169–202.

FIGURE 16–1

The cumulative survival in percent includes almost all the population until about age 75. At this time, the population declines greatly. Very few members are still alive at age 85. (From Fries, J., and Crapo, L. M. 1981. *Vitality and Aging.* W. H. Freeman and Company, New York. Used with the permission of the publisher.)

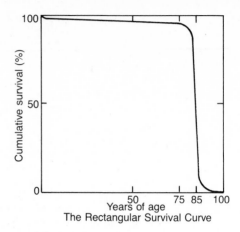

The Rectangular Survival Curve

reached age 75 is dead. A very few live to 113 years old, or older. It is observations like these that make the programmed aging hypothesis so intriguing.

HYPOTHESIS OF GENETICALLY PROGRAMMED AGING

The hypothesis of genetically programmed aging is provocative as well as intriguing. In the early 1960s, Hayflick reported that **fibroblasts** taken from human embryo tissue and cultured *in vitro* will undergo approximately 50 (35 to 65) cell divisions before they become senescent, and ultimately die. Fibroblasts are the irregularly shaped, highly branched connective tissue cells shown in Figure 16-2. Cell divisions result in doublings of cell population that are construed as **cell generations**. A cell generation is systematically assigned each time the number of cells in the population doubles. Figure 16-3 provides a diagram of this process. The $50\pm$ limit on cell doublings is known as the "doubling potential" or the **Hayflick limit**.

Hayflick and his colleagues subcultured the fibroblast cells continuously and kept careful records of the number of cell doublings. Without subculturing, the quantity of the tissue culture would become astronomical—if all the cells remained viable and continued to divide throughout the experiment, the number of cells after 50 doublings would be 2^{50} or 1,125,899,600,000,000 (1,125 trillion). The 50 doublings are therefore quite enough to yield a full-grown human adult. Scientists are aware, however, that cell populations under these condi-

356

FIGURE 16–2

Human fibroblast cells as observed with ordinary microscopy. (Courtesy of Gary Shipley, Mayo Foundation, Rochester, Minn.)

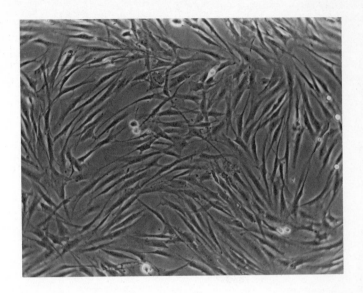

tions are heterogeneous. The cells differ in form, and while some of the cells divide, others do not.

Figure 16-4 illustrates the essence of the Hayflick technique. To test various hypotheses, the researchers froze many of their subcultures in liquid nitrogen. The freezing process does not seem to interrupt the cellular "clock" that keeps track of age as the number of cell dou-

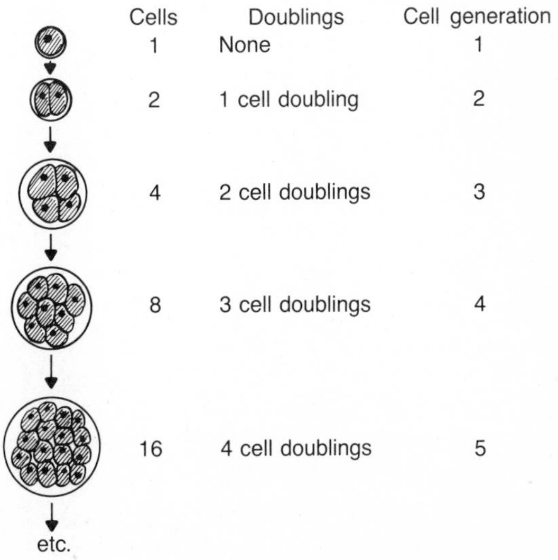

FIGURE 16–3

The cell doubling concept.

	Cells	Doublings	Cell generation
	1	None	1
	2	1 cell doubling	2
	4	2 cell doublings	3
	8	3 cell doublings	4
	16	4 cell doublings	5

etc.

357

FIGURE 16–4

The technique for subculturing human fibroblasts. The first step of the procedure is to break down the tissue into individual cells with the enzyme, trypsin. The cells are grown in glass vessels until they cover the entire surface of the growth medium. At this point, the cell population is divided in half, and the halves are subcultured again in the same manner. Subcultures can also be stored in liquid nitrogen ($-196°$ C). This subculturing process can be repeated about 50 times with human embryonic cells before the cells lose their ability to divide and then die. (From The cell biology of human aging by L. Hayflick. Copyright © 1980 by *Scientific American*, Inc. All rights reserved.)

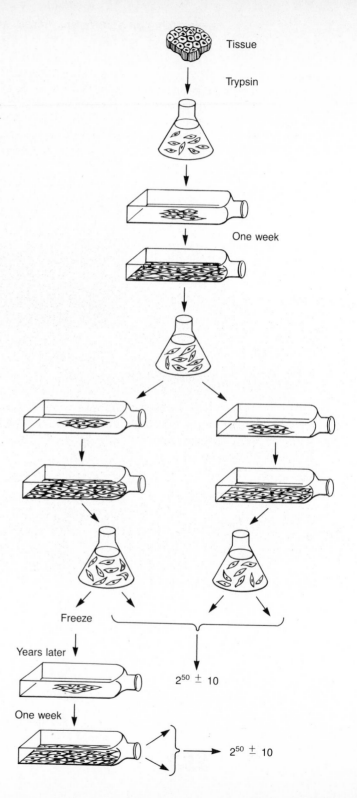

Tissue

Trypsin

One week

Freeze

Years later

One week

$2^{50} \pm 10$

$2^{50} \pm 10$

blings. Hayflick and his colleagues have frozen cells for as long as 13 years and have found that the numbers of doublings before freezing and after thawing always add up to approximately 50. That is, if the cells are frozen after 10 doublings, then they will double about 40 more times after thawing. The duration of freezing does not seem to make any difference. The cell counts time not as we usually understand it, but as cell generations.

Many scientists interpret the limit on cell doublings as evidence for cellular aging and believe that aging is an intrinsic characteristic of cells. The Hayflick technique eliminates any possibility that the limit is simply due to a lack of nutrients or an accumulation of waste products. Some scientists believe that the cell contains a genetic program that effects a gradual decline in cell function, ultimately bringing about cell death. If such a program is, as they hypothesize, regulated by genes, then *we are genetically programmed to die*.

The mechanism of cellular aging is unknown, and the idea that it occurs is by no means universally accepted. Some of those opposing the maximum cell doubling hypothesis claim that the results are nothing more than an artifact of normal cell culturing procedures. Nevertheless, the experiments are very significant contributions to our information about cellular aging. Even more controversial is the question of whether *in vitro* cellular aging effects aging at the organismal level. In other words, do organisms age because of cellular aging? There are some data indicating that the two phenomena are related.

Cell Senescence Experiments

Cellular aging is probably not exactly equivalent to the aging of the organism, but the possible relationship between them is certainly intriguing. The evidence for this relationship includes the observation that other types of human embryonic cells besides fibroblasts (skin, heart, and lung cells) show the same 50± population doublings. In addition, when cells from persons of various ages are cultured in the same way, cells from older donors show less doubling capacity than those from younger donors. Hayflick observed a range of only 14 to 29 doublings among eight adult donors. In another study, 23 individuals between the ages of 21 and 36 averaged 44.6 cell doublings, while 24 individuals between the ages of 63 and 92 averaged only 29.8 doublings. A definite correlation exists between the remaining life span of an individual and the doubling potential of his or her fibroblasts *in vitro*.

The maximum life span among different species of organisms varies tremendously—more than 50-fold among the placental mammals

TABLE 16-3

Population doublings for cultured embryonic fibroblasts from several species having a wide range of maximum life spans

Species	Range of Cell Doublings	Approximate Maximum Life Span (yrs)
Galapagos tortoise	90–125[a]	175
Human	40–60	110
Mouse	14–28	3.5

Note: [a]Cells taken from young donors rather than from embryos.

alone. Thus, it is worthwhile to compare the maximum life spans of different species and their embryonic cell doubling potentials. Table 16-3 shows such comparisons for several organisms. The correlation between cell doublings and life span is not perfect, but there is a general association of maximum life span with embryonic fibroblast doubling potential.

Other studies have investigated the doubling numbers for cells from individuals with pathological problems, genetic defects, and cytogenetic abnormalities. For example, the cells of individuals with xeroderma pigmentosum, which are deficient in DNA repair mechanisms, double just as many times as cells of normal individuals. On the other hand, fewer cell doublings are observed in persons with diabetes, Down syndrome, and several other genetic disorders than in normal individuals of the same ages. Several strange aging diseases are in a class by themselves.

Progeria and Other Aging Diseases

These unique diseases result in premature aging. Cells from their victims show a much reduced growth potential when compared with normal cells. This is especially true for cells taken from individuals with **progeria** and **Werner's syndrome**, which have sometimes been able to double only a couple of times. In progeria, the premature aging begins early in life, while in Werner's syndrome, it appears by about 20 to 30 years of age.

Progeria, also known as **Hutchinson-Gilford syndrome**, is a mysterious disease whose first symptoms appear several months after birth. All the aging symptoms occur during what are supposed to be the childhood years. Fortunately, progeria is a rare disease, probably occurring approximately once in 8,000,000 live births. At this time, only about ten progeria victims are known in the United States.

Affected persons, males and females, tend to resemble each other remarkably. Within two years of birth, their skin hardens, their hair falls out, and their growth is very limited. They tend to stop growing completely at a weight of about 28 to 30 pounds and a height of about 3 feet. In addition to being small and bald, they have prominent eyes, bent toes, other skeletal abnormalities, a large cranium, small face and chin, and strongly protruding veins. Affected children do not mature sexually, but they do have normal intelligence.

Children with progeria become wrinkled early in their lives, appear very old, and usually die in their teens. Death is often due to the same afflictions (such as heart disease and strokes) that strike elderly people. In 18 cases, the age at death ranged from 7 to 27.5 years, with a mean of 13.4 years.

The genetics of progeria is still unclear, but some scientists believe the disease is due to autosomal recessive inheritance. Firm information on this point has been difficult to obtain because so few cases are available for study. In several cases, however, victims of the disease have had one or more sibs with progeria.

Werner's syndrome has been called adult progeria because of its later onset. The hair begins to gray at about 20 years of age. Cataracts, joint deformities, skin problems, and a senile appearance soon follow. Victims are usually fertile, however, and they also have normal I.Q.'s. Many persons having Werner's syndrome develop diabetes mellitus with maturity. Death occurs at a mean age of 47 years, with a range of 31 to 63 years. Pedigree studies have prompted the conclusion that Werner's syndrome is also due to autosomal recessive inheritance.

Another disease that shortens the life span of its victims is **Cockayne syndrome**. This condition is characterized by dwarfism, mental retardation, and a senile appearance. Individuals with Down syndrome, Turner syndrome, and other genetic anomalies also have shorter than normal life spans, but it is questionable whether these genetic syndromes actually accelerate aging.

Does a cellular aging clock actually exist? Is the control of its pace affected in some of the premature aging diseases? The suffering connected with progeria, which is such a dramatic display of premature aging, gives special urgency to these questions. If a "death clock" does exist, we must determine its exact location in the cell.

Some experiments have demonstrated that the general location of the control for cellular aging is in the nucleus, not the cytoplasm. In

361

FIGURE 16–5

A cell fusion experiment designed and performed to determine the general location of the clock that regulates cellular aging. Nuclei from cells allowed to undergo 10 population doublings (young) were inserted into cytoplasts derived from cells that had been allowed to undergo 30 population doublings (old). Also, nuclei from cells allowed to undergo 30 population doublings (old) were inserted into cytoplasts derived from cells that had been allowed to undergo 10 population doublings (young). The results suggested that the nucleus was primarily responsible for cellular aging. (From The cell biology of human aging by L. Hayflick. Copyright © 1980 by *Scientific American*, Inc. All rights reserved.)

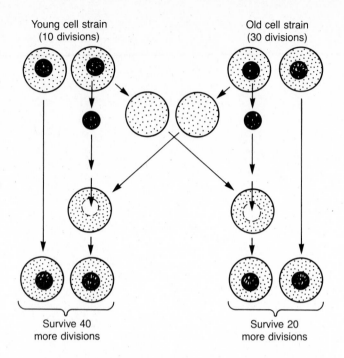

Young cell strain
(10 divisions)

Old cell strain
(30 divisions)

Survive 40
more divisions

Survive 20
more divisions

one well-planned study, researchers fused old nuclei with young **cytoplasts** (cells stripped of their nuclei) and young nuclei with old cytoplasts. They found that reconstructed cells consisting of young nuclei in old cytoplasts survived more cell population doublings than reconstructed cells consisting of old nuclei in young cytoplasts. This experiment is presented diagrammatically in Figure 16-5. They concluded that the ability to undergo a specific number of cell doublings before senescence depended on the nucleus, not the cytoplasm. In other words, the nucleus appears to be the location of the clock that determines cellular aging and programmed death.

THE FOUNTAIN OF YOUTH AND SOCIETAL PROBLEMS

According to the Hayflick model, cellular aging and death are perfectly normal events. Other data show that cell death is even essential to the normal development of many organisms, for it is cell death that helps to define the shape of the body and limbs in early development. Some cells in higher plants, such as the vessels for conducting water in stems and leaves, are not even functional until they die.

If the Hayflick model is correct, will we ever be able to modify the clock and extend the maximum life span? Cellular aging must be an extremely complex process. However, many cell biologists are researching the cell, studying cell cloning, carcinogenesis, development, **transformation**, and many other cellular processes. Transformation is the change that takes place in some cells enabling them to avert senescence even after the approximate 50 population doublings. (The HeLa cell discussed in Chapter 6 is a good example of a transformed cell.) With this great amount of scientific attention being devoted to this problem, some researcher could well learn how to alter the death clock.

However, we must ask whether the quality of life would improve with a lengthening of the life span. To many people enhancing the quality of life during our present life span is far more urgent than any increase in the length of the life span. Length without quality, they say, might raise more problems than it is worth. If the life span jumps to 200 or 300 years, with a proportionate increase in life expectancy, when should we retire? What will happen to social security? How will a longer life span affect unemployment, insurance, and various social programs? Overpopulation would become an even greater problem than it is now. Overpopulation leads to starvation, and starvation leads to other social problems. The entire subject should be of deep and lasting concern to society.

SUMMARY

Aging is an inevitable, slow decline in the physiological functioning of the body. Although difficult to study, its biological basis is now being probed vigorously. Recent hypotheses of aging include the somatic mutation theory, based on the fact that somatic mutations accumulate over the years. Another hypothesis proposes a degradation of the cell's DNA repair mechanisms. The error catastrophe hypothesis requires not mutations, but an accumulation and compounding of errors during protein synthesis. Free radicals, highly reactive cell components, may also accumulate and disrupt normal cell function over time. The body's immunological abilities, hormonal system, and central nervous system have also been implicated in the aging process.

A novel hypothesis connects the ability of cells to divide with the aging process at the level of the organism. Generally, populations of normal fibroblast cells from human embryos will only divide *in vitro* about 50 times before growing old and dying. This may be due to a genetic program for aging that is inherent in the cell. Support for this hypothesis has come from many observations and experiments. Although human life expectancy has increased considerably, the maximum life span has not. All species appear to have a specific maximum life span. The general pattern is that the longer a species' life span, the more times its embryonic fibroblasts can divide in culture. In humans, older persons have less potential for cell population doublings. Cells frozen at any point during the doubling count will resume doubling upon thawing and they will always reach $50 \pm$ doublings before undergoing senescence. The potential for cell population doubling is greatly reduced when the cells are taken from persons with premature aging diseases such as progeria, Werner's syndrome, and certain other genetic abnormalities. Cell fusion experiments suggest that the aging and death clock, if it truly exists, is located in the nucleus of the cell.

If the hypothesis of genetically programmed aging proves correct, scientists may some day discover the clock's exact mechanism and even determine how to control it to lengthen the life span. Many people doubt that such an event would really improve human well-being, for it would create great problems for our present societal practices and programs.

DISCUSSION QUESTIONS

1. What is the difference between "life expectancy" and "maximum life span"?

2. What is the critical difference between the somatic mutation and error catastrophe hypotheses of aging?

3. Many short-lived organisms, such as insects, are much more resistant to ionizing radiation than organisms with long life spans, such as humans. In terms of the somatic mutation theory, why does this present a paradox?

4. Why can we be sure that senescence and cell death in the Hayflick system of monitoring cell aging are not due to the accumulation of wastes or the lack of necessary nutrients?

5. Cancer cells are often considered a biological paradox because of their growth characteristics and their effects upon the organism. What do you suppose is meant by paradox in this context?

6. In what way could the deaths of individuals be advantageous to a species?

7. Can the life expectancy exceed the maximum life span? Theoretically, could the life expectancy equal the maximum life span? If so, under what conditions?

8. Do you think it is possible to test the cells from newborn human infants for their doubling capacity in order to determine their individual life-span potential? Why or why not?

9. If the aging process is genetically programmed, could we (or any other species) evolve a longer maximum life span? Under what conditions?

10. Should we actively seek to find and manipulate the so-called death clock?

11. Would you like to know your individual life-span potential, if possible? Why or why not?

12. The genetically programmed aging hypothesis suggests that advanced medicine cannot change the maximum life span. What role *can* medical progress play in the aging phenomenon?

13. To find the general location of the aging clock, researchers placed old nuclei in young cytoplasts and young nuclei in old cytoplasts. In addition, they placed young nuclei in young cytoplasts. What is the purpose of this latter part of the experiment? What would the various possible outcomes mean?

14. What direct connection might there be between cellular aging and death and organismic aging and death?

15. Assume that the aging clock and a means to slow it are discovered, so that the maximum life span becomes 400 years, with a life expectancy of 300 years. Choose one or more of the following societal factors, discuss the possible impact of the increased life span on it, and give your recommendations for dealing with the ensuing problems.
    a. life insurance,
    b. social security,
    c. retirement,
    d. overpopulation, and
    e. financial inheritance.

16. In a sense, everyone now living is a product of a long line of immortal cells. Explain this reasoning.

PROBLEMS

1. The average adult body contains an estimated 75 trillion cells. How many cell population doublings are necessary to amass that many cells, beginning with a one-celled zygote?

2. In the following human survival curves:
   a. What is the approximate average age of death for each of the years represented?
   b. Only the 1950 curve is completely correct. Why are the others wrong?

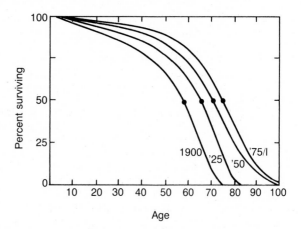

3. Suppose that we label exactly half of a population of cells with a radioactive substance; that each time a radioactive cell divides, the two daughter cells are also radioactive; and that our radioactive marker can be determined and scored. If, after five population doublings, 65 percent of the cells are radioactively labeled, what conclusion(s) can we draw?

4. The survival curves of two different species (a and b) appear in the graph below. Which species has the longer average life expectancy? Which has the longer maximum life span?

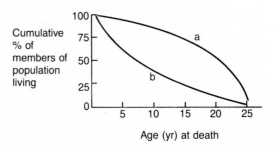

366

5. In the graph below, sketch the survival curve for a species whose every member has an equal chance of dying on any day, throughout the life span, due to predation. Assume that the maximum life span of this species is 20 years.

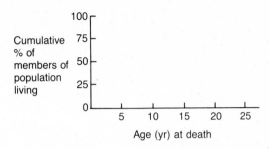

Cumulative % of members of population living

Age (yr) at death

REFERENCES

Cutler, R. G. 1976. Evolution of longevity in primates. *Journal of Human Evolution* 5:169–202.

Cutler, R. G. 1978. The origin, evolution, nature, and causes of aging. In *The biology of aging*, ed. J. A. Behnke, C. E. Finch, and G. B. Moment. New York: Plenum Press.

Cutler, R. G. 1979. Evolution of longevity in ungulates and carnivores. *Gerontology* 25:69–86.

DeBusk, F. L. 1972. The Hutchinson-Gilford progeria syndrome. *Journal of Pediatrics* 80:697–724.

Epstein, C. J., G. M. Martin, A. L. Schultz, and A. G. Motulsky. 1966. Werner's syndrome: A review of its symptomatology, natural history, pathologic features, genetics and relationship to the aging process. *Medicine* (Baltimore) 45:177–221.

Fries, J., and L. M. Crapo. 1981. *Vitality and aging*. San Francisco: W. H. Freeman and Co. Publishers.

Goldstein, S. 1969. Lifespan of cultured cells in progeria. *Lancet* 1:424.

Goldstein, S. 1971. The biology of aging. *The New England Journal of Medicine* 285:1120–1129.

Goldstein, S. 1974. Aging *in vitro*: Growth of cultured cells from the Galapagos tortoise. *Experimental Cell Research* 83:297–302.

Hayflick, L. 1965. The limited *in vitro* lifetime of human diploid cell strains. *Experimental Cell Research* 37:614–636.

Hayflick, L. 1975. Cell biology of aging. *BioScience* 25:629–637.

Hayflick, L. 1976. The cell biology of human aging. *The New England Journal of Medicine* 295:1302–1308.

Hayflick, L. 1980. The cell biology of human aging. *Scientific American* 242:58–65.

367

Hayflick, L., and R. S. Moorhead. 1961. The serial cultivation of human diploid cell strains. *Experimental Cell Research* 25:585–621.

Mitsui, Y., K. Matsuoka, S. Aizawa, and K. Noda. 1980. New approaches to characterization of aging human fibroblasts at individual cell level. In *Aging phenomena*, ed. K. Oota, T. Makinodan, M. Iriki, and L. S. Baker. New York: Plenum Press.

Muggleton-Harris, A. L., and L. Hayflick. 1976. Cellular aging studied by the reconstruction of replicating cells from nuclei and cytoplasms isolated from normal human diploid cells. *Experimental Cell Research* 103:321–330.

Orgel, L. E. 1963. The maintenance of the accuracy of protein synthesis and its relevance to aging. *Proceedings of the National Academy of Sciences* 49:517–521.

Pryor, W. A. 1970. Free radicals in biological systems. *Scientific American* 223:70–83.

Salk, D., E. Bryant, K. Au, H. Hoehn, and G. M. Martin. 1981. Systematic growth studies, cocultivation, and cell hybridization studies of Werner syndrome cultured skin fibroblasts. *Human Genetics* 58:310–316.

Schneider, E. L., and Y. Mitsui. 1976. The relationship between *in vitro* cellular aging and *in vivo* human age. *Proceedings of the National Academy of Sciences* 73:3584–3588.

Todaro, G. J., and H. Green. 1963. Quantitative studies of the growth of mouse embryo cells in culture and their development into established lines. *Journal of Cell Biology* 17:299–313.

Wright, W. E., and L. Hayflick. 1975. Contributions of cytoplasmic factors to *in vitro* cellular senescence. *Federation Proceedings* 34:76–79.

Wright, W. E., and L. Hayflick. 1975. Nuclear control of cellular aging demonstrated by hybridization of anucleate and whole cultured normal human fibroblasts. *Experimental Cell Research* 96:113–121.

17
Genetics and
Behavior

Biologists, psychologists, sociologists, and others have long debated the effects of heredity and environment on human behavior. Now most scientists believe that heredity and environment interact to determine many human behavioral traits. However, a great deal of argument remains over the relative importance of the two causes. Those who favor heredity and those who favor environment can each point to a few clearcut cases that support their views. Neither group can adequately explain the many behavioral traits that are not clearly controlled solely by heredity or environment.

The debate is vigorous because of its close connection to social issues, for whether behavioral traits, such as criminality and intelligence, are due to heredity or environment can profoundly affect how society approaches them. Yet some proponents have pushed their cases far beyond what the data justify. They think that if a trait has any hereditary basis at all, it must be inherited in a strictly deterministic way, or that if a trait has any environmental basis, it must be totally malleable. Neither view is accurate, especially within the realm of behavior. Even if a trait does have a genetic basis, environment will always affect it, sometimes very noticeably. Environment can modify the effects of genes, and combinations of genes may only determine the limits of the range of phenotypic expression. This concept is diagrammed in Figure 17-1. Human potentialities are not boundless.

We can see many examples where abnormal behavior is due to specific molecules, or to a biochemical deficiency. The most dramatic

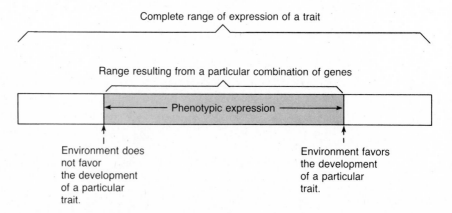

FIGURE 17–1
Individual genes are discrete, but their combinations might determine the limits of the range through which multigenic traits are expressed.

370

are the inborn errors of metabolism—Tay-Sachs and Huntington's diseases, phenylketonuria, Lesch-Nyhan syndrome—where defective genes and enzymes lead to mental retardation, depression, irritability, and even violence. Some forms of epilepsy seem to have a genetic basis. And there are many more examples, including those that may relate to chromosome aberrations.

STUDYING HUMAN BEHAVIOR

No one would deny that it is very difficult to study the inheritance of behavior in humans. Much of what we think we know comes from extending conclusions reached by means of animal studies to people. Such extrapolations are accompanied by inevitable uncertainties, but using animals to study the inheritance of behavior does have many advantages. It is relatively inexpensive to conduct studies using large numbers of rats, mice, cats, and dogs. They often produce numerous offspring, their environments can be controlled, and inbred (genetically homogeneous) lines are available. If animals with different genotypes are reared in identical environments, their behavioral differences can be attributed to genetic causes. If inbred animals are reared in different environments, their behavioral differences must have environmental causes. Naturally, the same observations might be made for humans, if we could manipulate them in the same ways.

Studies of both animals and humans rely on the concept of **heritability**, the ratio of genetic variance to total variance. With animals, experiments can be designed to reveal this ratio. With humans, we use various statistical tools, each of which requires assumptions and permits only cautious interpretations.

One of the classic ways to study human heredity is the pedigree, which we have already discussed. This is especially useful with single gene traits. However, most behavioral traits probably have a multigenic mode of inheritance, if they have a genetic basis at all. If a pedigree shows that a trait "runs in the family," this certainly points toward a hereditary link, but researchers require more definitive data. After all, family members do share an environment as well as genes. Studies of twins and adoptees have been used to separate these factors.

Twins and Concordance

Twin studies have long been a favored method of studying behavior, among other traits. The twin method capitalizes on the fact that while **dizygotic**, or fraternal, twins share half their genes on the average (as do all siblings), **monozygotic**, or identical, twins share *all* their genes. Monozygotic twins make up a human clone of two. The two types of twins develop in very different ways, as illustrated in Figure 17-2.

Twins make good research subjects because their genetic situation often allows us to distinguish hereditary from environmental influences. Identical twins can be reared together or apart, in the same or different environments, as can fraternal twins. Researchers can learn much by comparing the two types of twins in the two situations.

Finding a certain trait in two monozygotic twins does not by itself indicate that it was caused by heredity and not environment. Such

FIGURE 17–2
Monozygotic (identical) twins are formed when a very early developing cell mass divides into two embryos. Dizygotic (fraternal) twins form when two independently released eggs are fertilized by two different spermatozoa.

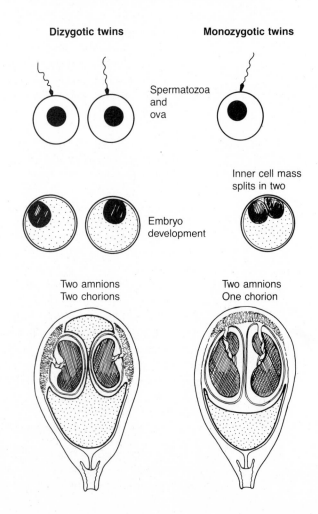

twins living together have the same heredity and virtually the same environment (although their environments may have subtle differences, starting with their positions in the uterus). The advantage of comparing monozygotic and dizygotic twins is that although the two environmental situations are usually comparable, their hereditary constitutions are quite different.

In genetic studies, it is better to use dizygotic twins of the same sex than any other two siblings of the same sex. Like monozygotic twins, dizygotic twins are born at the same time. They are the same age, reared at the same time, and generally exposed to the same environment. We can assume that the environments for two dizygotic twins are just as similar as those for two monozygotic twins. However, this assumption is sometimes criticized, because monozygotic twins may sometimes be placed in environments that are even more alike than those experienced by dizygotic twins. Notwithstanding such subtle environmental influences, the differences between monozygotic and dizygotic twins are often striking.

Researchers usually express their twin data in terms of **concordance rates**. The concordance rate is the percentage of cases in which both members of a twin pair have a particular discrete (yes or no) trait when it is known that one twin has the trait. If the second twin does not have the trait, the pair is **discordant**.

$$\text{concordance rate} = \frac{\text{concordant twins}}{\text{concordant twins} + \text{discordant twins}} \times 100\%$$

Table 17-1 gives hypothetical examples of high concordance and low concordance.

It is not enough to observe only the concordance rate for monozygotic twins. We must also observe the concordance rate for dizygotic twins, for they do not have identical genotypes. If a trait is highly hereditary, we would expect the concordance rate to be significantly lower for dizygotic than for monozygotic twins. If a trait were completely environmental, we would expect the rate to be similar for both kinds of twins. Table 17-2 lists dizygotic and monozygotic concordance rates for several traits. Some of these traits very clearly seem to have genetic causes, while others just as clearly seem to have environmental causes.

Many twin studies are impressive, and they probably deserve more attention than they receive. They indicate the heritability of specific traits, though they say very little about the mode of inheritance. If twin studies do not prompt further study, it may be because we are too cautious. Yet the caution may be justified. One of the more serious prob-

TABLE 17–1

Two hypothetical examples of concordance calculations

Twin Member Having the Trait	Other Twin Member[a]	Twin Member Having the Trait	Other Twin Member[a]
1	C	1	D
2	C	2	D
3	C	3	C
4	C	4	D
5	D	5	D
6	C	6	D
7	D	7	D
8	C	8	D
9	C	9	D
10	C	10	D

$$\text{Concordance} = \frac{8}{8+2} = 80\% \qquad \text{Concordance} = \frac{1}{1+9} = 10\%$$

Note: [a]C is concordance and D is discordance.

TABLE 17–2

Concordances of monozygotic and dizygotic twins for several traits

Trait	Concordance in Monozygotic Twins	Concordance in Dizygotic Twins
Schizophrenia	34% (203)	12% (222)
Tuberculosis	37% (135)	15% (513)
Manic-depressive psychosis	67% (15)	5% (40)
Hypertension	25% (80)	7% (212)
Mental deficiency	67% (18)	0% (49)
Rheumatic fever	20% (148)	6% (428)
Rheumatoid arthritis	34% (47)	7% (141)
Bronchial asthma	47% (64)	24% (192)
Epilepsy	37% (27)	10% (100)
Diabetes mellitus	47% (76)	10% (238)
Smoking habits (females only)	83% (53)	50% (18)
Cancer (at same site)	7% (207)	3% (767)
Cancer (at any site)	16% (207)	13% (212)
Death from acute infection	8% (127)	9% (454)

Source: Hartl, D. 1977. *Our Uncertain Heritage. Genetics and Human Diversity.* Philadelphia: J. B. Lippincott Co.

lems associated with twin studies is **ascertainment bias**. That is, con-cordance may be *reported* more often for monozygotic than for dizygotic twins simply because it is noticed more often.

Correlation

Whenever researchers wish to relate two paired variables, they calculate the **correlation coefficient**. This number measures the associ-ation between the two variables; it makes no assumptions about cause and effect.

Let us suppose that we can measure each variable and express it as a number. Then, if we find that high values of one variable tend to be associated with high values of the other and low values with low values, we see a **positive correlation**. If we graph the variables against each other, as in Figure 17-3, the line slopes upward to the right. The maximum correlation coefficient and the maximum slope of the line are each 1.00.

If high values of one variable are associated with low values of the other and vice versa, the line slopes downward to the right. Here we see a **negative correlation**, with a minimum value of -1.00. If the

FIGURE 17–3
The correlation coefficient is the slope of the line expressing how two vari-ables depend on each other. The maximum positive correlation is $+1.0$ (a). The maximum negative correlation is -1.0 (b). If the two variables are not correlated at all, the line slope and correlation coefficient are both 0.00 (c).

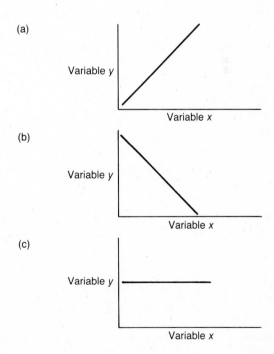

(a)

Variable *y*

Variable *x*

(b)

Variable *y*

Variable *x*

(c)

Variable *y*

Variable *x*

values of one variable are not correlated with any particular values of the other, we see a slope and correlation coefficient of 0.00.

The formula for calculating the correlation coefficient (r) is:

$$r = \frac{\Sigma XY - \dfrac{(\Sigma X)(\Sigma Y)}{n}}{\sqrt{\left[\Sigma X^2 - \dfrac{(\Sigma X)^2}{n}\right]\left[\Sigma Y^2 - \dfrac{(\Sigma Y)^2}{n}\right]}}$$

where X is one characteristic, Y the other, n the total number of variable pairs, and Σ the summation sign. Table 17-3 shows brief examples of correlation calculations; note the patterns of data that yield r values of 1.00, -1.00, and 0.00.

TABLE 17–3
Hypothetical examples of data resulting in perfect positive and negative correlations and no correlation

Example 1: Perfect positive correlation.

Variable X	Variable Y	XY	X^2	Y^2	
4	4	16	16	16	$r = \dfrac{40 - \dfrac{(12)(12)}{4}}{\sqrt{\left[40 - \dfrac{(12)^2}{4}\right]\left[40 - \dfrac{(12)^2}{4}\right]}}$
2	2	4	4	4	
4	4	16	16	16	
2	2	4	4	4	
$\Sigma = 12$	12	40	40	40	$r = 1.0$

Example 2: Perfect negative correlation.

4	2	8	16	4	$r = \dfrac{32 - \dfrac{(12)(12)}{4}}{\sqrt{\left[40 - \dfrac{(12)^2}{4}\right]\left[40 - \dfrac{(12)^2}{4}\right]}}$
2	4	8	4	16	
4	2	8	16	4	
2	4	8	4	16	
$\Sigma = 12$	12	32	40	40	$r = -1.0$

Example 3: No correlation.

2	4	8	4	16	$r = \dfrac{36 - \dfrac{(12)(12)}{4}}{\sqrt{\left[40 - \dfrac{(12)^2}{4}\right]\left[40 - \dfrac{(12)^2}{4}\right]}}$
4	2	8	16	4	
4	4	16	16	16	
2	2	4	4	4	
$\Sigma = 12$	12	36	40	40	$r = 0$

Perfect correlations are rare. When associations exist, the correlation values are usually between 0 and 1.00 or between 0 and −1.00. Strong associations yield values near the extremes (1.00 or −1.00). The magnitude of the values given to the variables and the size of the differences between the variables do not affect the final correlation coefficient. The correlation coefficient depends only upon how the variables are paired. As we will see throughout this chapter, the correlation coefficient is an invaluable statistical tool. It allows researchers to compare correlations for certain traits among pairs of persons, such as relatives or nonrelatives, and individuals living together or apart.

SCHIZOPHRENIA AND MANIC DEPRESSION

Human behavior is immensely variable, because of the tremendous diversity of human genotypes and environments, both of which play strong roles in molding behavioral phenotypes. These phenotypes include more than 100 kinds of mental deficiencies. One of these, **schizophrenia**, is a mental illness estimated to have a worldwide frequency of about 1 percent.

Schizophrenia is characterized by delusions, hallucinations, passivity, withdrawal, suspiciousness, and chaotic thoughts. The disorder can be mild, moderate, or severe, but it is often difficult to draw such lines, or even to distinguish schizophrenia from other types of mental disorders. The hypothetical causes suggested for schizophrenia include both structural and chemical brain abnormalities. One hypothesis calls for an overabundance of the neurotransmitter **dopamine**, which passes signals from nerve cell to nerve cell in certain parts of the brain. Another posits an overabundance of the cell membrane **receptors** through which nerve cells respond to dopamine. Still another proposes that a virus or viruslike agent is the cause of schizophrenia. Many investigators are convinced that schizophrenia has a genetic component as well, but the evidence is strictly statistical.

Schizophrenia definitely runs in families, and a person's risk of schizophrenia goes up with his or her degree of genetic relatedness to a diagnosed schizophrenic, as shown in Table 17-4. This may be because the relative's environment, which includes one or more schizophrenic persons, can itself be a significant cause. However, the concordance data for monozygotic and dizygotic twins in many countries, which appear in Tables 17-5 and 17-6, suggest a genetic component. Although the diagnostic criteria for schizophrenia have changed over the years, the general concordance patterns have not changed. Monozy-

377

TABLE 17–4

The probability of having schizophrenia under various conditions of relatedness to schizophrenic persons

Relationship to the Schizophrenic Person	Probability of Having Schizophrenia (%)[a]
Child of schizophrenic parents (both parents)	36.6
Sib (with one parent being schizophrenic)	13.79
Sib (with neither parent being schizophrenic)	8.24
Half-sib	3.22
First cousin	2.91
Aunt or uncle	2.01

Source: Slater, E., and Cowie, V. 1971. *The Genetics of Mental Disorders.* London: Oxford University Press.

Note: [a]The percentages are based upon diagnostically confirmed cases; that is, the study omits the probable schizophrenics.

gotic twins are more than three times more likely than dizygotic twins to both have schizophrenia.

Adoption studies have shown similar patterns. Schizophrenia is significantly more prevalent among persons born to schizophrenic mothers and raised from very early life by normal adoptive parents than among control groups of individuals born to nonschizophrenic mothers and also adopted by normal parents very early in their lives.

TABLE 17–5

Concordance values for schizophrenia among monozygotic and dizygotic twins in a series of studies conducted from 1928 to 1961

Country	Date	Monozygotic Twin Pairs			Dizygotic Twin Pairs		
		Number of Pairs	Concordant Pairs	Concordance (%)	Number of Pairs	Concordant Pairs	Concordance (%)
Germany	1928	19	11	58	13	0	0
United States	1934	41	25	61	53	7	13
Sweden	1941	11	7	64	27	4	15
United States	1946	174	120	69	296	34	11
England	1953	37	24	65	58	8	14
Japan	1961	55	33	60	11	2	18

Source: Gottesman, I. I., and Shields, J. 1966. *British Journal of Psychiatry,* 112:809–818.

TABLE 17-6
Concordance values for schizophrenia among monozygotic and dizygotic twins in a series
of studies conducted from 1967 to 1972

Country	Date	Monozygotic Twin Pairs			Dizygotic Twin Pairs		
		Number of Pairs	Concordant Pairs	Concordance (%)	Number of Pairs	Concordant Pairs	Concordance (%)
Norway	1967	69	31	45	96	14	15
Denmark	1969	25	14	56	45	12	26
Finland	1971	20	7	35	23	3	13
United States	1972	121	52	43	131	12	9
United Kingdom	1972	26	15	58	34	4	12

Source: Gottesman, I. I., and Shields, J. 1973. *British Journal of Psychiatry,* 122:15-30.

The role of the environment is also evident in such data. In the more recent studies, the concordance rate for monozygotic twins averaged 45.6 percent. If the trait were completely hereditary, the concordance should be 100 percent. Unfortunately, the specific environmental factors involved in schizophrenia have not yet been identified.

Many behavioral scientists agree that a genetic component exists for schizophrenia, but they can only speculate on the specific mode of inheritance. Some favor a **dominant gene model** with **incomplete penetrance**, which attributes the disorder to a dominant gene that is not always expressed. Others prefer the **polygenic model**, which says schizophrenia must be regulated by many gene pairs. The latter is probably the more popular mode. A third model is the **genetic heterogeneity model**, which says that different schizophrenics may have inherited their disorders in different ways. Finally, some researchers think there may be two different types of schizophrenia, one with and one without a strong genetic basis.

Manic-depressive illness occurs in 1 to 2 percent of the population of the United States, and its causes may be similar to those of schizophrenia. There is even evidence that it may be an alternate expression of schizophrenia genes. Environmental factors may play some role, but a strong genetic component can also be demonstrated. The overall concordance for monozygotic twins is 69 percent, but for dizygotic twins it is only 13 percent. In addition, comparisons of adoptees and their biological parents with control groups show a threefold greater incidence of manic depression among people with manic-depressive biological parents. Stress and other environmental factors may also be necessary for the disorder to be expressed. Some human geneticists have

379

concluded that manic depression is an autosomal dominant trait, but this has not been proved.

ALCOHOLISM

Some scientists assert that the link between genetics and **alcoholism** is well documented. Alcoholism is a complex dependence on alcohol with so much variability that it is sometimes difficult to distinguish alcoholics from nonalcoholics reliably. The disorder affects 3 to 5 percent of all adult males and almost 1 percent of all adult females in the United States. It affects every socioeconomic class and all levels of I.Q. and education. It exists among business people, politicians, professionals of all types, unskilled workers, the successful, and the "down and out."

Alcoholism clearly runs in families, but this may be because of genetic or environmental causes, or both. Heredity has been strongly implicated by studies of twins, half-sibs, and adoptees. In addition, alcohol preference can be bred into rats and mice, which indicates the trait is hereditary, at least for the animals tested. Biochemical investigations are contributing meaningful information about alcohol metabolism.

In one twin study, severe alcohol abuse among monozygotic twins yielded a concordance rate of 84.5 percent, while the rate for dizygotic twins was 66.7 percent. One adoption study compared men who were separated early from their biological parents, one of whom was an alcoholic, to a control group of adoptees whose biological parents were not alcoholics. The results showed almost four times as many alcoholics among the adoptees with alcoholic parents. In another study, sons of alcoholics adopted in early infancy by nonalcoholics showed alcoholism rates similar to those of their brothers raised by the alcoholic parents.

Genetic vulnerability to alcoholism seems to be a real phenomenon, but not everyone having this vulnerability will necessarily become an alcoholic. The condition is not one of **genetic determinism**. Instead, scientists suspect polygenic inheritance, with many environmental factors shifting the phenotype one way or the other. We now need to work out the biochemical differences that are inherited; the link between the chemical substance, alcohol, and the addiction remains elusive. The data are limited, but since alcoholism is responsible for so much human and social tragedy, the trait should certainly receive continued scientific study.

380

CRIMINALITY

Monozygotic twins show a 72 percent concordance rate for criminal behavior, while dizygotic twins show a 34 percent rate. Is there then any possibility of a genetic predisposition to criminality? Some evidence exists in favor of this possibility, and the idea of "bad seed" has been around for a long time. Yet the hypothesis regularly meets strong opposition.

Danish adoption studies have produced the striking results shown in Table 17-7. Children of noncriminal fathers adopted by criminal stepfathers were compared to children of noncriminal fathers adopted by noncriminal stepfathers. The crime rates were only average in both cases, suggesting that it makes little difference whether the children were reared in a criminal or noncriminal environment. However, children of criminal fathers adopted by noncriminal stepfathers had a higher crime rate than either of the above two groups. And children of criminal fathers adopted by criminal stepfathers showed the highest crime rate of all. Heredity may play an important role, but there does seem to be a strong interaction between heredity and environment.

Many studies on the possible inheritance of criminality have focused on persons with the aneuploid conditions XXY, XYY, and XXYY. Early data indicated the XYY persons were 30 to 40 times more likely to be placed in penal institutions than XY persons. Those initial studies

TABLE 17–7

Crime rates of progeny from criminal and noncriminal fathers reared by criminal and noncriminal stepfathers

Children of noncriminal fathers	adopted by	Criminal stepfathers	produce	Average crime rate
Children of noncriminal fathers	adopted by	Noncriminal stepfathers	produce	Average crime rate
Children of criminal fathers	adopted by	Noncriminal stepfathers	produce	Higher crime rate
Children of criminal fathers	adopted by	Criminal stepfathers	produce	Highest crime rate

Source: Criminality in adoptees and their adoptive and biological parents: a pilot study. In Mednick, S. A., and Christiansen, K. O. (eds.), 1977. *Biosocial Bases of Criminal Behavior.* New York: Gardner Press.

were criticized for lacking good controls (among other reasons), but the fact remains that a greater proportion of XYY persons than XY individuals are found in penal institutions. More recent studies reveal that about 2 percent of all male prisoners have the XYY karyotype, while the XYY frequency among all male newborns is only 0.1 percent. Therefore, the XYY genotype is 20 times more common in prisons than in the population as a whole. At first, researchers thought the extra Y chromosome might be the cause of criminal behavior. Additional data, however, indicated that individuals with the XXY karyotype have similar behavioral tendencies. Such males have an extra X chromosome, not an extra Y. The proportion of XXYY individuals that is eventually committed to prisons is even more dramatic, although this is a very rare karyotype, providing us with few data.

There can be little doubt about the frequencies of XXY and XYY persons found in mental and penal institutions. Studies have shown that persons with those aneuploid conditions are much more likely to be institutionalized than persons with the normal XY karyotype. Nonetheless, the cause of the relationship is by no means certain. No one has ever discovered a gene for crime. The data must be evaluated in the light of other observations. Persons with aneuploid conditions often show average I.Q.'s significantly lower than normal. Also, the mean I.Q. of institutionalized inmates is significantly lower than the mean I.Q. of the population as a whole. Further, people with below-average I.Q.'s exhibit above-average amounts of antisocial behavior. Coupling these observations leads some scientists to conclude that the lower intelligence of inmates is the main cause of the karyotype-prison association. However, we should note that persons with XXY or XYY karyotypes are mostly guilty of crimes against property, not crimes of violence.

The situation presents difficulties for society. Certainly, the link between crime and certain karyotypes cannot be regarded as trivial. On the other hand, we must be careful not to foster unwarranted prejudices by stigmatizing people. Most XXY and XYY persons live normal, law-abiding lives, and many people believe that little would be gained by labeling them at birth or later, before they have actually committed any crimes.

INTELLIGENCE

Is human intelligence controlled by heredity? Some aspects of the question are extremely sensitive, but very important, because their answers affect critical policy decisions at all levels of government.

The problems with this question begin with the definition of intelligence. Broadly speaking, intelligence is the ability to learn, reason, and create ideas, but the phenomenon is more complex than this broad definition suggests, and no definition satisfies everyone.

The second problem—assuming that we can resolve the first—lies in how to measure intelligence. Most testing schemes ask people to answer questions and solve problems that will produce a score in terms of **I.Q. (intelligence quotient)** points. The average score for American whites at any given age is 100, by definition. The group of scores at any age forms a **normal distribution** curve. On this curve, one **standard deviation** is 15 I.Q. points, and two-thirds of the population have I.Q.'s within 15 points of 100, or between 85 and 115. This distribution appears in Figure 17-4. Not all the experts agree that I.Q. scores really measure intelligence. Nevertheless, I.Q. scores do correlate well with other yardsticks of intelligence, such as school performance.

The heritability of I.Q. (if not of intelligence) is quite high, sometimes reported as high as 80 percent. However, critics have pointed out possible problems in using heritability calculations, especially with human populations. The statistic should be treated cautiously.

Many researchers have calculated correlation coefficients for the I.Q.'s of monozygotic and dizygotic twins living together and apart, for sibs living together and apart, and for various other related pairs. These are shown in Table 17-8. The correlations are strongest for monozygotic twins living together ($r = 0.87$) and apart ($r = 0.75$) and weakest for unrelated persons living apart ($r = 0.01$). If correlation coefficients are reliable, and if the studies summarized in the table were done carefully, these data reveal strong roles for both heredity and environment. After all, for unrelated persons living together and sharing only an environ-

FIGURE 17–4
The distribution of intelligence quotients among the white population in the United States.

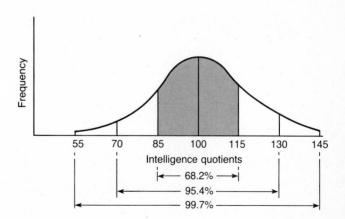

ment, r equals 0.24, much greater than the "no association" r of 0.01 for unrelated persons living apart.

Of the many other comparisons that can be made using data in Table 17-8, some stress the hereditary component of I.Q. scores, while others stress the environmental component. The importance of *both* factors—and their interaction—in determining how we perform on I.Q. tests seems obvious to most persons.

Giving infant animals and humans a more stimulating environment boosts later performance on intelligence tests. With humans, education has such an effect on I.Q. scores. The more stimulating environments and higher I.Q. performance both correlate with higher socioeconomic levels of families, but socioeconomic difference can be rooted in either heredity or environment. Generally, we should not be asking whether I.Q. test scores result from either nature or nurture. We should be asking how the two components interact, as they clearly do.

Much of the debate over intelligence stems from assertions by people such as Arthur Jensen that American blacks, who average some 15 points less on I.Q. tests than American whites, actually have significantly less intelligence. Jensen's study concludes that the difference is

TABLE 17–8

A summary of the correlation coefficients that compare the I.Q. values of persons having varying relatedness

Relationship	Correlation Coefficient	Genes Shared by the Two Persons (%)
Monozygotic twins living together	0.87	100
Monozygotic twins living apart	0.75	100
Dizygotic twins living together (same sex)	0.56	50
Dizygotic twins living together (opposite sex)	0.49	50
Sibs living together	0.55	50
Sibs living apart	0.47	50
Parent and child	0.50	50
Grandparent and grandchild	0.27	25
Uncle/nephew or aunt/niece	0.34	25
First cousin	0.26	12
Second cousin	0.16	3
Unrelated living together	0.24	0
Unrelated living apart	0.01	0

Source: Karlsson, J. L. 1978. *Inheritance of Creative Intelligence.* Chicago: Nelson-Hall.

due to heredity. Critics of this idea prefer an environmental explanation, attributable to socioeconomic and cultural differences. Some blame the apparent difference on pro-white biases built into the I.Q. tests. Others feel there is absolutely no scientific evidence for a hereditary component in racial differences relative to intelligence, and they have the official support of a Resolution of the members of the Genetics Society of America on heredity, race, and I.Q.

We must make a fundamental point about this hotly contested issue. Assume for the sake of argument that performance on I.Q. tests has a high heritability value in whites and that many visible character differences between the races are largely due to heredity. Given the above, we still cannot conclude that differences in I.Q. measurements between the races have a genetic origin. Such differences could still be environmental and are essentially unexplored.

Some people are convinced that we should not even pursue research on this question. They state that the information is too easily misinterpreted and misused, resulting in social injustices. Others disagree just as strongly, pointing to such things as the recent United States Office of Education survey testing the verbal and mathematical literacy of the American public. One-fifth of those surveyed did not know the meaning of "we are an equal opportunity employer." Fourteen percent could not write a check correctly. Twenty-nine percent could not calculate a week's pay if it included some overtime paid at time-and-a-half. Twenty-seven percent did not know the average body temperature. And so on. The difficulties of the respondents may be the result of a poor education rather than a lack of intellectual ability, but the data certainly support the need for additional research in this important area.

Most geneticists realize the dangers of applying the techniques of quantitative inheritance to the analysis of behavioral traits such as intellectual ability. Still, well-designed research on the genetic and environmental factors affecting behavioral traits and their variation within the human population may yield results useful to society. Therefore, such research efforts should not be discouraged. On the other hand, we must guard against the possible misuse of the data by those who think and act as if strict biological determinism exists.

MENTAL RETARDATION

Mental retardation is a serious human problem. The affected 3 percent of the population of the United States suffer from several different types of mental deficiencies; the term "mentally retarded" applies to all who have I.Q.'s below 70. Subclasses of the mentally retarded are defined in Table 17-9. Conditions that are caused by a single pair of recessive alleles include various inborn errors of metabolism: phenylketonuria, galactosemia, and Tay-Sachs disease. Many other conditions with more complicated genetic bases still need to be elucidated. Undoubtedly, hundreds of genes are involved in human mental processes. And because the environment plays a significant role as well, understanding mental retardation is a challenging task.

Mental retardation tends to affect more than one family member. Table 17-10, which summarizes a study reported in 1930, provides evidence of this fact. A more recent study concluded that 29 percent of all mental retardation cases were primarily genetic, 19 percent were probably genetic, 9 percent were probably environmental, and in 43 percent of cases the causes were unknown. Other studies have reached similar conclusions. The case for heredity is particularly strengthened by data such as those presented in Table 17-11. Retarded parents tend to have retarded children. Even if the parents are normal, but one of them has a retarded sib, their chances of having retarded children are five times greater than those of normal parents from normal families. This is a critical comparison, because in both cases the environment consists of two normal parents.

TABLE 17–9
Classification of the levels of mental retardation

Classification	I.Q.
Mildly retarded	55–69
Moderately retarded	40–54
Severely retarded	25–39
Profoundly retarded	0–24

TABLE 17–10
Percentage of mentally retarded siblings in families having different parental situations

When a Mentally Retarded Person Has:	% of the Siblings Who Are Also Retarded Is:
Normal parents	17.8
One retarded parent	41.3
Two retarded parents	93.2

Source: Brugger, C. 1930. *Zeitschrift für die Gesamte Neurologie und Psychiatrie Originale,* 130:66–130.

386

TABLE 17–11
Percentage of mentally retarded children in families from different parental unions

Parental Union	Retarded Children (%)
Retarded parent × retarded parent	39
Retarded parent × normal parent	13
Normal parent × normal parent (at least one of the parents has a retarded brother or sister)	2.5
Normal parent × normal parent (neither parent has a retarded brother or sister)	0.5

Source: Penrose, L. S. 1963. *The Biology of Mental Defect.* New York: Grune and Stratton.

In spite of these observations, most retarded children are born to normal parents. This is because: (1) many retarded persons do not have children; (2) there are many fewer retarded persons and married retarded persons than normal persons and marriages between normal persons; and (3) some mental retardations certainly have environmental causes.

Mentally retarded persons have significantly more gross chromosome aberrations than normal. Almost all aneuploid conditions and structural changes of chromosomes result in mental deficiency. Researchers have also found specific sites on several chromosomes where breakage or wide gaps tend to occur in a fraction of the cells cultured under certain conditions. One of these sites is located on the X chromosome, and it appears to be associated with a form of mental retardation. This curious chromosome is called the **fragile X chromosome**. In addition to specific chromosome aberrations, many metabolic defects are also associated with mental retardation. Still other forms of retardation are believed to have a polygenic basis. Because mental retardation is a widespread and serious disability, we need to support further research into its causes.

SOCIOBIOLOGY

Sociobiologists study the biological basis of how animals—including humans—interact with others of their own kind. Social behavior has evolved throughout the history of each species. And evolution, of course, is firmly based upon genetics.

Sociobiology considers how such human social behaviors as aggression, mating, parenthood, love, hate, altruism, and selfishness might be controlled by genes. Presumably, the exercise of these behaviors should improve the chances that those genes would be passed to new generations. This relationship can be seen clearly for mating and parenting behaviors, and less clearly for altruistic behavior, for example. The discipline is based on the idea of natural selection, and it proposes that even society and culture must ultimately have a genetic basis. Its hypotheses have met with much opposition and have generated vehement debate among biologists, sociologists, psychologists, philosophers, and others. Do society and culture emerge from heredity or environment? There are plenty of data to support the role of environment. Sociobiologists have some data indicating a genetic component at work within animal societies, but so far they have insufficient evidence for— or against—sociobiological concepts acting in human populations.

The opposition to human sociobiology should not surprise us. Certain sociobiological concepts tend to diminish the importance of cultural background, religious beliefs, free will, and equality of opportunity for women and minorities. Some scientists think sociobiology is another form of racism, sexism, or elitism.

As we have just noted, the evidence for the genetic roots of some human behavior is fairly strong. Nonetheless, data concerning social behavioral traits are lacking or somewhat tenuous, and we must be wary of premature conclusions. Since the human species has evolved rapidly in relative terms, it has a wide genetic diversity, presumably for both behavioral and physical traits. Much of the resistance to sociobiological hypotheses about human behavior stems from a fear of **biological determinism**. This fear is not unwarranted, but it should be tempered by the realization that few traits are solely genetic. Because most have environmental components, society can at least hope to manipulate many traits, such as the mental retardation of phenylketonuria or aggressiveness. We should not fear the implications of sociobiology, however uncomfortable they may be. Many geneticists and evolutionists advocate a continued effort to search out the truth with scientific objectivity. We should not sweep aside the unanswered questions.

SUMMARY

Most geneticists agree that human behavior has some genetic basis. The basis usually lies in complex multigenic systems, but some single-gene examples also exist. Some behavioral traits are linked to chromosomal aberrations.

The various methods used to study behavior in humans include investigations of similar traits in animals, heritability calculations, pedigree analyses, twin studies, adoption studies, and the statistical analysis of correlation. Selection for or against a trait in animals should be possible if the trait has a genetic basis, and such selection has been accomplished for some behavioral traits. Heritability measures the proportion of the total variance in a trait that is due to genotype alone. Twin and adoption studies offer some of the strongest evidence for the roles of both heredity and environment in certain behavioral traits. Correlation, which assesses the association between two variables, can be used whenever a trait can be measured.

Schizophrenia and manic-depressive illness seem to have similar causes, with twin and adoption studies giving the strongest evidence for a partial genetic influence. There are also strong indications that alcoholism has some genetic basis, and many studies have also raised the possibility of a genetic predisposition to criminality. Most of the criminal studies have focused on persons with abnormal numbers of sex chromosomes (XYY, XXY, and XXYY). Although persons with these karyotypes are much more likely to be institutionalized than XY persons, the reason may be their lower intelligence, rather than genetically based criminal tendencies.

Intelligence is a complex trait that is difficult to define, measure, and investigate. Numerous calculations for relatives and nonrelatives living together and apart have shown roles for both environment and heredity. The most intense controversy centers on the notion that intelligence differs among races as a result of hereditary causes. Several large studies of mental retardation offer convincing evidence that a significant proportion of mental deficiencies is due to genetic causes. Most scientists recognize that the interaction of environment with the individual's genotype plays a significant role in determining these behavioral traits.

The evolution of animal species is accompanied by a wide genetic diversity for behavioral and physical traits. Sociobiology is the scientific discipline that attempts to link genetics to social behavior. The application of sociobiological concepts to humans, however, has met with resistance.

DISCUSSION QUESTIONS

1. Is it possible for one monozygotic twin to be normal, while the other is mentally retarded? Briefly explain your answer.

2. Many genetic researchers believe that genetic factors contribute to many human behavioral differences. If they are right, does it follow that people's behaviors are determined by their genotypes? Explain.

3. How have studies of individuals with chromosome abnormalities contributed to our understanding of the genetics of behavior?

4. Why do sibs usually resemble each other behaviorally more than parents resemble their offspring?

5. The concordance rate for monozygotic twins is 55 percent for trait A. For trait B, it is 30 percent. Might trait B be more affected by heredity than trait A? Briefly explain why or why not.

6. The following concordance rates were obtained for a hypothetical trait; in each case, the twins were living together.
   Monozygotic concordance: 71%
   Dizygotic concordance: 34%
   Why do these data indicate a possible genetic role? Why do these data indicate an environmental role?

7. Could a trait have a heritability of 100 percent in a population at a particular time, and still be altered in the future by a change in the environment? Why or why not?

8. Geneticists strongly believe that schizophrenia, a fairly common mental disorder, is due to polygenic inheritance, although no one knows how many gene pairs might be involved. Discuss some of the data and the critical characteristics of the disorder that evidently led scientists to this conclusion.

9. Alcoholism has a frequency of about 5 percent among males in the United States and about 1 percent among females in the United States. What are some possible explanations for this difference?

10. Do you think the extra Y chromosome of an XYY individual accused of a crime should be an acceptable defense in court?

11. Assume that a genetic screening program shows that your four-day-old baby has an XYY karyotype. Would you want to know this? Would it affect how you would rear the child? If so, how?

12. Would you wish to adopt an XYY child? Why or why not?

13. What correlation data best indicate that intelligence (or I.Q. score) is not exclusively genetic in origin? Be specific and explain your reasoning.

390

14. The following graph displays the distribution of I.Q.'s in the American population. It is almost a normal curve of distribution, except for the hump to the left (somewhat exaggerated here). Explain the reason for the hump.

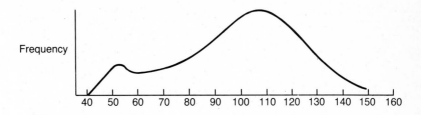

15. Suppose that one race has a mean I.Q. of 90 and a second has one of 105. The difference is 15 I.Q. points. Suppose that the heritability of I.Q. in the second race has been reliably calculated as 60 percent. Is it correct to say that 60 percent of the 15-point difference is due to genetic causes? Discuss your position.

16. Should prospective parents, not categorically opposed to abortion, selectively abort fetuses who will be mentally retarded, assuming that amniocentesis could allow us to acquire such information?

17. Two mentally retarded parents (I.Q.'s of 69 and 58) have a child with an I.Q. of 101. Should the child be removed from these parents and reared by foster parents with I.Q.'s comparable to or higher than the child's I.Q.? Defend your position.

18. Should we continue to study the basis of intelligence in humans? Why or why not?

19. A few scientists assert that our bodies are simply the vehicles by which genes perpetuate themselves. How does such a concept relate to sociobiology?

20. The usefulness of genetic studies of behavioral traits is sometimes questioned on the grounds that even if a genetic factor does exist, and it is measurable, nothing can be done about the situation. Do you agree or disagree with this thinking? Explain.

PROBLEMS

1. In the following data, S signifies "strain" and T signifies "treatment."
   a. Which case shows a strain effect, but not a treatment effect?
   b. Which case shows a treatment effect, but not a strain effect?
   c. Which case shows a strain-treatment interaction?

	case 1			case 2			case 3	
S	T1	T2	S	T1	T2	S	T1	T2
1	14	14	1	20	24	1	13	13
2	18	12	2	20	24	2	10	10
3	11	17	3	20	24	3	16	16

2. In two hypothetical cases, the environmental and genetic effects on a trait have been calculated. In the first case, the environmental effect is 50 percent. In the second, it is 40 percent. Which case shows the higher heritability?

3. Some laboratory strains of mice have been intensely inbred for behavioral and medical studies. Diagram the pedigree for the first three or four generations of such inbreeding. What are the advantages of such strains of mice?

4. An investigation of a certain population found 2000 pairs of twins, including 720 boy/boy pairs, 706 girl/girl pairs, and 574 boy/girl pairs. How many of the 2000 pairs of twins were probably monozygotic?

5. A study of a hypothetical behavioral trait in twins produced the following data:
   monozygotic twin with the trait: 180
   the other monozygotic twin with the trait: 90
   dizygotic twin with the trait: 200
   the other dizygotic twin with the trait: 16
   a. What is the concordance rate for the monozygotic twins?
   b. What is the concordance rate for the dizygotic twins?
   c. What do these concordance rates indicate?

6. What concordance rates for monozygotic and dizygotic twins would you expect for a genetic disease caused by a rare dominant gene? By a rare recessive gene?

7. What would you expect the concordance rate to be for Down syndrome in monozygotic twins? In dizygotic twins? Taken together, what do these two concordance rates indicate?

8. Show that it is *theoretically* possible to have genetically identical fraternal twins.

9. Why do dizygotic twins share an average of 50 percent of their genes? Show the relationship by deriving dizygotic twins from parents with *AABB* and *aabb* as their partial genotypes.

10. Monozygotic and dizygotic concordance rates for four different hypothetical traits are listed below. Arrange the traits in order, from the one with

the greatest genetic component to the one with the greatest environmental component.

Trait	Monozygotic concordance	Dizygotic concordance
A	2%	0.1%
B	92	90
C	61	61
D	16	9

11. Assume that you have selected and randomly paired a large number of Ph.D. chemists at major American universities. You then calculated a correlation coefficient for the paired I.Q. scores. For your control, you randomly selected, paired, and correlated many persons from a shopping center in a large city.
    a. What approximate correlation coefficient would you expect in each case?
    b. Would you expect the correlation to be significantly higher in one group than in the other? Why or why not?

12. Studying a certain human behavioral trait, you calculate correlation coefficients among various pairs of relatives and various pairs of nonrelatives. What correlation coefficients would you expect for:
    a. Monozygotic twins living apart if the trait's sole cause is heredity (that is, it has no environmental component)?
    b. Monozygotic twins living apart if the trait's sole cause is environmental (no genetic component)?
    c. Nonrelatives living together if the trait's sole cause is heredity?
    d. Nonrelatives living apart if the trait is caused by both heredity and environment?

13. The graphs below contain the I.Q. values of the following pairs of persons plotted against each other on the X and Y axes. Assign each set of persons to the correct graph.
    a. nonrelatives living together
    b. monozygotic twins
    c. sibs

14. Analyze the following data that indicate the percent of each group resorting to criminality and draw conclusions.

	Biological father a criminal	Biological father not a criminal
Adoptive father a criminal	36.2% (of 58)	11.5% (of 52)
Adoptive father not a criminal	22% (of 219)	10.5% (of 333)

Source: Rosenthal, D. 1971. *Genetics of Psychopathology*. New York: McGraw-Hill.

REFERENCES

Brugger, C. 1930. Genealogische Untersuchungen an Schwachsinnigen. *Zeitschrift fur die Gesamte Neurologie und Psychiatrie Originale* 130:66–130.

Gershon, E. S., W. E. Bunney, Jr., J. F. Leckman, M. Van Eerdewegh, B. A. DeBauche. 1976. The inheritance of affective disorders: A review of data and of hypotheses. *Behavioral Genetics* 6:227–261.

Goodwin, D. W. 1979. Alcoholism and heredity. *Archives of General Psychiatry* 36:57–61.

Goodwin, D. W., F. Schulsinger, L. Hermansen, S. B. Guze, and G. Winokur. 1973. Alcohol problems in adoptees raised apart from alcoholic biological parents. *Archives of General Psychiatry* 28:238–243.

Goodwin, D. W., F. Schulsinger, L. Hermansen, S. B. Guze, and G. Winokur. 1974. Drinking problems in adopted and non-adopted sons of alcoholics. *Archives of General Psychiatry* 31:164–169.

Gottesman, I. I., and J. Shields. 1966. Schizophrenia in twins: 16 years' consecutive admissions to a psychiatric clinic. *British Journal of Psychiatry* 112:809–818.

Gottesman, I. I., and J. Shields. 1973. Genetic theorizing and schizophrenia. *British Journal of Psychiatry* 122:15–30.

Hartl, D. 1977. *Our uncertain heritage: Genetics and human diversity*. Philadelphia: J. B. Lippincott Co.

Hook, E. B. 1973. Behavioral implications of the human XYY genotype. *Science* 179:139–150.

Jensen, A. R. 1969. How much can we boost I.Q. and scholastic achievement? *Harvard Educational Review* 39:1–123.

Karlsson, J. L. 1978. *Inheritance of creative intelligence*. Chicago: Nelson-Hall Publishers.

Mednick, S. A., and K. O. Christiansen, ed. 1977. *Biosocial bases of criminal behavior*. New York: Gardner Press Inc.

REFERENCES

Newman, H. H. 1932. Mental and physical traits of identical twins reared apart. *Journal of Heredity* 23:3–18.

Penrose, L. S. 1963. *The biology of mental defect*. New York: Grune & Stratton Inc.

Reed, E. W., and S. C. Reed. 1965. *Mental retardation: A family study*. Philadelphia: W. B. Saunders Co.

Rosenthal, D. 1971. *Genetics of psychopathology*. New York: McGraw-Hill Book Co.

Slater, E., and B. Cowie. 1971. *The genetics of mental disorders*. London: Oxford University Press.

Tsuang, M. T. 1978. Genetic counseling for psychiatric patients and their families. *The American Journal of Psychiatry* 135:1465–1475.

18
Evolution

Some people refuse to believe that human beings and other organisms are descended from ancestors unlike themselves. They prefer to believe that a supernatural power created the world all at once, including its wide range of life forms just as they exist today. These individuals cite various scriptures to support their beliefs.

There is no physical evidence whatsoever for "creationism," but there is substantial evidence for **evolution**. We have already considered some of this evidence in our study of heredity, which provides a means for inheriting traits, allows variations to arise and be maintained, and permits gene frequencies and traits to change in populations. There is the fossil record, too, which documents progressive change from the dawn of life some 3.5 billion years ago to today.

The idea of evolution is not new. It goes back, in one form or another, for more than 2000 years. **Abiogenesis** (life from nonliving components) dominated much of the thinking at that time, among other myths and legends. By the eighteenth century, some biologists viewed species as having a changeable rather than a fixed nature. They thought groups of species might have had a common ancestor. Contributing to their thoughts were observations of fossils, **vestigial** (no longer useful) parts of animals, and the development of new varieties of organisms through hybridization. Lamarck (1744–1829) added to these speculations with his idea that acquired characteristics could be inherited. The evolutionary concept was certainly in the minds of European biologists by the mid-1800s.

Scientific work at that time had shown that the world was much older than anyone had previously thought, and the fossilized remnants of long-vanished creatures were prodding people's imaginations. Yet no one had a very good idea of how evolution could work until Charles Darwin, shown in Figure 18-1, published his book, *On the Origin of Species by Means of Natural Selection*, in 1859.

CHARLES DARWIN AND THE ORIGIN OF SPECIES

At the age of 22, after losing his enthusiasm for studying both medicine and religion, Charles Darwin boarded the H.M.S. *Beagle* for a voyage that would continue from 1831 to 1836. The ship's mission was exploration, and Darwin was the ship's naturalist. Wherever it touched land, he collected samples of exotic plants and animals. He puzzled over them

398

FIGURE 18–1
Charles Darwin (1809–1882) in about
1881. (Used with permission of the
Library of Congress.)

fruitlessly until, in 1835, the *Beagle* reached the Galapagos Islands 600
miles west of Ecuador. There he heard the vice-governor of one of the
islands say that if all the animals from all the islands were lined up, he
could group them according to their islands of origin. This started Darwin
thinking. Why should the several Galapagos Islands, which shared the
same volcanic geological formations, soil, climate, and topography,
have different animals? His efforts to answer this question led directly
to the *Origin of Species*.

The animals that intrigued the vice-governor and Darwin were
varied. One island had 26 species of land birds but shared only one, a
bobolink, with other islands. The other 25 resembled South American
species. Another island had 16 species of snails all its own. As a group,
the islands had 13 species of finches, each with its own beak size and
shape, life style, and preferred environment. There were ground
finches and tree finches, seed eaters, insect eaters, and fruit eaters;
some of these are shown in Figure 18-2. It seemed clear that the beak
types suited the various feeding habits, and Darwin began to think
about the possibility that one original species had become modified to
form the various modern species, each pursuing different feeding strat-

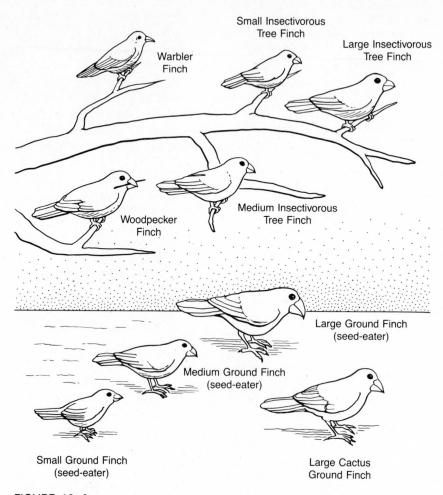

Small Insectivorous
Tree Finch

Large Insectivorous
Tree Finch

Warbler
Finch

Woodpecker
Finch

Medium Insectivorous
Tree Finch

Large Ground Finch
(seed-eater)

Medium Ground Finch
(seed-eater)

Small Ground Finch
(seed-eater)

Large Cactus
Ground Finch

FIGURE 18–2
Of the 14 related species of finches Darwin observed among the Galapagos
Islands, only 13 actually lived in the Galapagos. One inhabited another island in
the region. (From Volpe, E. Peter, UNDERSTANDING EVOLUTION. 4th ed. ©
1967, 1970, 1981 Wm. C. Brown Publishers, Dubuque, IA. All Rights Reserved. Reprinted by permission.)

egies for survival. His thoughts grew more definite as he reviewed the
specimens he had collected earlier in the voyage, and as he found similar patterns later on.

Once back in England, Darwin began to build the case for evolution. He studied **fossils**, noting that most differed from living species.

He saw how plants introduced into an area could spread quickly through the countryside. He studied how plant and animal breeders produce new breeds by **artificial selection** for favored traits, culling those that lack the traits and breeding those that have them. He argued that nature, too, must select for the traits that aid survival. He read Thomas Malthus' thoughts on the relationship of human population growth and food supply to famine and war. For more than 20 years, he diligently collected and assembled information on how new species might arise.

Although Darwin did not publish his work during those two decades, he did share it with his scientist friends. This was fortunate because in 1858 another naturalist, Alfred Russell Wallace, sent Darwin a copy of his paper explaining the origin of species by **natural selection**. Darwin's friends substantiated his priority in the field of evolution and convinced him to prepare a joint paper with Wallace for presentation at a scientific symposium later that year. At the same time, they also urged him to proceed with his book, which was published in 1859. Darwin and Wallace had reached the same conclusions about evolution quite independently, but the book—and Darwin's immense collection of supporting evidence—brought Darwin most of the recognition.

Darwin's theory of evolution had no real competition, but the public reaction to his book was immediate and controversial. Some thought he had unraveled the mystery of evolution, while others objected on religious, moral, philosophical, and aesthetic grounds. The debate was vigorous, but Darwin's work withstood the test, and his theory became widely accepted, at least among scientists. Darwin's insight had given evolution a plausible driving force—natural selection—and he had amassed a convincing array of supporting data. Unfortunately, Darwin was unaware of Mendel's classic studies of inheritance, and he did not have a good explanation for the maintenance of favorable variations. Eventually, the rediscovery of Mendel's work, the discovery of gene mutability, and the development of population genetics reinforced evolutionary thought.

NATURAL SELECTION

Darwin observed that plants and animals generally produce more offspring than ever survive to maturity. Such prolific reproduction must result in a competition for survival among the many offspring. He also observed that the members of a species differ from each other in various traits that affect survival. He reasoned that the individuals whose variations are best suited to survival will tend to survive. As this selection for survivors continues over many generations, the character of the species will gradually change to resemble those organisms having the most advantageous traits. This is natural selection. The scheme is essentially very simple, and its only shortcoming was one Darwin could not avoid—the mechanisms of heredity and variation were not yet known.

Present-day examples of natural selection are easy to find. Consider insect resistance to pesticides, which has caused serious problems in agriculture, and bacterial resistance to penicillin and other antibiotics, resulting from the widespread use and misuse of antibiotics. Both arise according to the simple version of natural selection sketched in Figure 18-3. Figure 18-4 represents a more complex example documented by H. B. D. Kettlewell in the 1950s. In about 1800, the tree trunks in wooded areas surrounding certain industrial centers of England were mostly light colored because of the lichens that covered them. At

FIGURE 18–3

A diagrammatic explanation of natural selection as the cause of change in the genetic composition of a population.

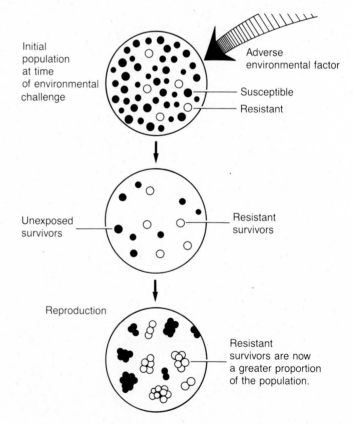

Initial population at time of environmental challenge

Adverse environmental factor

Susceptible

Resistant

Unexposed survivors

Resistant survivors

Reproduction

Resistant survivors are now a greater proportion of the population.

FIGURE 18–4

The appearance of dark and light moths upon trees with darkened bark and trees covered with the lighter lichens. (From Kettlewell, H. B. D. 1956. *Heredity*, 10: 287–301. Used with the permission of the Longworth Group Ltd.)

that time, most of the moths inhabiting the woods were also light in color, only a few were dark. The light and dark colorations were heritable characteristics. As the industrial revolution progressed, the tree trunks grew dark, colored by smoke and soot, and the frequency of the dark moths in the population increased greatly. In other words, the gene pool changed; *light* alleles became rare and *dark* alleles became common.

Kettlewell and others showed that the changes in the moth population were due to the environmental changes and bird predation. Moth coloration plays a protective, camouflaging role. Moth-eating birds can detect light-colored moths against dark tree trunks better than against light trunks. The light moths are thus selected against in a sooty environment. On the other hand, with the environmentalism of recent decades, air pollution in Great Britain has decreased, and the tree trunks have turned light again. Today, the dark moths show up more plainly, and the birds are eating them. Once more, moth populations in Great Britain are mostly light colored. The change has been thoroughly studied, even to the extent of filming the actual bird predation events, and the moth-bird interaction is now a classic example of natural selection.

We must stress that individual moths *did not* change from light to dark during their life times. Rather, the frequencies of the light and dark moths in the population changed from generation to generation, because of natural selection effected, in part, by bird predation.

Both dark and light moths belong to the same **species** or gene pool. All the organisms that can readily mate with each other and produce viable progeny under natural conditions make up a species. Theoretically, populations within a species can diverge from each other because of differences in selection pressures. They may diverge so greatly that they can no longer interbreed. The barriers to breeding are called **isolating mechanisms**, and they can differ widely. **Prezygotic** isolating mechanisms can involve differences in habitat, timing of sex-cell maturation, mating behavior, or even genital structure, making fertilization less likely. **Postzygotic** isolating mechanisms permit mating, but the resulting zygote or embryo may die, or the offspring may be sterile. If you think about it, you will understand why prezygotic mechanisms are less wasteful in biological terms than postzygotic mechanisms. Once isolating mechanisms exist, different groups cannot readily exchange genes. They can then diverge even more, until one species becomes two or more species.

The same mechanisms of selection apply to humans. We too are subject to natural selection. Genetic diseases offer some very clear examples of selection during the **prenatal** (between fertilization and

404

birth) and **postnatal** (between birth and reproduction) periods. In addition, some members of the human population leave more descendants than others, though it is hard to determine whether any genetic differences exist between these two groups.

Evolutionists have long debated what is the specific entity that is actually selected by natural selection, either to reproduce and survive or to succumb in the competition with other members of the population. Some models consider this entity to be the single allele. In other words, allele *A* may be more advantageous than allele *a*, regardless of the remainder of the genome. Other models consider the object of selection to be the entire individual, the phenotype molded by the organism's entire genome. Still others take the unit to be the entire species. One fairly common approach to evolution is to think of the gene as the unit of variation, the individual as the unit of selection, and the species as the unit that evolves. On the other hand, some biologists argue that natural selection is too simplistic, out of date, and not the only force involved in evolution.

CONTEMPORARY EVOLUTIONARY CONCEPTS

Evolution is one of the unifying concepts of all biology. Not all scientists agree on the precise mechanisms of evolution, but their disagreements should not be taken to mean that evolution does not happen. Evolution is change—not necessarily progress—and change happens over many generations. The debate is over mechanisms, and debate is the crux of the scientific method.

The classic overall view of evolution requires: (1) one or more sources of basic genetic variability to make change possible; and (2) one or more processes that can guide populations of organisms into particular channels.

The ultimate source of variation for biological evolution is **mutation**, changes in the genetic material. These changes may be small alterations in single genes or gross changes in chromosome structure or number. Variations also arise during **recombination** of genes by crossing over and meiotic segregation.

The guiding force, say **selectionists**, is natural selection, by which the environment screens the genetic variability, sorting organisms into winners and losers. On the other hand, some biologists believe that other forces, such as **genetic drift**, play an important part in evolution. Genetic drift is *random* genetic change in which the chance events of reproduction (especially in small populations) shift gene frequencies, sometimes drastically. Non-selectionist biologists describe their view as the **neutrality hypothesis** and support it with the observation that many mutations do not seem to affect the organism's ability to survive in its present environment. Therefore, selection cannot be the decisive factor. Instead, the laws of chance govern whether the neutral mutation will be established in the population. Although this hypothesis is attractive, especially to the molecular biologists, most evidence for and against neutrality is circumstantial at best.

Modern evolutionary biologists focus on two very different aspects of evolution. **Microevolutionists** study changes in the frequencies of alleles, particular chromosomes, and chromosome segments within a gene pool, as we observed in the example of the dark- and light-colored moths. Generally, microevolution leads to a population's **adaptation** to a changing environment, and even to speciation. **Macroevolution**, on the other hand, refers to the development of characteristics that distinguish the major groups of organisms (**genera, families, orders, classes,** and **phyla**).

Biologists argue over the relative roles of microevolution and macroevolution, but they usually agree that the two must interact. One facet of the debate concerns the rate of evolution. Those favoring **gradualism** say that evolution must be the consequence of mutations with small effects causing very gradual changes. (Darwin was a gradualist.) Those favoring **punctualism** believe the most important evolutionary changes occur as spasms of change alternating with long periods of evolutionary stasis. That is, millions of years might pass with very little change, followed by a host of new species springing up in a few millenia. Punctualism suggests that while the overall pace of evolution is slow, the fossil record should give the impression of a somewhat jerky evolution of life, and this is precisely what we observe.

EVIDENCE OF EVOLUTION

Microevolution has been observed in the laboratory and the field. Both microevolution and macroevolution have occurred in the past, and the signs remain for us to interpret. Some of the signs support gradualism, and some support punctualism, but none contradicts the Darwinian core.

Some indications of evolution have come from **biogeography**, the study of the geographic distribution of plants and animals. Adjacent populations often differ slightly from each other, while populations thousands of miles apart may be very different. **Taxonomy** as a discipline shows us that the numerous different kinds of organisms can be arranged in a hierarchical structure, as genera, orders, families, classes, and phyla, as shown in Figure 18-5. Clues have also come from **embryology**, the study of embryonic development. In the early stages of development, vertebrate embryos of different species closely resemble each other. Only later in development do they begin to diverge from each other in appearance. (In early stages, even the human embryo has a tail-like structure.)

Evolution requires that any group of species must stem from some single ancestral species. The more distant that ancestor is in time, the less they should resemble it and each other; and the less distant it is, the more they should resemble it and each other. We can see this relationship clearly in both anatomy (structure)—consider the similarities and differences of humans, apes, and monkeys—and in physiology (function). We can even see it when we examine chromosomes. For example, the chromosomes of the human, chimpanzee, gorilla, and orangutan are remarkably similar, even when observed with the high resolution afforded by the newer banding techniques, as illustrated in Figure 18-6. Such comparisons have allowed us to work out likely karyotypes for the common ancestor of humans and apes.

Recent investigations at the molecular level have added still more evidence for common origin and evolution. Proteins having the same function in different species have been shown to have very similar amino acid sequences, with the fewest differences between the proteins of closely related species, such as human and chimpanzee. The same is true for genes, as molecular biologists are finding with new techniques of determining the sequence of nucleotides in DNA.

FIGURE 18–5
The levels of taxonomic classification correspond to degrees of both similarity and relationship.

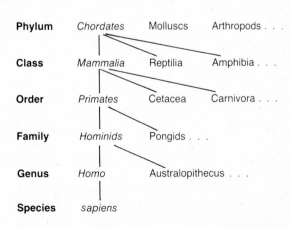

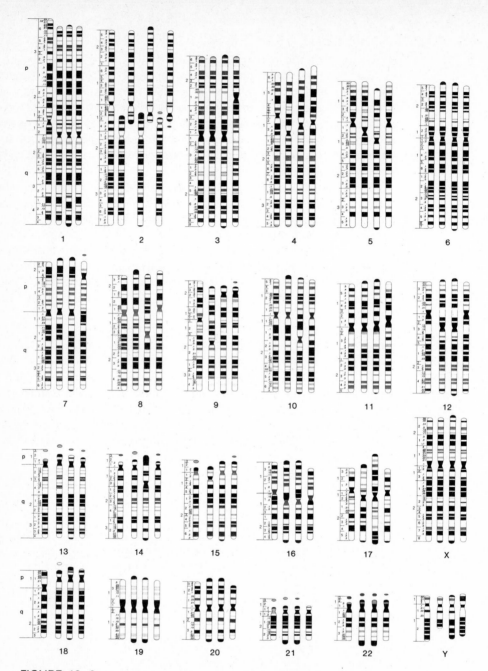

FIGURE 18–6
A diagrammatic comparison of the banding patterns and sizes of chromosomes of the human, chimpanzee, gorilla, and orangutan, arranged in that order from left to right. (From Yunis, J. J., and Prakash, O. 1982. *Science*, 215: 1525–1530. Copyright © 1982 by the American Association for the Advancement of Science. Used with the permission of the publisher and the author.)

Geology, paleontology, and the study of fossils have contributed much to our understanding of evolution. From the radioactivity in rocks, geologists have estimated the age of the earth to be approximately 4.6 billion years, a period long enough for the origin of life and its development to its present state. Paleontologists study early life by means of fossils, the remains or traces of once-living organisms. Although some evolutionists think the evidence for evolution does not have to depend on the fossil record, fossils nonetheless add an important dimension to the story. The oldest rocks are the deepest layers of any sedimentary formation. Often, paleontologists can trace a sequence of life forms up through layers of sedimentary rock and observe the actual sequence of evolutionary changes.

Because so many observations by scientists of many different disciplines have shown that evolution has taken place, scientists and nonscientists alike argue whether evolution should be called theory or fact. Exactly how life might have begun on this earth is more difficult to determine.

THE ORIGIN OF LIFE

We cannot know for sure just how life began, for this event is much too remote from our experiences and our methods of study. The earth is approximately 4.6 billion years old, and the oldest fossils—remains of single cells resembling bacteria and blue-green algae—are some 3.5 billion years old. Life appeared during the first billion years of the earth's existence, but we do not know how.

Some people believe in divine creation, but most scientists rule out miracles. A few suggest that life began when spores that originated in some other world drifted through space and reached the earth. Most look to conditions on earth itself for the answer, pointing out that abiogenesis, the origin of life from nonliving matter, had to occur at least once, sometime, somewhere.

The newly formed earth had an atmosphere of water vapor, carbon dioxide, ammonia, methane, hydrogen, and other gases, but not oxygen. When they were acted upon by lightning, heat, and other energies, gas molecules combined to form amino acids and other organic compounds. Rain washed these molecules into the seas to form a dilute organic "soup." Laboratory experiments have shown that this could have happened, for when mixtures of gases like those of the primordial earth are exposed to similar energies they give rise to the compounds needed to form a first living thing. Figure 18-7 shows how this was demon-

409

FIGURE 18-7

The apparatus devised by chemist Stanley Miller to produce amino acids and other organic compounds. (From *The origin of life* by G. Wald. Copyright © 1954 by *Scientific American*, Inc. All rights reserved.)

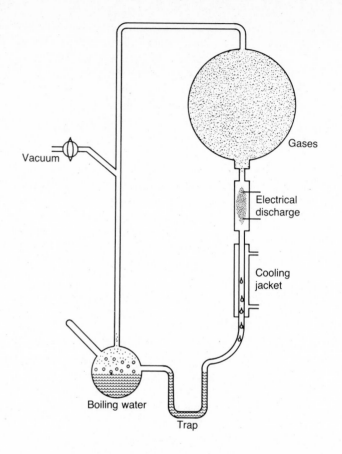

Vacuum

Gases

Electrical discharge

Cooling jacket

Boiling water

Trap

strated in the laboratory. Experiments have also shown that in the sea, small organic molecules could have reacted with each other to form larger ones. In fact, the reactions lead to small, bacteria-sized spheres. These spheres have many similarities to cells—they have membranes and hollow interiors, they grow and divide, they show enzymatic behavior, and they can contain nucleic acids. Researchers speculate that natural selection would be enough to turn these "protocells" into true cells, complete with genes, ready to evolve into the organisms we know today. The process of development was necessarily slow—it was only some 500 million years ago that life grew complex enough to move onto the land, and only 600 to 700 million years ago that life first became multicellular.

HUMAN EVOLUTION

In due course, the process of evolution gave rise to human beings. Exactly how this happened is far from clear, for the fossil record is incomplete and its interpretation changes from time to time. About 10 to 15 million years ago, the evolutionary line that led to humans diverged from those leading to apes and monkeys. Before that time, the separate lines were one. That is, humans are not descended from apes or monkeys, but humans, apes, and monkeys did have common ancestors.

Modern humans belong to the genus *Homo* and the species *sapiens*. They are sapient—they think, use tools, and communicate by speech. Yet they also share many features with other animals. As mammals, the group that developed from the reptiles some 300 million years ago, they have a diaphragm to aid in breathing; various mechanisms to maintain a constant, high body temperature; a large, well-developed brain; and hairs for insulation and protection. The young of the placental mammals, including humans, develop in a uterus and are nourished after birth by milk from mammary glands.

The placental mammals are split into 15 orders, one of which is the primates, a group of almost 200 species, including monkeys, apes, and humans. Primates are generally arboreal or terrestrial, quadrupedal, able to grasp, and very agile. They apparently evolved from tree-dwelling insect eaters resembling modern tree shrews. Today, there are two lines of monkeys, the Old and New World monkeys, and the line of **hominoids**—the gibbons, chimpanzees, other apes, and the humans.

Humans belong to their own subgroup, the Hominidae. The first member of this subgroup may have been Ramapithecus, whose fossilized teeth and jaw bones have been judged to be approximately 14 million years old. Three to four million years ago came the more human-like Australopithecines, whose several species walked upright, had small brains (compared with modern humans), and used crude tools. The evolution of the Hominidae is graphically presented in Figure 18-8. *Homo erectus*—the Java and Peking humans, or Pithecanthropines—appeared about 1.5 million years ago. They had larger brains, made fairly complex tools, used fire, hunted in groups, may have had speech and strong family organization, and spread widely across Africa and Asia.

The first members of *Homo sapiens* were the beetle-browed, large-brained Neanderthals, who disappeared about 40,000 years ago. Their successors, who may have evolved from them, were the Cro-Magnons. Physically indistinguishable from today's humans, they

411

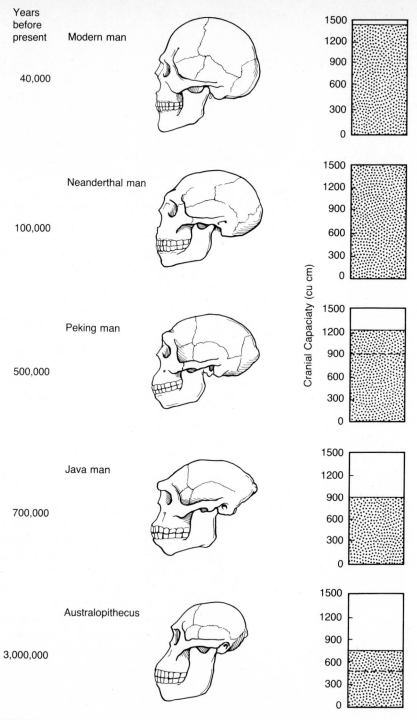

FIGURE 18–8

Human evolution as traced with fossil skulls and cranial capacities. (From Peking Man by N. Rukang, and L. Sheng Long, Copyright © 1983 by *Scientific American*, Inc. All rights reserved.)

shifted their way of life from hunting to agriculture about 10,000 years ago and spread widely around the world. Isolated regional groups developed their own idiosyncratic features and became the modern races of humanity.

CREATIONISM AND SOCIAL CONFLICT

The idea of evolution has been controversial ever since Darwin's day. The most reluctant to accept it have been those who take the Bible as their only source of truth, sometimes so literally that they believe Bishop Ussher's calculation from the Biblical begats that the earth was created at 9 A. M. on 23 October 4004 B. C.

In the United States for many years, these people were able to influence state legislatures, school boards, libraries, textbook publishers, and teachers to bar any discussion of evolution from both classrooms and textbooks. Their influence weakened after the famous Scopes trial of 1925 in Dayton, Tennessee. When Tennessee passed a law prohibiting the teaching of human evolution in public schools, John T. Scopes admitted that he taught the forbidden subject to his high school biology students. He was found guilty in court and fined, but the case was such an embarrassing spectacle that few states have wanted to ban evolution since then.

After 1925, the opposition to evolution was sporadic until the **creationist** revival of the 1960s, which was accompanied by a Creation Research Society (formed in 1963), a Creation-Science Research Center (1970), and an Institute for Creation Research (1972). The creationists claim to count some scientists among their numbers, but most of these people are physical scientists, not biologists.

The modern creationists do not seek to ban the teaching of evolution. Instead, they urge school boards and state legislatures to require that teachers and textbooks give equal attention to both creationism and evolution. They try to strengthen their case by calling their views "creation science" or "scientific creationism" and claiming that evolution is just as much religion as creationism, and creationism is just as much science as evolution. The Creation Research Society has even developed its own biology textbook.

One of the creationists' first great successes came when they convinced the California Board of Education to require textbooks to include creationism as an alternative to evolution. They also gained influence when Arkansas passed its Act 590, a law requiring "balanced treatment" for creationism and evolution. Excerpts from the Act are

413

reproduced in Figure 18-9. However, in 1981 the American Civil Liberties Union (ACLU) filed suit to invalidate Act 590, contending that it violated the separation of church and state and academic freedom; it was incredibly vague; and compliance with it was not possible. The presiding judge ruled in January 1982 that creationism is a religion, not a science, and that the Act was therefore unconstitutional. The state failed to show otherwise, and evolutionists carried the day. Some observers believe that creationists grossly underestimated the strength of scientists' convictions on this matter. Generally speaking, the public's understanding of the term *theory* does not parallel the scientist's use of the term. This contributes to some of the confusion between the thinking of scientists and lay persons. A scientific concept is elevated to a theory only after voluminous data that support the concept are already amassed and many contributing facts are available. A theory is not simply a guess.

Evolution *is* a scientific theory, and creationism *is not*. Evolution *is not* a religion, and creationism *is*. In addition, science is a lively activity, characterized by researchers doing experiments that test hypotheses. It is not *just* a body of facts. Creationism, on the other hand, has no problem-solving or research aspect at all. There is no part of it that is testable, for it is entirely a matter of miracles and faith. The debate is meaningless, and even creationists seem to know it. They spend most of their effort trying to discredit evolution. In fact, they become exuberant over the controversies among evolutionists, even though the

Section 1. Public schools within this State shall give balanced treatment to creation science and to evolution science. Balanced treatment to these two models shall be given in classroom lectures taken as a whole for each course.

Section 2. Treatment of either evolution science or creation science shall be limited to scientific evidences for each model and inferences from those scientific evidences, and must not include any religious instruction or references to religious writings.

Section 4. (a) "Creation science" means the scientific evidences for creation and inferences from those scientific evidences. Creation science includes . . . evidences and related inferences that indicate: (1) sudden creation of the universe, energy, and life from nothing; (2) the insufficiency of mutation and natural selection in bringing about development of all living kinds from a single organism; (3) changes only within fixed limits of originally created kinds of plants and animals; (4) separate ancestry for man and apes; (5) explanation of the earth's geology by catastrophism, including the occurrence of a worldwide flood; and (6) a relatively recent inception of the earth and living kinds.

FIGURE 18–9
Excerpts from Act 590, affecting the teaching of evolution and creationism in Arkansas.

controversies concern only the mechanisms, not the truth of evolution. Science is not rigid, and the theory of evolution continues to develop. The competing theories all fall within a certain framework, very much agreed upon by all evolutionists.

Creationists, however, tend to take things out of context. One of the so-called flaws of evolution that gives delight to creationists is the fossil record. They say the record holds no **transitional** forms, even though many fossils do show the transitions between differing organisms. *Archeopteryx*, shown in Figure 18-10, is a link between reptiles and birds—an excellent example of a transitional form. Furthermore, they say the record has too many gaps, even though the gaps are very well explained. Very few organisms have ever died under conditions that favored fossilization; many fossil-bearing sediments have been destroyed by erosion; many invertebrate organisms were not capable of fossilization; and paleontologists have yet to find many fossils. If the gaps are real, and not just artifacts of history, they may be due to the actual mode of the evolutionary process, for the concept of punctualism has been gaining support. Evolution may occur in large jumps, and there may be no real gaps. Finally, there is so much evidence for evolution from various disciplines that evolutionists think we hardly need a good

FIGURE 18-10
The fossil *Archeopteryx*, an extinct animal intermediate in form between dinosaurs and birds. (Courtesy of the Smithsonian Institution.)

fossil record. Gaps in the record cannot disprove evolution, despite the opinions of the creationists.

Most religions find no contradiction in evolution because they do not take the scriptures literally. Many people can reconcile their religion and evolutionary theory by believing that a Divine Creator created plants, animals, and humans by means of the evolutionary process. In their system of belief, evolution does not disprove the existence of a Creator.

Yet the issues will not go away. There are 40 million fundamentalists in America. While we respect these people's beliefs, we must insist that creationism should not be taught in public schools under the guise of science.

SUMMARY

Charles Darwin and Alfred Wallace had the essential insight that gave evolution a mechanism—natural selection—and made it one of the great themes of biology. Darwin is usually given most of the credit, for his 1859 book, *On the Origin of Species*, stated the case persuasively. His basic points were that organisms reproduce in excess, the progeny vary, a struggle for survival takes place, the fittest organisms tend to survive, and the traits that contributed to survival are passed on to the offspring. A classic example of natural selection is the changing of the population composition of dark and light moths in certain wooded areas as a result of bird predation in a polluted environment. When such changes are great enough, two diverging populations may eventually be unable to exchange genes. Such changes are called isolating mechanisms, and they can lead to speciation.

Mutation, chromosomal changes, and gene recombination produce genetic variability, and natural selection drives evolution in certain adaptive directions. In addition, some biologists believe adaptively neutral changes can be established in a population by chance. Microevolution refers to those changes that occur within populations and lead occasionally to speciation. Macroevolution refers to very large changes that distinguish major groups of organisms from each other. Gradualism is the concept of a slow, step-by-step evolution, while punctualism is the concept of evolution as big jumps alternating with slower gradual changes.

The evidence for evolution is now overwhelming. Such data have come from biogeography, taxonomy, embryology, physiology, geology, paleontology, genetics, and molecular biology. The origin of life is more difficult to explain, but several theories have been developed, and some data show the feasibility of abiogenesis occurring on the primitive earth.

Ever since Darwin's time, there have been religious and social controversies over evolution. Today, debates and lawsuits are both part of a climate of opposition between creationists and evolutionists over what should be taught in the public schools about our origins.

DISCUSSION QUESTIONS

1. Why is it wrong to credit Darwin with the theory of evolution? What did Darwin really give us?

2. Darwin and Wallace independently conceived and explained the idea of natural selection long before modern knowledge of genetics was available. List the basic tenets of natural selection as presented by Darwin, and then discuss what genetics has added to them in the past 50 years.

3. Although Darwin and Mendel worked at about the same time, Darwin was unaware of Mendel's accomplishments. Do you think his ideas would have been different if he had known of Mendel's work? If so, in what ways?

4. In what sense does evolution have a genetic basis?

5. *Genetic variability* exists when a population of organisms has many different genotypes. How are natural selection and genetic variability opposing forces in organic evolution?

6. Why are there so many species?

7. If we allow enough time, will the human species, *Homo sapiens*, eventually evolve into one or more new species? Why or why not?

8. Many people believe that the human population does not evolve as other species do. Others say that humans are just as subject to natural selection as the rest of the living world. Think this through, and discuss it as deeply as you can.

9. Give several good examples of natural selection in the human population and discuss them.

10. During the last century, "Mendelism" and "Darwinism" have entered the standard vocabulary of biology. Just what does "Darwinism" mean?

11. Old-timers often say, "Dogs do not chase cars as often as they did 50 years ago," or "Pheasants run more and fly up less when being hunted than they did 40 years ago." No matter whether they are right or wrong, to what concept are these old-timers referring?

417

12. What do you think is the fundamental unit of evolution? Explain.

13. Why do you think all organisms have the same genetic code? (See Chapter 4 to review the genetic code.)

14. What human traits do you think are due to biological evolution? What human traits do you think are due to cultural evolution?

15. Discuss the possibility that abiogenesis could occur again, just as evolutionists believe that it occurred once to begin organic evolution on earth.

16. Do you think that evolution and creationism should have equal time in the high school classroom? Defend your position.

PROBLEMS

1. A researcher puts 1,000,000 bacteria in a flask containing an appropriate liquid medium. She then adds an amount of antibiotic that would ordinarily kill all the bacteria. Only 10 bacteria survive this chemical challenge, and she then lets them grow into a second population of 1,000,000. Finally, she adds the same amount of antibiotic again.
   a. About how many bacteria will probably survive this second exposure to the antibiotic?
   b. Do these results provide evidence for the adaptation (resistance) to have taken place before the bacteria were exposed to the antibiotic or for the adaptation to have taken place after the bacteria were exposed to it?

2. Suppose that astronauts leave some fruit flies on another planet, where they develop into a viable population. Four centuries later, other astronauts return some of these flies to earth, where biologists find that they belong to a new species of fruit fly. In detail, how did the biologists probably test the flies for being a new species?

3. In 1955, H. B. D. Kettlewell released and recaptured both dark (*carbonaria*) and light (*typica*) moths in both polluted and unpolluted woodlands. Analyze the data below and state any conclusions that may be appropriate.

		typica	*carbonaria*
unpolluted woodland	released	393	407
	recaptured	54	19
polluted woodland	released	64	154
	recaptured	16	82

418

REFERENCES

Darwin, C. R. 1859. *On the origin of species by means of natural selection, or the preservation of favored races in the struggle for life.* London: John Murray.

Gallup, G. 1978. Gallup youth poll. *Riverside Press Enterprize*, Nov. 23 (cited by Moore, J.A., 1979. *American Biology Teacher* 41:544–547.)

Gurin, J. 1981. The creationist revival. *The Sciences* 21:16–19, 34.

Kettlewell, H. B. D. 1955. Selection experiments on industrial melanism in Lepidoptera. *Heredity* 9:323–342.

Kettlewell, H. B. D. 1956. Further selection experiments on industrial melanin in the Lepidoptera. *Heredity* 10:287–301.

Kettlewell, H. B. D. 1961. The phenomenon of industrial melanism in the Lepidoptera. *Annual Review of Entomology* 6:245–262.

Mayr, E. 1978. Evolution. *Scientific American* 239:46–55.

Miller, S. L. 1953. Production of amino acids under possible primitive earth conditions. *Science* 117:528–529.

Moore, J. N., and H. S. Slusher. 1970. *A search for order in complexity.* Grand Rapids, MI: Zondervan Publishing House.

Numbers, R. L. 1982. Creationism in 20th-century America. *Science* 218:538–544.

Raloff, J. 1982. Of God and Darwin. *Science News* 121:12–13.

Rukang, W., and L. Shenglong. 1983. Peking man. *Scientific American* 248:86–94.

Stanley, S. M. 1981. *The new evolutionary time table: Fossils, genes, and the origin of species.* New York: Basic Books Inc., Publishers.

Volpe, E. P. 1981. *Understanding evolution.* Dubuque, IA: Wm. C. Brown Group.

Wald, G. 1954. The origin of life. *Scientific American* 191:44–53.

Wells, H. G., J. S. Huxley, and G. P. Wells. 1970. The facts supporting evolution. In *Evolution of man*, ed. L. B. Young. New York: Oxford University Press.

Yunis, J. J., and O. Prakash. 1982. The origin of man: A chromosomal pictorial legacy. *Science* 215:1525–1530.

419

19

Genetic Engineering

*O*nly thirty years ago, we knew very little about the gene. Watson and Crick had just discovered the structure of DNA. The secrets of transcription and translation were undiscovered. In the years to follow, however, the genetic code was elucidated. Researchers next began to unravel some of the gene's control mechanisms and to better understand the nature of mutation and differentiation. By the 1970s, people were already recalling Aldous Huxley's *Brave New World*, with its mechanical wombs and chemical predestination. Were geneticists acquiring the power to play God? Some found such genetic progress exciting. Others became frightened and cautious of genetic advances.

What are the possibilities of the branch of biological technology we call **genetic engineering**? This branch of study includes the deliberate manipulation of the genetic material, repairing, modifying, deleting, adding, and exchanging the pieces of DNA called genes. The techniques are powerful, and they will certainly revolutionize biology.

GENETIC ENGINEERING METHODS

Many molecular biologists have concentrated on transferring segments of DNA from one organism to another. The process has four main steps:

1. The isolation, or synthesis, of the DNA segment or gene to be transferred;
2. The cloning or propagation of the DNA segment;
3. The transfer of the DNA segment to the host cell or organism; and
4. The stabilization of the DNA segment in its new surroundings.

Scientists have conquered the many problems associated with these tasks in work with one-celled bacteria. However, some technical difficulties must be overcome before this kind of technique can be routinely applied in humans and other higher forms of life.

DNA segments can be isolated and purified. Often, these segments comprise specific genes, and once their base sequences are known, they can be synthesized *in vitro*. Initially, the synthesis of a gene was so time-consuming that only very small genes could be assembled. H. G. Khorana and his colleagues first developed the procedures, synthesized a gene of 207 base pairs, and showed it to have the appropriate activity. Today, faster and more efficient techniques are

422

available, including computerized "gene machines" that automatically construct specified DNA segments. Researchers now have the biochemical tools for obtaining the particular DNA segments they need for their genetic engineering work.

Once a DNA segment has been isolated, it can be replicated to give the researcher enough copies for the additional genetic engineering steps. This replication or propagation is called **gene cloning**. It depends on **recombinant DNA** techniques, which allow DNA from various sources to be joined. Researchers often join DNA segments to specific bacterial DNA structures called **plasmids**. When they return the recombinant plasmids to the bacteria and allow the bacteria to multiply, the plasmids and their inserted passengers multiply too. Later, they can recover many copies of the specific DNA segment. We will consider the technology of recombinant DNA in more detail below.

In bacteria, plants, or animals, genetic engineering requires some way to transfer genes into cells. Current methods of gene transfer include:

1. Incubation of host cells with the purified DNA segments to be transferred;
2. Injection of the purified DNA directly into the nuclei of the host cells;
3. Injection of the whole chromosomes, in some cases, or their transfer by **cell fusion** techniques; and
4. The use of **vectors** to carry DNA segments into the host cells. The chief vectors are plasmids and viruses. Plasmids are small, circular, self-replicating DNA molecules often found in bacteria; they are not part of the regular bacterial chromosome. Figure 19-1 shows the position of plasmids in a bacterial cell.

FIGURE 19–1
Bacterial genes lie not only on their chromosome, but also on plasmids. Four such plasmids are shown here.

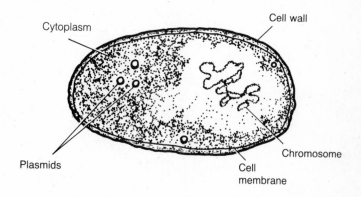

Cytoplasm

Cell wall

Plasmids

Cell membrane

Chromosome

423

The introduction and incorporation of foreign DNA into a recipient cell by exposing the cells to purified DNA and its subsequent expression in those cells is called **transformation**. The basic concept of transformation by this method is shown diagrammatically in Figure 19-2. This type of transformation event is usually rare. When a phenotypic change does occur in the host cells, it is often difficult to demonstrate that the foreign DNA has actually been incorporated into the host DNA. Transformation has been most successful with one-celled organisms or cell suspensions, although scattered success with injecting the DNA into whole higher organisms has been reported. Adding calcium phosphate to the mixture of host cells and donor DNA considerably improves the chances for successful transformation. With plant cells, it helps to remove the cell walls first. In a few cases, whole metaphase chromosomes have been effectively transferred from donor cells to host cells.

One can also inject donor DNA directly into host cell nuclei, using an extremely small pipette, or **micropipette**. Some cells will survive the treatment, and the efficiency of DNA transfer is usually better than

FIGURE 19-2
The basic mechanism of transformation through DNA-mediated gene transfer.

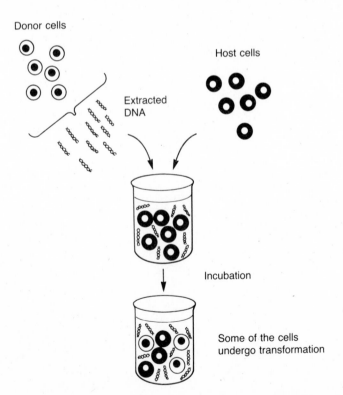

Donor cells

Host cells

Extracted DNA

Incubation

Some of the cells undergo transformation

the methods described above. This method has proved quite successful in mouse experiments.

Cell fusion techniques have provided still another way to transfer genes. The techniques now permit almost any two types of somatic cells, plant or animal, to be fused *in vitro*. The resulting hybrid cell contains the nuclei and the chromosomal materials of both parental cells. Generally, one set of chromosomes degrades in the cytoplasm. When the nuclei fuse, subsequent cell divisions will lose most of the chromosomes of one set. Nonetheless, one or more chromosomes of the lost set often remain in the hybrid cell; or in some cases, their fragments are integrated into the whole chromosomes of the surviving set. This sequence of events is illustrated in Figure 19-3. The technique has proved most useful in mapping genes to particular chromosomes. It has not yet provided any therapeutic applications in humans, but it is certainly promising enough to warrant further study.

Viruses can also be used to carry DNA from a donor cell to a host cell. Unlike the **virulent** viruses that kill the cells they infect, some

FIGURE 19–3
Cell fusion as a possible means of effecting gene transfer.

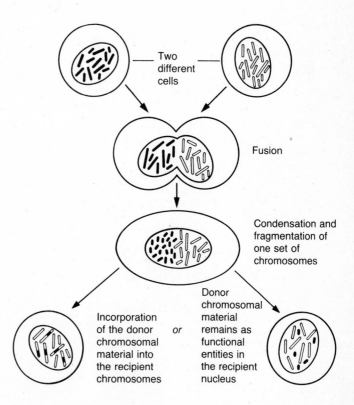

425

viruses can infect a host cell without killing it. All viruses insert their DNA into the host cell, where it causes the host cell to manufacture new virus particles. Ultimately, the new viruses are released by bursting or **lysing** the host cell. They can then infect other cells. Occasionally, viral DNA incorporates host-cell DNA during the infection. After lysis, the virus can carry the host DNA to a new host cell, as shown in Figure 19-4. If the virus does not kill the second host, it can give it a new functional gene. Molecular biologists have found such viruses useful for gene transfers.

The final requirement for successful genetic engineering is that the newly inserted gene or genes function in a stable manner. That is, they must be expressed properly. Many genetic engineering techniques allow little control over (1) what genes will be transferred, (2) where they will integrate in the host DNA, or (3) how they will behave after they are transferred. A gene may be expressed differently depending on where it is located on the host cell's chromosomes. This differential expression is the **position effect**. A second problem is that when a gene is placed in the host cell with all its control regions intact, the newly inserted control regions may regulate other adjacent host genes in an abnormal way. Gene control is still one of the substantial unknowns in this kind of work. The systems involved appear to be very complex and sensitive.

Recombinant DNA techniques used with bacterial plasmids offer so effective a means of gene transfer that they have become practically routine, in spite of the controversies surrounding them. To some extent, researchers have even overcome the instability problem. Other techniques are nearing this level of reliability, and we are gradually becoming more effective in placing genes into cells from diverse sources and making them work.

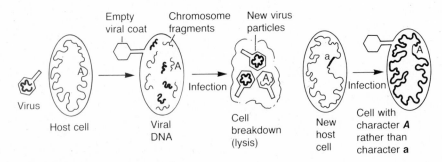

FIGURE 19–4
Gene transfer by viral infection of host cells.

THE BASICS OF RECOMBINANT DNA

Recombinant DNA techniques promise the revolutionary ability to move genes from any organism to any other organism and to synthesize new genes as desired. These techniques began to be developed in the early 1970s, when researchers discovered the enzymes known as **restriction endonucleases**. Restriction enzymes are produced by bacteria, from which they can be extracted for use *in vitro*. They cut DNA molecules into smaller pieces, but the divisions are not made at random. Each specific restriction enzyme cuts DNA wherever there is a particular symmetrical sequence of base pairs. Figure 19-5 shows the sequences recognized and cut by four different restriction enzymes. In each case, the broken ends of the DNA are "sticky" (chemically cohesive); they are short segments of single-stranded DNA which, because of base complementarity, can bind to each other or to the other ends generated when the same enzyme cuts a different DNA molecule at the same base sequence. For example, the restriction endonuclease ECO RI from the bacterium *Escherichia coli* recognizes and cuts DNA to leave the following ends:

The specific sequence cut by a restriction enzyme may occur in many places throughout the DNA of an organism; therefore, the DNA can be cut into many fragments. When DNA from two different sources is fragmented in this way and mixed together, the sticky ends can undergo complementary base pairing. Adding the **polynucleotide ligase** enzyme then seals the single-stranded nicks that remain between the joined ends. The different DNAs are thus spliced together as hybrid DNA molecules.

FIGURE 19-5
Several examples of restriction endonucleases that cut DNA molecules at points having specific base pair sequences.

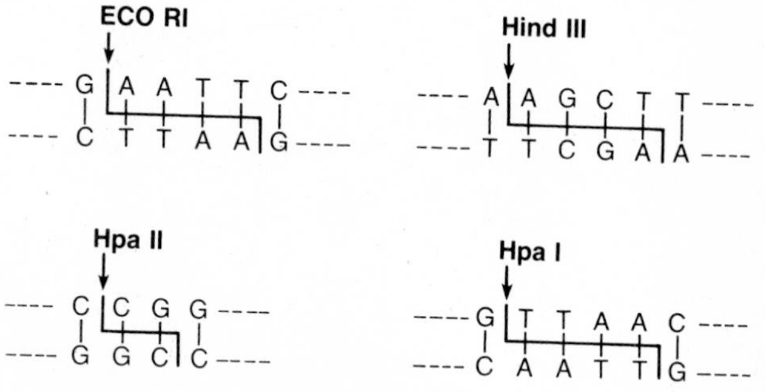

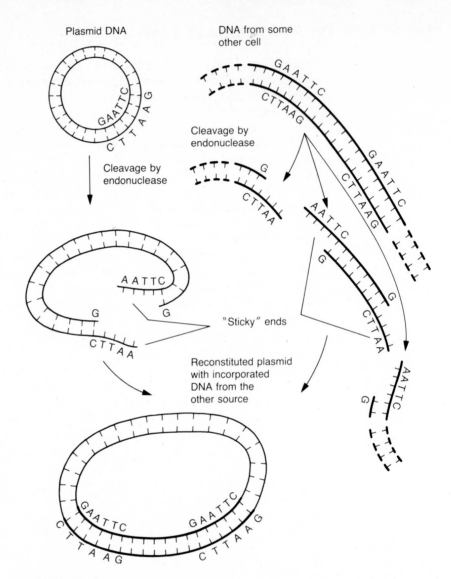

FIGURE 19-6
The formation of a hybrid DNA molecule.

Plasmids can be extracted from bacteria, hybridized by this splicing technique with DNA fragments from any other source, and introduced into bacterial cells with the aid of certain salts and a rapid temperature change. This process appears in Figure 19-6. As the transformed bacteria multiply, so does the plasmid and its inserted DNA sequence. In this way, researchers can synthesize thousands and even millions of copies of a particular DNA segment. The copies can then be extracted and purified for study. If the bacterium can be induced to synthesize the gene product, the DNA can be left in place as a true gene transplant. Such transplants of foreign genes into bacteria have created tremendous excitement in genetics, medicine, and business, for they suggest new ways to manufacture important substances. However, they have also stirred fears of disaster and prompted government regulation of recombinant DNA experiments.

REGULATING RECOMBINANT DNA RESEARCH

Almost as soon as it became possible to do recombinant DNA research, some of the scientists involved expressed their concern over possible hazards. Could it result in new infectious organisms and new plagues or epidemics? Could transplanting tumor-causing genes into a virus or bacterium produce an infectious cancer? Could modified organisms overrun whole ecosystems? No one could answer these questions. No one had any relevant experience. But most thought that the questions deserved serious consideration. Recombinant DNA work requires joining bits of DNA from different species, and even different phyla and kingdoms. The favorite organism for this work is *E. coli*, a bacterium that thrives in the human gut and elsewhere. Because of its intimate association with human beings, it was believed to be the perfect vehicle for an epidemic or plague. Some scientists even thought it might become a tool for terrorists.

Other questions addressed inhumane experiments on humans and the propriety of ''playing God.'' Scientists and others—philosophers, ethicists, lawyers, educators, and laymen—recognized that this was a time for scientific responsibility and ethics.

Feelings ran high from the start, for some people felt that the research should proceed unhampered by restrictions of any sort, while others argued for total prohibition. Interested scientists took up the issue at the 1973 Gordon Research Conference on nucleic acids and formed a committee to study the problems. Eventually, the committee issued an

429

international appeal for a moratorium on certain kinds of recombinant DNA experiments. In 1975, the now-famous Asilomar (California) Conference developed guidelines for conducting recombinant DNA research safely. The National Institutes of Health soon adopted the guidelines, with modifications, and used them to regulate recombinant DNA research. A summary of the regulatory requirements appears in Table 19-1. Congressional hearings were also held, but legislation on the matter has not yet appeared.

Although most scientists were willing to establish and abide by regulatory guidelines, a few spoke for complete prohibition of recombinant DNA research. Of course, the possible benefits of the research were stated over and over, while some scientists pointed out that recombinant DNA was as significant in the history of science as the splitting of the atom, that this was a tremendous opportunity to probe deeply into the unknown aspects of heredity, and that precipitous action would curtail basic research. A few scientists, including James Watson, argued that the whole issue of possible risks was exaggerated and that there was no need for alarm. During the debates, it was pointed out that evolution's way of sorting out genes without human intervention had given us bubonic plague, yellow fever, polio, cancer, and a host of other devastating problems. Scientists acknowledged the international scope of the situation, admitting that the experiments would ultimately be accomplished with or without controls, in one country or another.

Some scientists took advantage of this lively debate to point out the need for a scientifically informed public. In fact, the debate was somewhat sensationalized, attracting members of society as both spec-

TABLE 19-1

Regulatory requirements for recombinant DNA research.

Risk Level	Laboratory Access	Safety Measures
P1 Minimal	Free	Standard laboratory practice
P2 Low	Limited	Autoclaving (sterilization) of equipment and materials
P3 Moderate	Restricted	Lab specially engineered for containment of organisms used; air locks
P4 High	Tightly restricted	Special lab and use of organisms unable to survive outside lab

tators and participants. The participating laymen joined the fray in Cambridge, Massachusetts, home of Harvard University and the Massachusetts Institute of Technology (M.I.T.). In the summer of 1976, the Cambridge City Council placed a three-month moratorium on recombinant DNA research within the city. By January 1977, a newly organized Cambridge Experimentation Review Board had drafted recommendations for the control of recombinant DNA research in Cambridge. The recommendations included compliance with the NIH guidelines and additional restrictions. A few other cities soon adopted or considered similar measures. Local politics had entered the discussion.

Scientists voluntarily imposed a moratorium on their own recombinant DNA research—an initiative unprecedented in the history of science. However, it alerted government and society in general to the possible implications of recombinant DNA work. Many scientists now feel they created a "monster." The biohazards of recombinant DNA now seem much less than they did at first. The NIH guidelines have been relaxed recently, but they are still in effect, perhaps in an effort to prevent the enactment of new laws. Some people, both inside and outside of science, do not want to see the regulations relaxed too much. They believe it is still too early to be sure that recombinant DNA experiments are safe. Undoubtedly, the arguments will continue.

APPLICATIONS OF GENETIC ENGINEERING

Genetic engineers have already performed a number of gene transplants, from rats and toads to *E. coli*, from plants to yeast, beans to sunflowers, rabbits to monkeys and mice, humans to rats, and more. The genes have coded for insulin, nitrogen fixation enzymes, seed storage proteins, growth hormone, and so on. And in many cases, the genes have been expressed in their new hosts. When the gene for rat growth hormone was transferred into the fertilized eggs of mice, the recipient mice grew an average of 50 percent larger than normal. Some genes have worked well enough in bacteria that in 1982, human insulin produced by bacteria went on sale to the public.

There are numerous success stories. The chemicals synthesized through genetic engineering techniques and now undergoing clinical and animal tests are human **somatotropin**, human albumin, bovine growth hormone, and a vaccine for foot and mouth disease in cattle. Human insulin, under the name Humulin, was the first product generated by recombinant DNA techniques to reach the market. Somatotropin is a

431

human growth hormone that maintains its biological activity even when it is synthesized by engineered *E. coli*. Figure 19-7 shows the key steps in the process of forming substances by means of engineered bacteria.

Interferon is an antiviral substance normally secreted by human white blood cells. Scientists hoped that this substance would prove effective against cancer. The gene for one type of human interferon has been totally synthesized. In another experimental approach, the DNA sequence for an interferon has been linked to a bacterial plasmid that functions in both bacteria and yeast. Yeast cells with the plasmid can synthesize up to a million molecules of active interferon apiece. One liter of bacterial culture can produce 20–40 milligrams of interferon. Conventional methods use approximately 20,000 pints of blood to yield as much. Before genetic engineering, the annual output of interferon by a

FIGURE 19–7
An overall view of the scheme for obtaining a desired substance by using recombinant DNA procedures and bacteria.

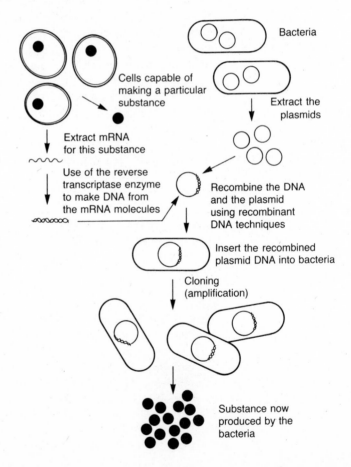

major supplier could treat about 600 persons per year at a cost of roughly $10,000 each. Because the chemical is very difficult to harvest from blood cells, at one time the cost of interferon was $22 billion per pound. Soon the cost may be as low as $10 per treatment. The new availability of interferon has allowed scientists to test its effectiveness adequately for the first time. Unfortunately, the anticancer power of interferon seems limited, but it is useful against viral infections, such as the common cold.

In a similar way, genetic engineering has improved the prospects for obtaining somatostatin. Approximately 500,000 sheep brains used to be necessary to supply only 5 milligrams of the hormone. Now it takes only two gallons of engineered bacteria culture.

Extraordinary social benefits could follow from the genetically engineered production of large volumes of antibodies, hormones, a more effective polio vaccine, a vaccine for herpes, enzymes, food supplements, flavorings, pharmaceuticals, and numerous other specialty chemicals. Genetic engineering has already been used to detect certain genetic disorders in humans, although this application is still in the early stages of development. Successes have been reported, but mostly *in vitro*. Quite some time ago, for example, cells from a patient with galactosemia were changed to normally functioning cells by virus-mediated gene transfer. In other words, the deficient cells were given the normal gene and as a result they developed the ability to make the appropriate enzyme. Obviously, curing a human being of a genetic disorder will be a much more formidable task. The gene transfer will have to take place in many cells, or in a whole organ, or even in the entire body. Such complex applications of genetic engineering must await major breakthroughs. So far, genetic engineering is most effective in terms of combating genetic disorders when it is used in prenatal diagnoses of certain blood conditions, such as sickle cell anemia and thalassemia. These disorders are identified by using restriction enzymes in the analysis of the fetal DNA.

Other applications may be generated by the behavioral scientists, who are developing methods of using gene-splicing techniques to isolate genes for behavioral anomalies. Still other possible uses loom in agriculture, affecting both plants and animals. Applications to plants are coming more slowly than those to animals, but genetic engineering in that area is gaining momentum. Plant research now includes tissue culturing, cell fusion techniques, and recombinant DNA technology. Investigators are carrying out intensive work on seed proteins, nitrogen fixation, photosynthesis, pest resistance, pathogen resistance, and stress tolerance. The major limiting factors are the lack of good vector systems

433

and the complexity of genetic systems in higher plants. Some of the more cautious scientists feel that the promises of genetic engineering in agriculture have been oversold; still other indications are that agricultural applications are moving faster than expected.

The ideas and the discoveries keep coming steadily. In the making are projects to develop bacteria that degrade oil spills, separate metal from low-grade ore, and produce industrial chemicals. How many of the ideas will be feasible remains to be seen, but one thing is certain: by means of genetic engineering research, scientists are learning exactly how genes work at the molecular level. Gene regulation is still one of biology's "black boxes," but our knowledge of this mystery is growing rapidly as a result of the new capabilities of molecular biology.

INDUSTRY, MARKETS, AND PATENTS

Biotechnology is the area where biology and industry meet. Its workers produce prosthetic devices, pharmaceutical products, industrial chemicals, and more. Today, the most active branch of biotechnology is genetic engineering, which promises entirely new products as well as new ways of obtaining old products. Consequently, genetic engineering has become extremely appealing to investors. In the last few years, more than 150 companies have been founded to deal with the biotechnology of recombinant DNA. Cetus Corporation, Genentech, Inc., Biogen, Molecular Genetics, Inc., Genex, and Zoecon have all been prominent in the daily news. Genentech, for instance, produced and marketed the first bacteria-grown vaccine, developed against foot and mouth disease. This company is one of the biggest of its kind, and it now produces a dozen or more genetically engineered products.

Scientists, investors, and business people are actively seeking profitable products that can be developed by means of recombinant DNA techniques. However, after the initial flurry of rapid development, investors became more cautious. Some observers believe the business world greeted recombinant DNA and its industrial applications too enthusiastically. Such skeptics believe that many of the new companies will eventually go out of business. And indeed, some already have. Yet for other companies, the future appears to be promising.

Most of the early excitement centered on the production of therapeutic agents such as insulin, somatostatin, and interferon, among others. Scientists have also been interested in engineering bacteria to digest pollutants, perform particular industrial fermentations, separate

434

metal from ore, and degrade various other substances. More recently, interest has focused on agriculture.

Heading the list of new problems stemming from biotechnology is the question of awarding patents on genetically engineered organisms. The problem of patent awards first arose when microbiologist Ananda Chakrabarty filed for a patent on a bacterium capable of digesting the components of crude oil. Without using recombinant DNA techniques, he supplied the *Pseudomonas* bacterium with plasmids bearing genes for oil-degrading enzymes. Each enzyme can degrade one component of crude oil, and each one is found in a different bacterium. Chakrabarty combined them all in his special variety of *Pseudomonas*. His patent request was initially denied, with the arguments going all the way to the Supreme Court. The government argued that living things are products of nature and thus are not subject to patents. But in a landmark decision on 16 June 1980, the Supreme Court ruled by 5 to 4 that human-engineered organisms are patentable under federal patent law. The controversy continues, but several universities now hold patents on biomedical products developed through genetic engineering, and many more applications for patents have been submitted since the Supreme Court decision in 1980.

Another situation causing concern arises because the appeal of genetic engineering to investors has increased the interactions of scientists, universities, and industry. Molecular biology has become a business, and some scientists have become entrepreneurs, executives, and consultants for biotechnological companies. Some have left academia for jobs in industry, and others are collaborating with industrial scientists. A few companies have provided university researchers and departments with large sums of money in return for exclusive rights to discoveries.

Will such university–industry arrangements have a detrimental or a beneficial influence on academic research? As always, observers disagree. Critics point out that strong ties to business may compromise academic freedom, independence, and objectivity. Research and training priorities may change, the flow of information may be curtailed, and researchers may become reluctant to exchange substances and organisms. Many good scientists are already leaving universities to enter the business world, and in the future we may face a severe shortage of capable researchers and teachers.

On the other hand, there is a great deal of optimism being expressed in academia, industry, and by interested laymen. Some feel that the new professional ties will be good for academia, research, industry, and the economy in general. Many companies have donated vast amounts of money to support university research, sometimes in exchange for collaboration or product rights, but often without strings. Some people view the ties as a way to accelerate our progress in molecular biology. They are not concerned about companies profiting from academic research, for, they say, industry performs an essential role in turning research successes into products useful to society.

One conclusion seems inescapable: the new arrangements between universities and industry will bring about changes. Quite possibly, we may be approaching a whole new way of doing research, both basic and applied.

"PLAYING GOD"?

Some people fear genetic engineering. They feel that the genetic engineers are "playing God," and that their efforts will soon make it possible for parents to design their children or governments to design their citizens. They look toward a "quality controlled" society, where the popularity of certain genes and the unpopularity of others may blot out the population's essential variability.

Various religious groups are sensitive to advances in human biology, and genetic engineering troubles many theologians. They—and others—are opposed to intervening in human evolution with tailored genes, and we hear much concern about whether gene splicers have appropriate ethics.

Yet many people feel that there is entirely too much gloom surrounding genetic engineering, and that much of the fear is unjustified. They believe that we hear too much about legless humans for space travel, humans with gills for underwater activities, and humans with one pair of hands for heavy work and a second pair for more delicate work. They say that genetic engineering offers us great hope that outweighs the philosophical objections, and that we must proceed with it. Advocates say that genetic engineering is one expression of the essential creativity of humans, and considering the pain and suffering that exist in the world, it is time for us to explore this means of enriching people's lives.

Certainly scientists must do the work carefully and cautiously, and keep society informed at every step. Special problems are bound to arise, for scientists still know too little about how genes interact, and the necessary new bioethics still has to develop. There are no precedents, and society will have to show sound judgment. Genetic engineering is just beginning, but it has a strong and promising future.

SUMMARY

Genetic engineering means adding, deleting, or exchanging genes to alter an organism's heredity. It requires the isolation or synthesis of a gene, the cloning of a gene, the transfer of a gene into the target cells, and the proper function of the gene in its new setting. None of these steps is insurmountable, although the key operation, gene transfer, is more difficult in higher forms of life than in bacteria. The various gene transfer methods include incubation in DNA, microinjection of DNA, cell fusion, and the use of vectors (vehicles such as viruses and plasmids for carrying the foreign DNA).

The powerful recombinant DNA technique allows DNA from different sources to be combined with the use of restriction enzymes that cut DNA at specific sites and leave "sticky ends." DNA recombined with plasmid DNA can be placed in bacteria, and sometimes the foreign genes will function in the bacterial cell. Fears that such recombinant organisms could present hazards have led to various precautionary regulations for recombinant DNA work. Many members of the scientific community now minimize the dangers, but the regulations remain in effect.

A number of mammalian genes have been successfully placed in bacteria, and in a few cases genes have been transferred from one mammal to another. Human cells have also been transformed in various ways, at least *in vitro*. Most of the contemplated applications of genetic engineering involve using engineered bacteria to produce specialty chemicals and degrade pollutants and other compounds, developing human gene therapy, and improving domestic plants and animals. Because of the commercial possibilities, industry has become very involved in molecular biology and numerous genetic engineering firms have been founded. However, many of the projects are long term, and only a few products have reached the marketplace so far.

The biotechnological era has definitely arrived, and assuredly there will be benefits. But society may also have new problems to face as science moves closer and closer to being able to control human destiny.

DISCUSSION QUESTIONS

1. Why does "recombinant DNA" describe genetic engineering technology well?

2. How could one claim that genetic engineering had been performed as early as 1928? 1944? (Review Chapter 2 if necessary.)

3. What is the difference between "genetic technology" and bona fide "genetic engineering"?

4. Which of the following constitute true genetic engineering as we have defined it? (a) cloning; (b) in vitro fertilization; (c) artificial insemination; (d) embryo transfer from one organism to another; (e) cell fusion; (f) transformation; and (g) recombinant DNA technology.

5. Someday it may be possible to surgically remove a person's genetically defective organ, such as the pancreas, transfer nondefective genes into most of the organ's cells, and implant the organ back into the person, with the new genes properly functioning. Would this accomplishment be true genetic engineering as we have defined it? Would you consider it a complete "cure" for the individual? Can you think of any genetic consequences?

6. Many scientists are involved in "sequencing" DNA—determining the order of its nucleotides. What benefits might such a laborious task have?

7. Some scientists have advocated sequencing all the DNA of the human species. Would it matter whose DNA we sequenced? If so, whose DNA should we sequence?

8. Briefly, why might genetic engineering have its greatest impact on agriculture, rather than upon human genetic therapy?

9. What do you think of the statement "knowledge is always good"?

10. To what extent should the public be involved in determining policy related to genetic engineering?

11. What role should legislators play in setting genetic engineering policy?

12. Is regulating recombinant DNA research a matter of society interfering with scientific freedom?

13. A molecular biologist with both a Ph.D. and an M.D. was reprimanded for doing recombinant DNA experiments on human subjects without the full approval of his superiors at his university. He claimed to have acted as a physician, not just as a researcher, and therefore to have had every right to do the experiments. Do you agree with this argument? Why or why not?

14. What steps should be taken, if any, to encourage the public to be better informed about genetic engineering?

15. Describe and explain any connection that might exist between genetic engineering and human evolution.

16. Is genetic engineering "playing God" or "playing human"?

17. Should scientists be able to patent their genetically engineered organisms?

18. If you had an *extra* $10,000 that you did not need for any specific purpose, how much of it would you be willing to invest in the stocks of a genetic engineering company? Explain.

REFERENCES

Edge, M. D., A. R. Greene, C. R. Heathcliffe, P. A. Meacock, W. Schuch, D. B. Scanlon, T. A. Atkinson, C. R. Newton, and A. F. Markham. 1981. Total synthesis of a human leukocyte interferon gene. *Nature* 292:756–762.

Grobstein, C. 1979. *A double image of the double helix: The recombinant-DNA debate*. San Francisco: W. H. Freeman & Co. Publishers.

Hitzeman, R. A., F. E. Hagie, J. L. Levine, D. V. Goedell, G. Ammerer, and B. D. Hall. 1981. Expression of a human gene for interferons in yeast. *Nature* 293:717–722.

Huxley, A. 1932. *Brave new world*. New York: Harper & Row, Publishers Inc.

Khorana, H. G. 1979. Total synthesis of a gene. *Science* 203:614–625.

Merril, C. R., M. R. Geier, and J. C. Petricciani. 1971. Bacterial virus gene expression in human cells. *Nature* 233:398–400.

Mertens, T. R. 1975. *Human genetics: Readings on the implications of genetic engineering*. New York: John Wiley & Sons Inc.

Olson, K. C., J. F. Fenno, N. Lin, R. N. Harkins, C. Snider, W. H. Kohr, N. J. Ross, D. Fodge, G. Prender, and N. Stebbing. 1981. Purified human growth hormone from *E. coli* is biologically active. *Nature* 293:408–411.

Palmiter, R. D., R. L. Brinster, R. E. Hammer, M. E. Trumbauer, M. G. Rosenfeld, N. C. Birnberg, and R. M. Evans. 1982. Dramatic growth of mice that develop from eggs microinjected with metallothionein growth hormone fusion genes. *Nature* 300:611–615.

20
Agriculture and Applied Genetics

In 1970, plant breeder Norman E. Borlaug, shown in Figure 20-1, won the Nobel Peace Prize for his role in initiating the "Green Revolution." He was instrumental in breeding high-yield wheat strains for developing countries. These strains have shorter and stronger stems and are much better adapted to the environments of the countries for which they were bred. The conformations of these wheat varieties are compared in Figure 20-2. Given plentiful fertilizer, these plants produce dramatically increased wheat yields. Using them, Mexico changed from being an importer of wheat to one of the world's largest wheat exporters. India increased its wheat yield an unbelievable 700 percent. Other countries have also used the semi-dwarf wheat with success.

These new varieties represent a tremendous agricultural achievement and a great contribution to world food supplies. Because food is essential for both economic growth and political stability, these varieties have aided the cause of world peace. Since 1970, Borlaug has irritated many environmentalists with his controversial opinions about the use of land and forests, but his contributions to developing countries have certainly illuminated the important role of the agricultural scientist in the world today.

Genetics has contributed immensely to the increases in crop yields that have occurred over the years, summarized in Table 20-1,

FIGURE 20–1
Norman Borlaug received the Nobel Peace Prize in 1970 for improving wheat and increasing food production in third-world countries. (Photograph courtesy of the Rockefeller Foundation.)

FIGURE 20–2
Bundles of different varieties of wheat displayed to show their large variation in height. (U.S. Congress, Office of Technology Assessment, *Impacts of Applied Genetics: Micro-organisms, Plants, and Animals*, GPO stock no. 052-003-00805-0. Washington, DC: U.S. Government Printing Office, April 1981.)

but the magnitude of its contribution is hard to estimate. Other factors have also been involved in the increases. Farm management, pest controls, fertilizers, and better farming techniques may be responsible for half of all the increases in yield over the past 50 years, with the other half due to genetics. Geneticists alone cannot feed the world. This endless task requires the efforts of plant and animal breeders and other agriculturalists as well.

Yet the Green Revolution had limitations, and it fell short of its promises. People concerned about the environment criticize the occasional ecological carelessness that accompanied the Green Revolution. They point, for example, to the serious erosion caused by some overzealous agricultural practices. They further assert that many agricultural scientists do not give appropriate attention to the nutritive quality of crops, focusing instead on yield, color, shelf life, time of ripening, and how well the product can withstand shipment. One of the most crucial concerns is that current agricultural practices greatly restrict the

443

TABLE 20-1

Average crop yields per acre in the United States for the years of 1930 and 1975

Crop	1930	1975	Unit	Increase (%)
Wheat	14.2	30.6	Bushels	115
Rye	12.4	22.0	Bushels	77
Rice	46.5	101.0	Bushels	117
Corn	20.5	86.2	Bushels	320
Oats	32.0	48.1	Bushels	50
Barley	23.8	44.0	Bushels	85
Grain sorghum	10.7	49.0	Bushels	358
Cotton	157.1	453.0	Pounds	188
Sugar beets	11.9	19.3	Tons	62
Sugar cane	15.5	37.4	Tons	141
Tobacco	775.9	2011.0	Pounds	159
Peanuts	649.9	2565.0	Pounds	295
Soybeans	13.4	28.4	Bushels	112
Snap beans	27.9	37.0	Cwt	33
Potatoes	61.0	251.0	Cwt	129
Onions	159.0	306.0	Cwt	92
Tomatoes:				
(Fresh market)	61.0	166.0	Cwt	172
(Processing)	4.3	22.1	Tons	413
Hops	1202.0	1742.0	Pounds	45

Source: The National Plant Genetic Resources Board, U. S. Department of Agriculture. 1978. *Plant genetic resources. Conservation and use.* Washington, DC: U. S. Government Printing Office.

genetic pool; that is, too many farmers grow the same varieties of crop plants. The resulting lack of diversity can cause serious problems, as we will see later in this chapter.

Although some countries have been able to control population increases, many have not. World population is now growing by several hundred thousand persons each day. The United Nations has identified 43 countries with serious food deficits and inadequate diets. Many persons close to the agricultural scene have predicted severe world food shortages in the 1990s.

Society and its agricultural scientists must plan for the future. Most of these scientists believe that the genetic limits of crop improvement have not been reached. Classical breeding has already been very successful in increasing plant yields, insect and disease resistance, and in some cases nutritive value. The same techniques can and will do more. Using such revolutionary techniques as genetic engineering may add a whole new dimension to agriculture.

This chapter focuses on the applications of genetics to problems of plant production. Animal foods are also vital, but for most people of the world—and their domestic animals—plants are the basic source of food.

TRADITIONAL PLANT BREEDING

Crop plants were first domesticated and bred by prehistoric people, who selected plants with the most desirable traits from the population of wild plants. Today's plant breeders, too, are constantly seeking improved plant varieties. With the major crop plants, they need only an average of ten years to produce a new and satisfactory variety.

Plant breeders exploit the reproductive and other biological characteristics of plant species. In one basic technique, they cross different strains of the species they are attempting to improve. They then produce F_2 progeny by self-fertilization to create new combinations of genes. Using further generations produced by self-fertilization, they stabilize the many resultant lines among the F_2 progeny so they breed true. Finally, the breeders select the progeny lines whose gene combinations they judge to be best for one reason or another and test them, usually in several different environments or locations in the country or world. Favorable test results lead to the release of the new variety to farmers. The main steps in this process appear in Figure 20-3.

Despite their past successes, plant breeders cannot be complacent. Pathogens and insects are constantly evolving and as a result they attack crops in new ways. Consequently, plant breeders must continually develop new varieties resistant to natural threats. The breeders are involved in two races, one against the natural enemies of crops, and another against world food shortage. In addition to resistance and yield, they must seek improved nutritive quality.

Most breeders believe that even better results can be achieved. Some think we should make concerted efforts to select plants that are better adapted to environments that are considered unsuitable for present agricultural uses. Such schemes might be very successful, for artificial selection can be a powerful influence on the development of organisms. It can even adapt plants to salty water; California researchers already have a variety of oats that can be irrigated with seawater.

445

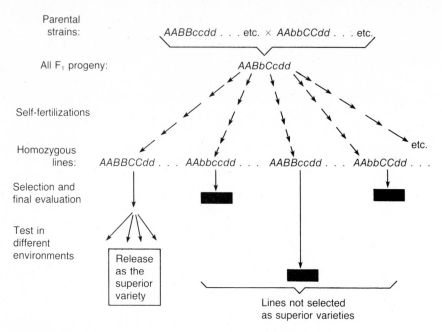

Parental strains:

All F₁ progeny:

Self-fertilizations

Homozygous lines:

Selection and final evaluation

Test in different environments

FIGURE 20–3
One simple method of breeding for new crop varieties.

Hybrid Corn

Corn is the world's third most important crop. Together with wheat, rice, and sorghum, it supplies more than half of the world's calories. This is partly due to the acreage planted and partly to enormous improvements in yield—from 20 bushels per acre in the United States at the beginning of this century to 35 to 40 bushels per acre in 1950 to 109 bushels per acre in 1981. This progression is shown graphically in Figure 20-4. This spectacular increase in yield resulted from new planting methods, use of fertilizers, better equipment, the introduction of herbicides, and the development and wide use of **hybrid corn**. The success of hybrid corn prompted the hybridization of many other domestic plants.

The introduction of high-yield hybrid corn was a major event in the agriculture of the United States. During World War II, corn was essential for food, alcohol production, and other materials needed for the war effort. Corn remained just as important during the post-war years, both in the United States and in much of the war-torn world.

446

FIGURE 20–4
Corn yields during the era of hybrids up to 1981. (From D.N. Duvick. 1984. Genetic diversity in major farm crops on the farm and in reserve. *Economic Botany*, 38: 161–178. Used with permission.)

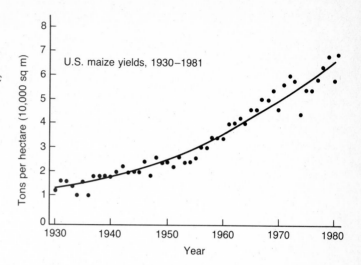

The term hybrid means different things to different people. To many geneticists, a hybrid is the progeny of a cross between two genetically different parents. In the simplest case, the difference exists at one or only a few gene pairs, as in:

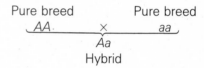

Many plant breeders think of hybrids as resulting from a cross between two different **strains**. Strains are usually morphologically different from each other in one or more ways, but they do belong to the same species. Some people, including many evolutionary biologists, reserve the term hybrid for the progeny of a cross between parents of two different species, which occurs only rarely in natural environments. Hybrid corn is the result of crossing different strains, all of the same species, *Zea mays*.

At first, much hybrid corn was produced by the **double cross** method described by D. F. Jones in 1918 and shown in Figure 20-5. In the first step, **inbred** (homozygous) strains are developed by self-fertilization over many generations. The inbred strains are genetically uniform and can produce only one kind of gamete. Theoretically, they contain absolutely no variation, unless a mutation occurs.

447

FIGURE 20-5

A diagrammatic explanation of producing hybrid corn by the double-cross method. Inbred lines *A* and *B* are crossed to obtain the hybrid *AB*, and inbreds *C* and *D* are crossed to obtain the hybrid *CD*. The progeny from the single crosses are then intercrossed to obtain the double-cross hybrid, *ABCD*. (Reprinted with permission of Macmillan Publishing Company from *Genetics* by Monroe W. Strickberger. Copyright © 1976 by Monroe W. Strickberger.)

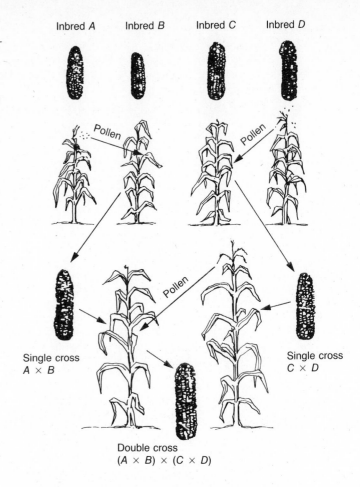

Inbred *A* Inbred *B* Inbred *C* Inbred *D*

Pollen   Pollen

Single cross
A × *B*

Pollen

Single cross
C × *D*

Double cross
(*A* × *B*) × (*C* × *D*)

homozygous inbred strain: *AAbbCCddEEFFgghh* . . . etc.
gametes: all*AbCdEFgh* . . . etc.

Superior inbred strains, or those with certain advantageous traits, are then crossed to give single hybrids. The *double cross* arises when, after stabilization, two different single hybrids are crossed with each other, or **intercrossed**. Such crossing is also called **outbreeding**. In the commercial production of hybrid corn seed, plant breeders carefully **cross-fertilize** their single hybrids by transferring pollen from the tassels of one hybrid to the ear-silk of the other.

Today, the double-cross method produces about 20 percent of hybrid corn. Many variations, such as single crosses and three-way crosses between single crosses and inbreds, produce most of the rest. Single crosses yield hybrid plants with tremendous uniformity. However, the seeds harvested from the hybrid plants produce highly variable progeny, due to the many gene combinations that can segregate from the highly heterozygous hybrid plant. Farmers must therefore buy more seed from the commercial source each year, rather than using some of their own harvested seed from the previous season.

The greater yields of hybrids are due to **hybrid vigor** or **heterosis**, an increase in vigor and growth capacity often seen in crossed plants and animals. The precise genetic mechanism underlying heterosis has not been defined, but we do know that the combinations of genes in crosses often do better than the combinations in inbred strains. This phenomenon is known as **combining ability**.

The Loss of Genetic Variability

In 1970, an epidemic of **corn leaf blight**—*Helminthosporium maydis*, as shown in Figure 20-6—struck the corn crop in the United States. Before that year, most strains of corn had been quite resistant to this disease. However, the fungus had apparently mutated enough to make the heavily used strains of corn vulnerable. The blight destroyed

FIGURE 20–6
Southern corn leaf blight (*Helminthosporium maydis*). (Photo courtesy of A. J. Ullstrup.)

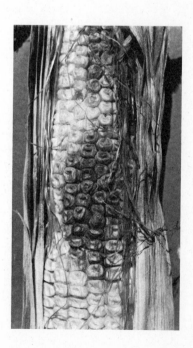

449

15 percent of the nation's overall corn crop, and half of the crop in some heavily infested areas.

Fortunately, because the United States has vast resources to buffer shortages, the blight caused no great hardship. However, even slight reductions in crop yields can be devastating in other countries, and history records numerous famines resulting from crop loss. Probably the most famous—or infamous—is the Irish potato famine of 1845–1848, when a fungus destroyed the potato crop and one million Irish people died of starvation or directly related diseases. Another million-and-a-half Irish people escaped that fate by emigrating to other countries. Such consequences of crop loss are staggering.

The causes of the 1970 corn loss in the United States, the 1845–1848 Irish potato famine, and the high losses attributable to **stem rust** of United States wheat in 1916, 1953, and 1954 are fundamentally the same. In each case, the basic problem was a lack of genetic variability in the crop. Too many of the nation's farmers had planted varieties genetically similar to each other, and when a disease developed to which they were not resistant, the nation suffered a massive crop loss.

In 1970, 80 to 90 percent of the United States' corn crop was varieties with Texas male-sterility cytoplasm, a trait that greatly eases the labor of hybridizing. These varieties were excellent yielders, and the seed was economical. Farmers preferred them, but they were susceptible to the new race of corn blight. Because of this vulnerability, these varieties brought us to the brink of a major crisis. United States' science, industry, and agriculture deserve credit for the rapid recovery from the disaster, but the corn blight mutant also caused some long-term changes. The country is now less fond of genetic uniformity, and new corn varieties are available. Nevertheless, many people worry that the changes may not be enough to avert future agricultural disasters.

Genetic uniformity remains a worldwide problem. Nine-tenths of the world's total calories come from only 30 crops. Only six varieties make up 71 percent of the corn crop in the United States; only four varieties make up 65 percent of the rice crop; and only nine varieties make up 50 percent of the wheat crop. Many geneticists and agriculturalists are concerned, and some are frankly alarmed.

We can see a closely related problem in the clearing of tropical jungles for farming. Massive land clearing threatens and even eliminates many species of wild plants and animals. Once lost, these gene combinations, or **germ plasm**, cannot be replaced. The loss of this material is especially frightening because tropical areas have produced most of the world's genetic diversity and may contain many biological characteristics that humans could find useful, including genes for resistance es-

450

sential to future agriculture. Although it took nature eons to select those successful gene combinations, we can eliminate them almost overnight. When we do, we replace the rich wildlife with a handful of domestic species that are all very similar to each other. We become vulnerable, and we deprive ourselves of an invaluable genetic resource.

There is a clear need for germ plasm preservation, but meeting that need is a difficult task complicated by policies, politics, lack of funds, and international problems. Periodically planting and harvesting the seeds of many different strains to maintain the resource, and then properly storing them in cool, dry facilities is a massive, expensive undertaking. In addition, there has not been a great deal of public concern about maintaining genetic diversity, especially while we continue to have bumper crops. However, knowledgeable persons know that things can change very quickly. With this in mind, scientists are taking some steps to preserve germ plasm. Even commercial companies are beginning to aid in this important effort. Yet we must realize that while genetic diversity is essential for human and wildlife survival, by itself it is no guarantee against disaster.

UNCONVENTIONAL BREEDING

Technology has brought artificial insemination, embryo transfer, and *in vitro* fertilization to the aid of animal breeders. The same techniques cannot be used with plants, but there are ways to reproduce plants *in vitro*. Researchers have regenerated several plant species from single cells in laboratory culture. The first major success came in 1958, with the growth of a whole carrot plant from a single root cell (this was one of the first demonstrations of the phenomenon of **totipotency**). The process, which depends on manipulating plant hormones and nutrients in a synthetic medium, is now almost routine for tobacco, potatoes, rice, sugar cane, orchids, and other plants. Before long, it should prove useful with other major crop plants.

The ability to grow plant cells and tissues *in vitro* makes it possible to harvest special plant products—pharmaceuticals, pesticide components, flavors, and aromatics—from cells cultured in large vats or tanks. Researchers need only coax the cells into expressing the appropriate genes. They can then extract and purify the desired material from the vats, with great savings in time and space. However, several barriers remain to be overcome before the process will be practical for widespread use.

451

Tissue culture techniques also permit the production of **plant-lets** (young plants). Work conducted in a single lab for a few weeks can then replace acres of nursery space and months of time, for one gram of tissue may be enough to clone 500 plants. Ultimately, the plantlets are placed in soil and grown conventionally.

Plant tissue culture also permits the selection of mutant cells. Scientists can treat cell suspensions using methods very much like those used to mutate and select bacteria. Millions of cells can be grown in glass vessels, irradiated, and artificially selected in order to recover advantageous mutants. The process, which is shown in Figure 20-7, has not yet proved extremely successful with most major crop plants, but it has yielded a few mutants resistant to certain drugs. Researchers speculate that such techniques may someday become routine with crop plants. The biggest drawback at this time is the difficulty of regenerating functional plants from the acquired mutant cell or cells.

Still another *in vitro* plant technology involves cell fusion or **somatic cell hybridization**. Cultured plant cells are enzymatically stripped of their **cellulose cell walls** to yield **protoplasts**. The protoplasts—even if they are from plants of different species—can then be made to merge or fuse. Hybrid cells can be formed and grown into hybrid plants even

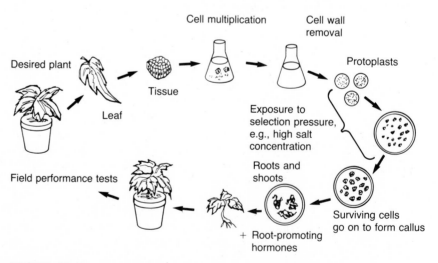

FIGURE 20–7
The technique of regenerating plants from single cells in culture. U.S. Congress, Office of Technology Assessment, *Impacts of Applied Genetics: Microorganisms, Plants, and Animals*, GPO stock no. 052-003-00805-0. (Washington, DC: U.S. Government Printing Office, April, 1981.)

452

when natural incompatibilities prevent hybridization by more normal reproductive routes.

The first demonstration of protoplast fusion occurred in 1972, when Peter Carlson and his colleagues combined somatic cells from two species of tobacco, *Nicotiana glauca* and *N. langsdorffi*, and successfully regenerated **somatic hybrids**. These two species can also hybridize through normal sexual reproduction, although success requires the doubling of the hybrid zygote's chromosomes after fertilization. The resulting plant is **amphiploid**, as shown in Figure 20-8, and it can produce viable gametes. These sexually produced plants and the somatic hybrid plants have the same chromosome constitution, and their characteristics are nearly identical.

Although plant protoplast technology needs further development to be economically useful, it has already yielded some interesting ideas and experiments. Plant biologists would like to induce mutations in cultured plant cells to supplement naturally occurring variation. They also would like to use haploid cells for this purpose, so that recessive mutations could be detected more easily. In such cases, the chromosomes would have to be doubled later to establish the desired diploid cells. This **diploidization** step seems to be no obstacle; effective methods already exist for the purpose. Some researchers believe that even more flexible and promising plant breeding techniques may result from genetic engineering.

GENETIC ENGINEERING AND AGRICULTURE

Genetic engineering, especially by recombinant DNA techniques, has excited not only medical and pharmaceutical researchers, but also agricultural scientists. We even hear predictions of a second Green Revolution, for if genetic engineering techniques can effectively increase food production and improve crop quality, the genetic engineering impact may be larger in agriculture than in medicine. Relatively few people have incurable diseases or require expensive drugs, but everyone eats—and often.

So far the application of genetic engineering to agriculture is only experimental. There is much promise, but there are also difficulties. We still need a good way to incorporate exogenous DNA into plant cells; this is the process of **transformation**, which has worked much better with bacteria, and even with animal cells, than with plants and plant protoplasts. We also need to learn how to regenerate a wider variety of

453

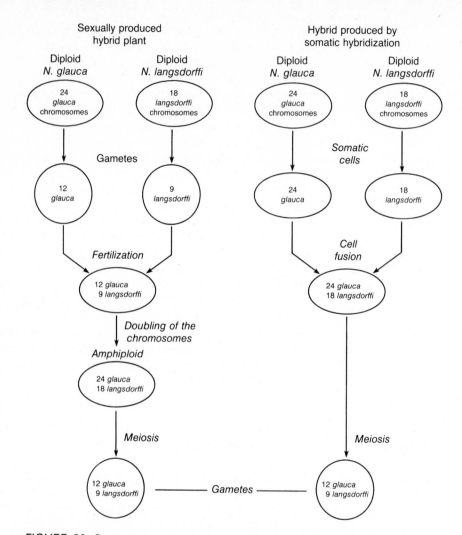

FIGURE 20-8
The schemes for producing hybrid *N. glauca* × *N. langsdorffi* plants with identical chromosome constitutions by both sexual reproduction and somatic hybridization.

plants from single cells; how to coax many agronomic traits, normally expressed in specific tissues of the whole plant, to express in cells grown *in vitro*; and how to manipulate polygenic traits, which include many valuable agronomic traits. Because of these and other gaps in our knowledge, many agricultural scientists are skeptical or even pessimistic about the applications of genetic engineering to agriculture. Others are sure that the advances will come—eventually.

So far, most genetic engineering has taken the form of placing genes from various organisms into bacteria. Nevertheless, even genetic engineering of bacteria can have agricultural applications. For example, one commercial company has placed the gene for bovine (cattle) growth hormone into bacteria. The bacteria produce the hormone, which is then harvested, purified, and fed to cattle to speed their fattening. Similar gene transplants in mammals have induced mice to produce rabbit hemoglobin and rat and human growth hormone. In the latter case, the mice grew half again as large as normal. Such genetic engineering experiments buoy our hopes of eventually accomplishing similar feats with agriculturally important animals, such as cattle and other livestock. Other achievements—such as the successful transfer of a gene for storage protein from a French bean to laboratory-cultured sunflower tissue—hold still more promise for the valuable contribution of genetic engineering to plant agriculture.

Most agricultural scientists point out that classical plant and animal breeding is also genetic engineering from a technical standpoint. But so far, genetic engineering in plants has accomplished relatively little using the newer revolutionary techniques. Instead, scientists have been focusing on the basic research that will eventually bring some of their ideas to fruition. Plant scientists would like to splice genes for improved characteristics into the DNA of plant cells, and to do so rapidly and precisely. They would like to convert some annual plants—corn, for example—into **perennials**, which renew their growth each year from roots, not seeds. They would like to engineer plants to be more tolerant of adverse conditions.

Agricultural scientists are committed to developing crops that can convert atmospheric nitrogen into usable form. Only certain plants, mostly **legumes**, are capable of fixing nitrogen; they do so with the aid of **symbiotic** bacteria (*Rhizobium*). There are some algae and bacteria that can fix nitrogen without the symbiotic relationship with another organism. Because most crop plants lack this capability, they require nitrogen in the form of fertilizers, whose production is costly in terms of energy and money. Nitrogen-fixing crop plants would be very valuable; unfortunately, we may not be able to engineer them for some years because many genes are probably involved, making the task extremely difficult.

Biotechnology in agriculture, even if it is successful beyond our aspirations, will not replace traditional animal and plant breeding. It will be a supplement, with the traditional practices being essential to the

455

success of the new technology. However, biotechnology will be a vital supplement, for without it we may be unable to keep pace with the world's need for nutrition and calories.

THE PROBLEM OF PESTS AND RESISTANCE

Insect pests have always been one of agriculture's most serious problems. Incredibly, they consume an estimated one-third of all the world's crops. The study of these pests is economic **entomology**, one of whose objectives is to control pest damage. For some years, chemical insecticides have served this purpose fairly well, but they have drawbacks such as toxicity to nontarget (nonpest) organisms and carcinogenicity to humans. Because of these disadvantages, some effective pesticides—DDT, for example—have been banned from ordinary use.

Some crop growers are convinced that they need the banned pesticides to protect their crops. One cotton grower presented the data in Table 20-2 in the mid-1970s to support this necessity. However, environmental organizations have vehemently rejected these cries for help. They point out that crop growers receive most of their information and advice on pest control from sales people representing chemical companies, not from agricultural scientists. They also stress that using chemical pesticides only increases farmers' dependence on these substances. The insect pests rapidly develop resistance to the pesticides as a result of mutations. As a result, new pesticides are required. Table 20-3 shows how real this problem is by documenting the rapid buildup of resistance to methyl parathion by the tobacco budworm. Many other examples could also be cited.

We need high crop production to satisfy the world's food requirements, but many effective pesticides that enhance crop yield are

Table 20–2
Relative efficacies and costs of DDT and other pesticides on cotton

	1972 (using DDT)	1975 (using non-DDT pesticides)
Yield	800 lb/acre	400 lb/acre
Cost for pest control	$29/acre	$70/acre
Number of pesticide applications during growing season	12	18

TABLE 20–3
Comparative resistance to the insecticide, methyl parathion, in tobacco bud-worms collected from various locations from 1964 to 1970

Location and Year	LD_{50} (mg/kg larvae)[a]	Fold Increase in Resistance Over 1964, College Station
College Station, Texas		
1964	0.01	1X
1965	0.03	3X
1966	0.02	2X
1967	0.02	2X
1968	0.04	4X
1969	0.06	6X
1970	0.26	26X
Brownsville, Texas		
1968	0.09	9X
Weslaco, Texas		
1968	0.05	5X
1969	0.11	11X
1970	0.54	54X
Tampico, Mexico		
1968	0.03	31X
1969	1.69	169X

Note:[a] LD_{50} is the milligrams of methyl parathion per kilogram of the insect larvae required to kill 50 percent of them effectively.
Source: P. L. Adkisson. 1969. How insects damage crops. *Connecticut Agricultural Experiment Station Bulletin,* 708:155–164.

highly toxic, harmful to wildlife, and dangerous to humans. The target organisms quickly become resistant. Fortunately, other approaches to controlling pests are available. We can use **varietal control** by breeding crops that are resistant to pests. We can use **pheromones**, chemical messengers produced by individual insects that control their behavior under natural conditions. The pheromones most useful for pest control are sex attractants, which can be extracted from the insects or made synthetically. They can then be used to attract insects to a cage or trap, to a concentrated pesticide source, or to chemical sterilizing agents. Pheromones have the advantage of being nontoxic and very species specific.

Biological pest-control methods have become very attractive in recent years. They include the use of predators and parasites to control pest populations. Ladybird beetles can be very effective against aphids. Parasitic wasps and flies are also very useful. Bacterial and virus dis-

457

eases have been tested with encouraging results, especially in combination with some chemicals. Unfortunately, only about one-tenth of all
attempts at biological control have succeeded, even partially. Nothing
works against some pests, such as the codling moth that plagues apple
orchards.

The **sterile male** or **autocidal** method is an ingenious technique.
In the 1950s and 1960s, researchers learned how to treat large numbers of male insects, usually with radiation, so that they are unable to
produce viable sperm. When they are released among wild, untreated
insects, sterile males can greatly reduce the rate of reproduction and
the size of the insect population.

The principal factors that make the sterile-male technique effective are:

1. The sterile males must be able to compete with normal
   males for mates;
2. Large numbers of sterile males must be used to increase
   the chances that they, rather than fertile, wild males, will
   mate with the wild females;
3. Optimally, the females should mate only once and then successfully shun all other males thereafter; and
4. It must be possible to rear and process the insects in great
   quantity.

If we fully understand the biology of the target pest, the technique can
be very effective, as shown in Table 20-4.

TABLE 20-4
The theoretical decline in successive generations when a constant number of
sterile males is released into a population of one million fertile females and one
million fertile males

Generation	Number of Virgin Females in the Area	Number of Sterile Males Released Each Generation	Ratio of Sterile to Fertile Males Competing for Each Virgin Female	Females Mated to Sterile Males (%)	Theoretical Population of Fertile Females in Each Subsequent Generation
F_1	1,000,000	2,000,000	2:1	66.7	333,333
F_2	333,333	2,000,000	6:1	85.7	47,619
F_3	47,619	2,000,000	42:1	97.7	1,107
F_4	1,107	2,000,000	1807:1	99.95	Less than 1

Source: E. F. Knipling. 1955. *Journal of Economic Entomology,* 48:460.

The attack upon the blowfly, *Cochliomyia hominivorax*, exemplifies a massive, long-term attempt to control an insect by the sterile-male technique. This pest's larval stage, the screwworm, feeds within lesions and cuts of cattle and sheep, resulting in ruined hides, unhealthy livestock, and even death of newborns. It caused large economic losses in the southern United States until the United States Department of Agriculture and various livestock organizations began to rear blowflies by the millions and sterilize them with gamma rays. The first batches of sterile males were released on an island, and because migration was impossible, the local screwworm population was quickly eradicated. The method also succeeded in Florida, and eventually it was used in a large part of the southwestern United States. Between 100 and 150 million sterile flies were released per week. In most years, infestations were reduced and the economic savings made the practice worthwhile, but there was only limited success in 1968, 1972, 1975, and 1976, and a severe outbreak occurred in 1978. The mass-reared flies had apparently lost their mating competitiveness, one of the requirements for success. Researchers have been seeking ways to overcome this problem.

For many insects, scientists have not been able to develop a sterile-male method of pest control. The boll weevil, for example, completely loses its reproductive competitiveness when irradiated. With this problem in mind, researchers have sought other, genetic means to make pest organisms sterile. For example, chromosomal abnormalities can reduce fertility up to 100 percent and have been tested as possible insect control measures. Other novel ideas include genetic manipulation to ensure that all progeny are males, altering growth patterns so that insects develop at an inopportune time of year, and crossing different species to create sterile hybrids for release into wild populations.

The system of insect control most favored at present is **integrated management**. This system is based upon mixed, multiple controls, ranging from chemical pesticides to the sterile-male technique, with each control measure applied at a critical time during the year. To be effective, this approach requires a good understanding of the pest's biology and ecology.

The practicality of various possible control methods is being investigated for many insect pests, including the screwworm, mosquito, fruit fly, house fly, codling moth, tsetse fly, and boll weevil. The war against insects is by no means over, and we can be sure the cost to control some insects will be high, even with the aid of genetic knowledge.

459

SUMMARY

Norman Borlaug's 1970 Nobel Peace Prize is but one indication of the importance of agriculture to world well-being. He developed a high-yielding dwarf-like strain of wheat and was instrumental in starting the Green Revolution. According to some observers, the Revolution fell short of its aims because less attention was given to nutritive quality than to storage life and other commercial qualities, and intensive agriculture often led to erosion. Most countries in the world still suffer from food shortages, requiring agricultural scientists to face many challenges.

Traditional crop breeding relies upon sexual reproduction, hybridization, and selection for desirable traits. Developing and testing new strains can take many years before seeds reach the farmer. Yet the process can improve nutrition, yield, and resistance to pests and disease. The new genetic technologies may be able to deliver results much more quickly.

Corn is a very important crop worldwide, and the development of hybrid varieties and better agricultural practices have greatly increased yields. Initially, most hybrid corn resulted from double crosses; much of today's hybrid corn comes from single or three-way crosses. In all cases, cross-fertilization of strains promotes hybrid vigor (heterosis) and results in better yields. At the same time, the plant breeder's success leads to the loss of genetic variability. Since many farmers plant the same varieties, a nation's food supply can be vulnerable to disease epidemics. This is precisely what happened in the United States in the early 1970s, when corn with the Texas male-sterile cytoplasm encountered a new race of corn leaf blight. This epidemic nearly caused a serious food shortage. We have good reason to preserve genetic variability in our crops, for once lost, the same variability cannot be replaced.

Unconventional breeding methods include the techniques of *in vitro* fertilization and embryo transfer for animals and the *in vitro* culturing of plant cells, the fusion of plant protoplasts, and the growth of some of the resulting somatic hybrids into functioning adult plants. Genetic engineering of plants—transferring genes from one plant to another by genetic manipulation—also holds great promise, although so far it has accomplished relatively little. Scientists hope someday to change plants from annuals to perennials, to give plants genes for nitrogen fixation, and to perform other useful transformations.

Major agricultural problems are insect pests and their resistance to chemical pesticides. New methods to control insects include the sterile-male technique, which has shown mixed success against some pests, notably the screwworm of the southern United States. However, the radiation used to sterilize the male insects can reduce reproductive competitiveness in some species, making the system useless. Cytogenetic sterilization methods, such as the use of chromosomal abnormalities, have also been investigated. Other methods of insect control include the use of varietal control (genetic resistance), pheromones (chemicals to attract and capture insects), and biological controls (natural enemies of pests). Integrated management stresses the use of mixed strategies against a specific pest. Genetics, ecology, and biology in general will undoubtedly continue to be important disciplines in the urgent task of preventing insects from devouring the world food supply.

DISCUSSION QUESTIONS

1. Discuss some of the ways an adequate food supply can help stabilize a nation and promote international peace.

2. How could shortening the wheat plant and thickening its stem, as Norman Borlaug did with crops in Mexico, have anything to do with increasing its yield?

3. Although the United States produces some foods in surplus, many agricultural scientists are steadily producing new varieties and developing new breeding techniques, some quite unconventional and novel. Discuss a few of the possible reasons for continuing this intensive agricultural research in the face of surpluses.

4. Some people have said that hybrid corn won World War II. How could this be possible?

5. Use your knowledge of genetics to explain why farmers must buy their hybrid seed corn each year, rather than using some of the corn from their previous harvest.

6. Plant breeders often discuss the "combining ability" of the different strains they use in their crosses. Might combining ability play a role in human crosses? Discuss the hypothetical consequences of such a factor in humans.

7. Why are plant biologists so excited by the success and potential of protoplast fusion?

8. Do you think there should be an international effort to preserve plant gene pools, or should each nation take on this responsibility in its own way? Give reasons.

461

9. Explain why it will be very difficult to use genetic engineering to stabilize multigenic traits in plants.

10. In the past, southern cotton growers sprayed their cotton fields two to three times each summer to protect against boll weevils. Years later, they sprayed 12 to 30 times each summer, and still their yields were less than in previous years. What was the problem?

11. The sterile-male technique to control insects is the result of scientific ingenuity, and it has worked fairly well in a few cases. What are some of the key factors necessary to make this insect control program effective?

12. A great deal of ecology is involved in any insect control measure. Elaborate.

PROBLEMS

1. A single-cross hybrid results from crossing the two inbreds:

   AABBccddeeFF × *aaBBccDDEEFF*

   How many different lines could theoretically be selected from among the F_2 progeny and inbred over several generations to a homozygous state?

2. Diagram a three-way cross in the production of hybrid corn.

3. Using letters for genes, show that hybrids gained by the single-cross method will result in a population of extremely uniform plants. In the same way, show that hybrids gained by the double-cross method will not necessarily display uniformity.

4. Somatic cells from two tobacco species have been fused and grown into a somatic hybrid plant. Since this plant is nearly identical to the hybrid plant obtained by means of a sexual cross, the two tobacco species must be fairly homozygous. Show why this relationship exists by representing the genes with letters.

5. Draw a diagram to show that a mature somatic hybrid plant should be capable of normal meiosis. Also show with a diagram that a mature sexually produced hybrid plant will usually be sterile. How can this sterility be avoided?

6. Given the release of 40 million sterile males into a population of 20 million female insects and 20 million male insects, and assuming equal competitiveness, what is the chance that any female will mate with a sterile male? What would the chance of mating with a sterile male be if 180 million of them were released?

REFERENCES

Adkisson, P. L. 1969. How insects damage crops. *Connecticut Agricultural Experiment Station Bulletin* 708:155–164.

Carlson, P. S., H. H. Smith, and R. D. Dearing. 1972. Parasexual interspecific plant hybridization. *Proceedings of the National Academy of Sciences* 69:2292–2294.

Herskowitz, I. H. 1973. *Principles of genetics.* New York: Macmillan Publishing Co. Inc.

Jones, D. F. 1918. The effects of inbreeding and crossbreeding on development. *Connecticut Agricultural Experiment Station* 207:1–100.

Knipling, E. F. 1955. Possibilities of insect control or eradication through the use of sexually sterile males. *Journal of Economic Entomology* 48:459–462.

Muhammed, A., R. Aksel, and R. C. von Borstel, ed. 1977. *Genetic diversity in plants.* New York: Plenum Press.

National Plant Genetic Resources Board, U.S. Department of Agriculture. 1978. *Plant genetic resources: Conservation and use.* Washington, DC: U.S. Government Printing Office.

National Science Foundation. 1978. Crops are ecosystems too. *Mosaic* (July/August) 24–31.

Office of Technology Assessment. 1981. *Impacts of applied genetics: Microorganisms, plants, and animals.* Washington, DC: U.S. Government Printing Office.

Palmiter, R. D., R. L. Brinster, R. E. Hammer, M. E. Trumbauer, M. G. Rosenfeld, N. C. Birnberg, and R. M. Evans. 1982. Dramatic growth of mice that develop from eggs microinjected with metallothionein growth hormone fusion genes. *Nature* 300:611–615.

Russell, W. A. 1974. Comparative performance for maize hybrids representing different eras of maize breedings. In *Proceedings of the 29th Annual Corn and Sorghum Research Conference.*

Serebrovsky, A. S. 1940. On the possibility of a new method for the control of insect pests. *Zoology Journal* 19:618–630.

Steward, F. C., M. O. Mapes, and K. Mears. 1958. Growth and organized development of cultured cells. II. Organization in cultures grown from freely suspended cells. *American Journal of Botany* 45:705–708.

Strickberger, M. W. 1976. *Genetics.* New York: Macmillan Publishing Co. Inc.

Ullstrup, A. J. 1977. Diseases of corn. In *Corn and corn improvement,* ed. G. F. Sprague. Madison, WI: American Society of Agronomy, Inc.

21
Genetics and Politics

Genetics makes a tremendous impact on medicine, agriculture, psychology, industry, and even philosophy. This textbook has demonstrated that impact. Anyone who needs further evidence of the importance of genetics has only to browse through the pages of such journals as *Science* or *Nature*. Even popular magazines, such as *Time*, *Newsweek*, and *Ladies Home Journal*, and daily newspapers frequently describe new developments and new applications of genetics.

It should surprise no one that genetics also affects politics, and that politics affects genetics. Genetics exerts so much influence on how people should live, the choices they must make, and how they must define themselves and each other—all issues traditionally in the domain of politics, philosophy, ideology, and religion—that it arouses strong feelings. Issues relating to genetics can affect the outcomes of elections, and the outcomes of elections can dictate what kind of genetic research receives public support and funding.

The number of politically sensitive genetic issues increases yearly. Programs such as genetic screening, genetic counseling, worker protection regulations, and eugenic measures, among others, have stirred controversy. Politicians have debated aspects of reproductive biology that involve abortion, birth control, *in vitro* fertilization, artificial insemination, embryo transfer, surrogate mothers, and the possibility of cloning. Educated and informed people disagree over biological determinism, intelligence quotients, other aspects of the nature-nurture controversy, and the application of Darwinian evolutionary concepts to society.

Several United States government agencies are directly concerned with science. These agencies seek legislation and funds to support their own activities and to support scientific research throughout the country. They include the National Science Foundation (NSF), National Institutes of Health (NIH), United States Department of Agriculture (USDA), United States Department of Health and Human Services (HHS), the Food and Drug Administration (FDA), and others. Many American scientists often feel that they are pressured to take political stances that agree with current government positions, simply because they depend on political agencies for their funding. Even scientists employed on campus and by industry depend at least indirectly on government funding, so they cannot be apolitical. Yet not all scientists accept the pressures to conform to the research direction emanating from Washington. A few run for office themselves, intent on reshaping the priorities more to their likings. Others lobby, trying to persuade members of Congress and agency officials to support unpopular or novel paths of research.

Researchers also influence the politics of science in other ways. Each year, 2000 to 3000 prestigious scientists are called upon by Congress and government agencies to give advice and various other forms of testimony on technical issues. As a result, scientists can have powerful effects on legislation. Scientific input, often relating to genetics, has influenced whether laws pass or fail.

Scientific input can also shape military plans (as it did in the case of the atomic bomb) and administration policies in health, education, and civil rights. Unfortunately, this input reaches government in a very unsystematic way. One proposal for a more systematic method of consultation called for a "science court" composed of specialists who could evaluate the evidence on controversial technical issues and decide which side of a given controversy was best supported. The court's decision would then serve to guide government policy making. However, many people have objected to the idea of a science court. Composed wholly of scientists, they say, it would be too elitist. They suggest drawing members from lay groups such as business people, the law, and the clergy.

Certainly we can point to cases where the lack of humanizing influences on the science-politics combination has led to disaster. More often, the disaster has occurred because ideologues collaborated with one or two scientists whose ideas matched the ideology too closely, and then closed their minds to other points of view. This is what happened in Nazi Germany, where National Socialist ideologues allowed hereditarian notions to take a central place in German policy. With all other information shut out, these views quickly led to the most drastic eugenics measures, and to the deaths of millions of people. Since the Holocaust, scientists have been more vocal, and less inclined to remain in their laboratories when they believe their governments are making mistakes.

Governments recognize this increased political activity. In many lands, and occasionally even in the United States, governments try to keep scientists quiet. Their techniques range from withdrawal of funding for research to outright oppression in some countries—including harassment, denial of privileges, imprisonment, and even official or unofficial execution. Other scientists are exiled.

Some countries consider their scientists to be dangerous people, for science does not necessarily respect ideology. Yet not all persecution of scientists is based on their science. Sometimes it is for their political views or activities, especially if they support political opponents or systems they consider more favorable to the values of science—truth and freedom.

Are scientists really dangerous? Consider the famous geneticist H. J. Muller, whom Barry Cohen recently acclaimed as a model for scientists concerned with society and politics. In his time, many did think him dangerous, for he was a controversial personality all his life. Muller served as faculty advisor to the Communist students at the University of Texas; he advocated sperm banks and other eugenic practices; and he was an active atheist. In addition, he openly opposed the Lysenko movement in the Soviet Union.

The Lysenko movement in the Soviet Union is a good example of how dangerous a science–politics mesh can be, though here the danger is that of error, not truth. There is no better illustration of how a program based on misunderstood genetics and politics could cause disaster, for Lysenkoism nearly destroyed Soviet agriculture in the years following World War II.

THE LYSENKO CASE

The incredible story of Lysenkoism lasted almost 20 years. It began in 1948, when Trofim D. Lysenko became director of the Soviet Academy's Institute of Genetics and president of the Lenin Agricultural Academy, with the help of theoretician I. I. Prezent and Stalin's absolute power.

Lysenko believed in **Lamarckian** evolution. That is, he believed that the characteristics an organism acquires during life can become heritable and pass to succeeding generations. Furthermore, he insisted on his "physiological theory" despite a long history of genetic experiments showing that it was false. Most of his supporting data were very weak, and some, it seems, had even been faked to appear more impressive.

How could Lysenko achieve his position of power? His extreme environmentalist views seemed more relevant to Marxism and Communist political doctrines than did the more generally accepted ideas. In fact, Lysenko boldy denounced classical genetics because it contradicted those doctrines, and his power quickly destroyed classical genetics in the Soviet Union. Textbooks were rewritten and curricula changed. Unbiased, pure genetic research practically disappeared.

Lysenko's power was enhanced by his promises to use his expertise in genetics to expand the productivity of Soviet agriculture greatly, just when the need for more crops was critical. Stalin believed

468

him, and the political apparatus supported Lysenko's control of biology and agriculture and his supression of other scientific ideas. Soviet scientists who dared to oppose the Lysenko movement were demoted, arbitrarily dismissed, jailed, exiled, and according to some Western observers, even executed. One famous plant geneticist, N. I. Vavilov, was accused of being a British spy and sent to a Siberian labor camp, where he eventually died.

Lysenkoism rejected the sound breeding practices and hybridization techniques being developed in the West. This was a drastic mistake, causing Soviet agricultural productivity to decline disastrously. Meanwhile, Western agriculture and Western genetics achieved great advances. DNA was identified as the genetic material, and its molecular structure was worked out. Molecular biology was born, and the pace of genetic discovery quickened.

The Soviet Union had a strong tradition in biological and agricultural development before Lysenkoism. Subsequently under Krushchev, the situation improved somewhat. Brezhnev and Kosygin finally restored genetic research to its original status of free inquiry. Lysenko was discredited and removed from control, leaving only the new word "Lysenkoism" to mark his place in history.

The Soviet Union's return to sound genetics has been welcomed by all geneticists. The Lysenko episode was a sad affair for science, politics, and human rights, but now Soviet scientists are striving to catch up with the rest of the world, especially in molecular genetics. Yet, Soviet agriculture still lags, and Lysenko's error is to some extent to blame for recent large sales of grain to the Soviet Union by the United States and Canada. These grain sales have fueled still more vigorous political debates.

THE CONTROL AND USE OF GENETIC RESEARCH

Genetics—and science in general—responds to political pressures in places other than the Soviet Union. In the United States, these pressures are usually expressed in legislation and funding, as after the 1957 launchings of the Soviet satellites, Sputniks I and II. The United States reacted by putting billions of dollars into science education, science instrumentation, space research, and other scientific and technological areas. In 1961, President John F. Kennedy committed the nation to putting a man on the moon within that decade. The result was an explosion of scientific advance in all areas—and man did, in fact, reach the moon.

469

Another example of how political commitment can support science comes from the "war on cancer" prompted by Congress and President Nixon in the early 1970s. The "war" put large amounts of money into cancer and related research, and this research has paid off in a much improved understanding of the cancer process, basic biology, and molecular genetics.

These examples illustrate what may be the only proper way for a government to control research. Certainly, governments have the right to decide what types of research to encourage through funding. And if they provide the money, they have both a right and a responsibility to monitor the research, at least in the sense of ensuring that public monies are well spent. They exert this control largely by screening research proposals, but also by channeling funds for training students into certain fields and supporting new construction at particular schools.

This kind of system often means that scientists themselves, who work for and advise government agencies, do the controlling, even though government is supposed to represent everyone, not just scientists or politicians or bureaucrats. The general public needs to be involved in the processes of choosing research avenues to follow and controlling the research funded. It needs to be heard through organizations, lobbies, political parties and platforms, political representatives, and the voting process. It *is* heard in all these ways, through the voices of special interest groups. Politicians have won and lost elections because of their stands on controversial issues such as abortion.

Today, public participation affects American research in several areas. For instance, a moratorium on "test-tube baby" research (*in vitro* fertilization) has been in effect since 1975. This means that the government will not fund this kind of work, and specific research must be funded by private sources. As another example, a massive study designed to detect XYY persons as children, which would then allow the monitoring of their behavior over a long period of time, was abandoned because of mounting pressures from both inside and outside the scientific community. Many of the criticisms centered on the dangers of stigmatizing people without any adequate justification.

Research in reproductive biology is always in the public eye, and genetic research on humans, especially in the realms of intelligence and race, genetic engineering, and cloning, has many vocal opponents. As a result, it is difficult to find much funding for studies of intelligence in races. The public strongly favors science in general, but many people object to anything that relates to the creation of new life forms or threatens traditional practices related to religion and one's philosophy of life.

470

Many people also object to research that appears to be useless or trivial. Wisconsin's Senator William Proxmire made his indignation public when he inaugurated his "Golden Fleece" award in 1975. He intended to dramatize examples of how the government wastes money on useless research, but some scientists believe his award to be flawed. All too often, the award goes to "useless" basic research that is not properly understood, but is nevertheless important. Basic research has always been important to the future. Faraday's "useless" work on electricity ultimately led to radio, radar, television, and pocket calculators, among other things. "Useless" work on rockets is paying off in communication and weather satellites, and more. We might wonder whether Mendel's gardening with the lowly sweet pea would have been funded, and if it were, whether it would have earned a Golden Fleece award.

Government seems much more reasonable when it reacts to possible hazards, as in the federal controls on recombinant DNA research. But we must note that it was neither the government nor the general public, but the scientists involved, who first drew public attention to the possible dangers inherent in this kind of research. In an almost unprecedented way, the scientists imposed a moratorium on their own work, devised safety guidelines, and called for regulation. This led to a great deal of debate, testimony, and lobbying in the mid-1970s. Legislation for liability and strict fines for researchers who ignored the guidelines were seriously proposed. The "Friends of DNA" argued for less control on the research. And cities such as Cambridge, Massachusetts, proposed their own controls. The Cambridge city council wanted to forbid scientists at Harvard University to do recombinant DNA work within the city without its approval.

Some opposition to genetic research is based on ideology or concerns for safety. There is also a faction of the public that is completely opposed to human genetic research because of religion or ethics. Many of these opponents of research worry that new discoveries could be twisted to hamper progress toward racial equality. Some say that it is not possible to resolve questions such as the relationship of intelligence to race, that only fragmentary, incomplete data are within human reach. They argue that imperfect data can lead only to abuses and difficulties in progress toward equal opportunity.

On the other hand, many geneticists argue that abuses arise not from genetic research, but from a lack of genetic research. Genetic research, not its absence, was what helped refute the Nazi concept of racial superiority. Proponents of the freedom to research such areas fur-

471

ther point out that we must acquire more genetic information regarding human potentials if we are ever to achieve equal opportunity for all. Biologically diverse people cannot develop to their optimum levels under the same environmental conditions.

Most scientists and politicians agree that politically motivated suppression of scientific research is not a productive route for society. We must guard against the misinterpretation of data, but we should not fail to seek the data. Research results are crucial to present and future decisions that will greatly affect society. And everyone in our society must participate in these decisions. To do this, everyone needs to be political, and to understand the issues.

SUMMARY

When genetics interacts with society, it becomes inescapably political. Political issues in genetics include many present programs and areas, as well as innumerable issues that we can barely imagine now. Many governmental agencies are directly or indirectly concerned with science. Numerous scientists advise Congress each year, science courts have been contemplated, and much legislation has been affected by science.

Unfortunately, the interaction of politics and science, especially genetics, has sometimes turned out badly for all concerned. For example, fanatical Nazis, obsessed with the false notion of the "master race," practiced extreme eugenics. Probably the most bizarre case is that of Lysenkoism in the Soviet Union from 1948 to the middle 1960s. Lysenko's dictatorial imposition of archaic genetics and his suppression of classical genetics and its advocates were disastrous. Politically motivated oppression of scientists continues today in some countries.

Politics and government are heavily involved in controlling and directing much research in the United States. Special interest groups, various organizations, lobbyists, funding priorities, and public sentiments all play important roles in directing research in genetics and biology in this country.

DISCUSSION QUESTIONS

1. Citizens of the United States enjoy freedom of speech, religion, and the press. Should we also have freedom of scientific inquiry? Discuss your response.

2. Do you think science courts should be established, made up wholly of scientists, to hear and judge controversies having scientific bases? What would be the advantages? The disadvantages?

3. Why is reproductive biology a political issue?

4. Why do you suppose scientists are so often singled out for harassment and even persecution?

5. Lysenkoism is gone, but the Soviet Union's progress toward the level of Western genetics has been very slow. What are some of the possible reasons for this slowness?

6. Do you think we should conduct sound scientific research on certain aspects of society, even if we fear that our results will be politically misused?

7. Do you think geneticists should have pursued their long-term study of XYY persons, or do you think the decision to end the study was appropriate? Give reasons.

8. Should we study the correlation between race and intelligence? Why or why not?

9. If you received unlimited government funding for human genetics research, what would you investigate? Why?

REFERENCES

Cohen, B. M. 1982. H. J. Muller: A memoir. *Journal of Heredity* 73:477.

Further Reading

Chapter 1

Baer, A. S., ed. 1977. *Heredity and society: Readings in social genetics*. New York: Macmillan Publishing Co. Inc.

Baer, A. S. 1977. *The genetic perspective*. Philadelphia: W. B. Saunders Co.

Bornstein, J., and S. Bornstein. 1979. *What is genetics?* New York: Julian Messner.

Carlson, E. A. 1966. *The gene: A critical history*. Philadelphia: W. B. Saunders Co.

Childs, B. 1974. A place for genetics in health education, and vice versa. *American Journal of Human Genetics* 26:120–135.

Childs, B. 1979. Education in genetics for the medical profession and the public. *The Biological Sciences Curriculum Study Journal* 2:7–8.

Childs, B. 1982. Genetics in the medical curriculum. *American Journal of Medical Genetics* 13:319–324.

Childs, B. 1983. Why study human genetics? *The American Biology Teacher* 45:42–46.

Childs, B., C. A. Huether, and E. A. Murphy. 1981. Human genetics teaching in U.S. medical schools. *American Journal of Human Genetics* 33:1–10.

Darby, W. J. 1980. Science, scientists, and society: The 1980's. *Federation Proceedings* 39:2943–2948.

Dunn, L. C. 1962. Cross currents in the history of human genetics. *American Journal of Human Genetics* 14:1–13.

475

Ehrlich, P. R., R. W. Holm, and I. L. Brown. 1976. *Biology and society*. New York: McGraw-Hill Book Co.

Gardner, E. J. 1983. *Human heredity*. New York: John Wiley & Sons Inc.

Goldschmidt, R. B. 1950. Fifty years of genetics. *American Naturalist* 86:313–340.

Hartl, D. L. 1977. *Our uncertain heritage. Genetics and human diversity*. Philadelphia: J. B. Lippincott Co.

Hartl, D. L. 1983. *Human genetics*. New York: Harper & Row, Publishers Inc.

Hurd, P. D. 1978. The historical/philosophical background of education in human genetics in the United States. *The Biological Science Curriculum Study Journal* 1:3–8.

Jenkins, J. B. 1983. *Human genetics*. Menlo Park, CA: The Benjamin/Cummings Publishing Co., Inc.

Judson, H. F. 1980. *The search for solutions*. New York: Holt, Rinehart & Winston General Book.

Lerner, I. M. 1968. *Heredity, evolution, and society*. San Francisco: W. H. Freeman & Co. Publishers.

Lipkin, M., Jr., and P. T. Rowley, ed. 1974. *Genetic responsibility*. New York: Plenum Press.

Ludmerer, K. M. 1972. *Genetics and American society. A historical appraisal*. Baltimore: The Johns Hopkins University Press.

Mange, A. P., and E. J. Mange. 1980. *Genetics: Human aspects*. Phildelphia: Saunders College.

Motulsky, A. G. 1983. Impact of genetic manipulation on society and medicine. *Science* 219:135–140.

Nagle, J. D. 1979. *Heredity and human affairs*. St. Louis: C. V. Mosby Co.

Novitski, E. 1982. *Human genetics*. New York: Macmillan Publishing Co. Inc.

Oosthuizen, G. C., H. A. Shapiro, and S. A. Strauss, ed. 1980. *Genetics and society*. Oxford: Oxford University Press.

Owen, R. D. 1983. Genetics in the 20th century. *The Journal of Heredity* 74:314–319.

Ravin, A. W. 1978. Genetics in America: A historical overview. *Perspectives in Biology and Medicine* 21:214–231.

Reed, S. C. 1979. A short history of human genetics in the U.S.A. *American Journal of Medical Genetics* 3:282–295.

Schull, W. J., and R. Chakraborty, ed. 1979. *Human genetics. A selection of insights*. Stroudsburg, PA: Dowden, Hutchinson & Ross, Inc.

Scriver, C. R., D. E. Scriver, C. L. Clow, and M. Schok. 1978. The education of citizens: Human genetics. *The American Biology Teacher* 40:280–284.

Stern, C. 1975. High points of human genetics. *The American Biology Teacher* 37:144–149.

Stern, C., and E. R. Sherwood, ed. 1966. *The origin of genetics*. San Francisco: W. H. Freeman & Co. Publishers.

Stine, G. J. 1977. *Biosocial genetics—human heredity and social issues*. New York: Macmillan Publishing Co. Inc.

476

Stubbe, H. 1972. *History of genetics*. Cambridge, MA: The MIT Press.

Sturtevant, A. H. 1954. Social implications of the genetics of man. *Science* 120:405–407.

Sturtevant, A. H. 1965. *A history of genetics*. New York: Harper & Row, Publishers Inc.

Weissmann, G., ed. 1979. *The biological revolution. Applications of cell biology to public welfare*. New York: Plenum Press.

Winchester, A. M., and T. R. Mertens. 1983. *Human genetics*. Columbus: Charles E. Merrill Publishing Co.

Chapter 2

Bauer, W. R., F. H. C. Crick, and J. H. White. 1980. Supercoiled DNA. *Scientific American* 243:118–133.

Bennett, J. 1980. Building a cell. *The American Biology Teacher* 42:518–524.

Crick, F. H. C. 1954. The structure of the hereditary material. *Scientific American* 191:54–61.

Crick, F. H. C. 1957. Nucleic acids. *Scientific American* 197:188–200.

Hoagland, M. 1981. *Discovery. The search for DNA's secrets*. Boston: Houghton Mifflin Co.

Hotchkiss, R. D., and E. Weiss. 1956. Transformed bacteria. *Scientific American* 195:48–53.

Hurwitz, J., and J. J. Furth. 1962. Messenger RNA. *Scientific American* 206:41–49.

Judson, H. F. 1979. *The eighth day of creation*. New York: Simon & Schuster.

Kornberg, A. 1968. The synthesis of DNA. *Scientific American* 291:64–78.

Lerner, I. M. 1968. *Heredity, evolution and society*. San Francisco: W. H. Freeman & Co. Publishers.

Miller, J. A. 1982. The gene idea. *Science News* 121:180–182.

Mirsky, A. E. 1968. The discovery of DNA. *Scientific American* 218:78–88.

Olby, R. 1974. *The path to the double helix*. Seattle: University of Washington Press.

Panati, C. 1980. *Breakthroughs*. Boston: Houghton Mifflin Co.

Portugal, F. H., and J. S. Cohen. 1977. *A century of DNA*. Cambridge, MA: The MIT Press.

Sayre, A. 1975. *Rosalind Franklin and DNA*. New York: W. W. Norton & Co., Inc.

Sinsheimer, R. L. 1962. Single-stranded DNA. *Scientific American* 207:109–116.

Watson, J. D. 1968. *The double helix*. New York: Atheneum Publishers.

Weisburd, S. 1984. Dissecting the dance in DNA. *Science News* 125:362–364.

Chapter 3

Blixt, S. 1975. Why didn't Gregor Mendel find linkage? *Nature* 256:206.

Crow, J. F. 1979. Genes that violate Mendel's rules. *Scientific American* 240:134–143, 146.

John, B., and K. R. Lewis. 1972. Somatic cell division. *Oxford Biology Reader 26*, ed. J. J. Head and O. E. Lowenstein. London: Oxford University Press.

John, B., and K. R. Lewis. 1976. The meiotic mechanism. *Oxford Biology Reader 65*, ed. J. J. Head. London: Oxford University Press.

Lerner, I. M. 1968. *Heredity, evolution and society*. San Francisco: W. H. Freeman & Co. Publishers.

Levine, L., ed. 1971. *Papers on genetics*. St. Louis: C. V. Mosby Co.

Lewis, K. R., and B. John. 1972. *The matter of Mendelian heredity*. New York: John Wiley & Sons Inc.

Mazia, D. 1961. How cells divide. *Scientific American* 205:100–120.

Mazia, D. 1974. The cell cycle. *Scientific American* 230: 54–64.

Novitski, E., and S. Blixt. 1978. Mendel, linkage, and synteny. *BioScience* 28:34–35.

Olby, R. C. 1966. *Origins of Mendelism*. London: Constable & Company.

Oldroyd, D. 1984. Gregor Mendel: Founding-father of modern genetics. *Endeavour* 8:29–31.

Orel, V. 1984. Mendel scientific legacy in 1984. *Biologia* 39:341.

Stern, C., and E. R. Sherwood, ed. 1966. *The origin of genetics: A Mendel source book*. San Francisco: W. H. Freeman & Co. Publishers.

Taylor, J. H. 1958. The duplication of chromosomes. *Scientific American* 198:36–42.

Therman, E. 1980. *Human chromosomes. Structure, behavior, effects*. New York: Springer-Verlag.

Chapter 4

Benzer, S. 1962. The fine structure of the gene. *Scientific American* 206:70–84.

Crick, F. H. C. 1962. The genetic code. *Scientific American* 207:66–74.

Crick, F. H. C. 1966. The genetic code III. *Scientific American* 215:55–62.

Crick, F. 1970. Central dogma of molecular biology. *Nature* 227:561–563.

Hurwitz, J., and J. J. Furth. 1962. Messenger RNA. *Scientific American* 206:41–49.

Ingram, V. M. 1958. How do genes act? *Scientific American* 198:68–74.

Koshland, D. E. 1973. Protein shape and biological control. *Scientific American* 229:52–64.

Nirenberg, M. W. 1963. The genetic code II. *Scientific American* 208:80–94.

Nomura, M. 1969. Ribosomes. *Scientific American* 221:28–35.

Phillips, D. C. 1966. The three-dimensional structure of an enzyme molecule. *Scientific American* 215:78–90.

Phillips, D. C., and A. C. T. North. 1973. Protein structure. *Oxford Biology Reader 34*, ed. J. J. Head. London: Oxford University Press.

Rich, A., and S. H. Kim. 1978. The three-dimensional structure of transfer RNA. *Scientific American* 238:52–62.

478

Scriver, C. R., and C. L. Clow. 1980. Phenylketonuria: Epitome of human biochemical genetics. *The New England Journal of Medicine* 303:1336–
1342.

Stone, I. 1974. Humans, the mammalian mutants. *American Laboratory* 7:32–
39.

Travers, A. A. 1974. Transcription of DNA. *Oxford Biology Reader 75*, ed. J. J.
Head. London: Oxford University Press.

Yanofsky, C. 1967. Gene structure and protein structure. *Scientific American*
216:80–94.

Chapter 5

Bodmer, W. F., and L. L. Cavalli-Sforza. 1976. *Genetics, evolution, and man.*
San Francisco: W. H. Freeman & Co. Publishers.

Boyd, W. C. 1950. *Genetics and the races of man.* Boston: Little, Brown and
Co.

Boyd, W. C. 1963. Genetics and the human race. *Science* 140:1057–1065.

Cavalli-Sforza, L. 1969. Genetic drift in an Italian population. *Scientific American*
221:30–37.

Cavalli-Sforza, L. L. 1973. Some current problems of human population genetics. *American Journal of Human Genetics* 25:82–104.

Cavalli-Sforza, L. L. 1974. The genetics of human populations. *Scientific American* 231:80–89.

Coon, C. S. 1965. *The living races of man.* New York: Alfred A. Knopf Inc.

Dobzhansky, T. 1973. *Genetic diversity and human equality.* New York: Basic
Books Inc., Publishers.

Ebling, F. J., ed. 1975. *Racial variations in man.* New York: John Wiley & Sons
Inc.

Glass, H. B. 1953. The genetics of the Dunkers. *Scientific American* 189:76–
81.

Goldsby, R. 1971. *Race and races.* New York: Macmillan Publishing Co. Inc.

Hartl, D. L. 1977. *Our uncertain heritage. Genetics and human diversity.* Philadelphia: J. B. Lippincott Co.

King, J. C. 1971. *The biology of race.* New York: Harcourt Brace Jovanovich
Inc.

Lerner, I. M. 1968. *Heredity, evolution and society.* San Francisco: W. H. Freeman & Co.

Mead, M., T. Dobzhansky, E. Tobach, and R. E. Light, ed. 1969. *Science and
the concept of race.* New York: Columbia University Press.

Provine, W. B. 1973. Geneticists and the biology of race crossing. *Science*
182:790–796.

Reed, T. E. 1969. Caucasian genes in American Negroes. *Science* 165:762–
768.

Wilson, E. O., and W. H. Bussert. 1971. *A primer of population biology.* Stam-
ford, CT: Sinauer Associates Inc.

479

Chapter 6

Beaconsfield, P., G. Birdwood, and R. Beaconsfield. 1980. The placenta. *Scientific American* 243:94–102.

Beck, F., D. B. Moffat, J. B. Lloyd. 1973. *Human embryology and genetics*. Oxford: Blackwell Scientific Publications.

Biggers, J. D. 1981. *In vitro* fertilization and embryo transfer in human beings. *The New England Journal of Medicine* 304:336–342.

Cates, W., Jr. 1981. The Hyde amendment in action. *Journal of the American Medical Association* 246:1109–1112.

Cherrenak, F. A., M. A. Farley, L. Walters, J. C. Hobbins, and M. J. Mahoney. 1984. When is termination of pregnancy during the third trimester morally justified? *The New England Journal of Medicine* 310:501.

Clarke, B. 1975. The causes of biological diversity. *Scientific American* 233:50–60.

Coale, A. 1974. The history of the human population. *Scientific American* 231:40–51.

Curie-Cohen, M., L. Luttrell, and S. Shapiro. 1979. Current practice of artificial insemination by donor in the United States. *The New England Journal of Medicine* 300:585–590.

Djerassi, C. 1979. *The politics of contraception*. New York: W. W. Norton & Co.

Edwards, R. G. 1981. Test-tube babies. *Carolina Biology Reader 89*. Burlington, NC: Carolina Biological Supply Co.

Edwards, R. G. 1981. The beginnings of human life. *Carolina Biology Reader 135*. Burlington, NC: Carolina Biological Supply Co.

Edwards, R. G., and R. E. Fowler. 1970. Human embryos in the laboratory. *Scientific American* 223:44–54.

Edwards, R. G., and J. M. Purdy. 1982. *Human conception* in vitro. New York: Academic Press Inc.

Edwards, R. G., and D. J. Sharpe. 1971. Social values and research in human embryology. *Nature* 231:87–91.

Facklam, M., and H. Facklam. 1979. *From cell to clone*. New York: Harcourt Brace Jovanovich Inc.

Fraser, F. C., and R. A. Forse. 1981. On genetic screening of donors for artificial insemination. *American Journal of Medical Genetics* 10:399–405.

Freedman, R., and B. Berelson. 1974. The human population. *Scientific American* 231:30–39.

Galston, A. 1975. Here come the clones. *Natural History* 84:72–75.

Greep, R. O., M. A. Koblinsky, and F. S. Jaffe. 1976. Reproduction and human welfare: A challenge to research. *BioScience* 26:677–684.

Grobstein, C. 1979. External human fertilization. *Scientific American* 240:57–67.

Grobstein, C. 1982. Coming to terms with test-tube babies. *New Scientist* 96:14–17.

Grobstein, C., M. Flower, and J. Mendeloff. 1983. External human fertilization: An evaluation of policy. *Science* 222:127–133.

Gurdon, J. B. 1968. Transplanted nuclei and cell differentiation. *Scientific American* 219:24–35.

Hardin, G. 1968. The tragedy of the commons. *Science* 162:1243–1248.

Hardin, G. 1982. Some biological insights into abortion. *BioScience* 32:720–727.

Henig, R. M. 1978. *In vitro* fertilization: A cautious move ahead. *BioScience* 28:685–688.

Holmes, H. B., B. B. Hoskins, and M. Gross, ed. 1980. *Birth control and controlling birth*. Clifton, NJ: Humana Press, Inc.

Holmes, H. B., B. B. Hoskins, and M. Gross, ed. 1981. *The custom-made child? Women-centered perspectives*. Clifton, NJ: Humana Press.

Humber, J. M., and R. F. Almeder, ed. 1979. *Biomedical ethics and the law*. New York: Plenum Press.

Huxley, A. 1946. *Brave new world*. New York: Harper and Brothers.

Johnson, M. H., and B. J. Everitt. 1980. *Essential reproduction*. Boston: Blackwell Scientific Publications.

Karp, L. E. 1981. Artificial insemination: A need for caution. *American Journal of Medical Genetics* 9:179–181.

Kieffer, G. H. 1979. *Bioethics: A textbook of issues*. Reading, MA: Addison-Wesley Publishing Co. Inc.

Kieffer, G. H. 1980. IVF-*In vitro* fertilization. *The American Biology Teacher* 42:211–218, 231.

Kolata, G. B. 1978. *In vitro* fertilization: Is it safe and repeatable? *Science* 201:698–699.

Kolata, G. B. 1983. *In vitro* fertilization goes commercial. *Science* 221:1160–1161.

Mader, S. S. 1980. *Human reproductive biology*. Dubuque, IA: Wm. C. Brown Group.

Mastroianni, L., Jr., and J. D. Biggers, ed. 1981. *Fertilization and embryonic development* in vitro. New York: Plenum Press.

McKinnell, R. G. 1979. *Cloning*. Minneapolis: University of Minnesota Press.

Motulsky, A. G. 1974. Brave new world. *Science* 185:653–663.

Neumann, M., ed. 1978. *The tricentennial people. Human applications of the new genetics*. Ames, IA: Iowa State University Press.

Nilsson, L., M. Furuhjelm, A. Ingleman-Sunberg, C. Wirsen, and B. Farsblad. 1976. *A child is born*. New York: Delacorte Press/Seymour Lawrence.

Oosthuizen, G. C., H. A. Shapiro, and S. A. Strauss, ed. 1980. *Genetics and society*. Cape Town: Oxford University Press.

Porter, I. H., and R. G. Skalko, ed. 1973. *Heredity and society*. New York: Academic Press Inc.

Reining, P., and I. Tinker, ed. 1975. *Population: Dynamics, ethics and policy*. Washington, DC: American Association for the Advancement of Science.

Revelle, R. 1974. Food and population. *Scientific American* 231:160–170.

Rhodes, P. 1981. Childbirth. *Carolina Biology Reader 111*. Burlington, NC: Carolina Biological Supply Co.

Robison, W. L., and M. S. Pritchard, ed. 1979. *Medical responsibility. Paternalism, informed consent, and euthanasia*. Clifton, NJ: Humana Press.

Samuels, A. 1982. Artificial insemination and genetic engineering: The legal problems. *Medical Science Law* 22:261–268.

Segal, S. J. 1974. The physiology of human reproduction. *Scientific American* 231:52–62.

Swanson, H. D. 1974. *Human reproduction*. New York: Oxford University Press.

Whittingham, D. G., S. P. Leibo, and P. Mazur. 1972. Survival of mouse embryos frozen to −196° and −269°C. *Science* 178:411–414.

Williams, P., and G. Stevens. 1982. What now for test tube babies? *New Scientist* 93:312–316.

Chapter 7

Arehart-Treichel, J. 1975. Correcting enzyme defects in the test tube. *Science News* 107:211–212.

Bloom, A. D., and S. James, ed. 1981. *The fetus and the newborn*. New York: Alan R. Liss Inc.

Childs, B. 1974. A place for genetics in health education, and vice versa. *American Journal of Human Genetics* 26:120–135.

Childs, B. 1978. A new view of health: Genetics and environment. *The Biological Sciences Curriculum Study Journal* 1:9–12.

Childs, B. 1979. Education in genetics for the medical profession and the public. *The Biological Sciences Curriculum Study* 2:7–8.

Childs, B., C. A. Huether, and E. A. Murphy. 1981. Human genetics teaching in the U.S. medical schools. *American Journal of Human Genetics* 33:1–10.

Crow, J. F. 1967. Genetics and medicine. In *Heritage from Mendel*, ed. R. A. Brink. Madison: University of Wisconsin Press.

Davis, J. W., B. Hoffmaster, and S. Shorten, ed. 1978. *Contemporary issues in biomedical ethics*. Clifton, NJ: Humana Press.

Desnick, R. J. 1980. *Enzyme therapy in genetic disease: 2*. New York: Alan R. Liss Inc.

Drimmer, F. 1973. *Very special people*. New York: Amjon Publishers Inc.

Frigoletto, F. D., Jr., and I. Umansky. 1979. In *Genetic disorders and the fetus*, ed. A Mulinsky. New York: Plenum Press.

Harris, A. 1981. On the track of cystic fibrosis. *New Scientist* 90:167–169.

Hook, E. B., D. T. Janerich, and I. H. Porter, ed. 1971. *Monitoring birth defects and environment. The problem of surveillance*. New York: Academic Press Inc.

Humber, J. M., and R. F. Almeder, ed. 1979. *Biomedical ethics and the law*. New York: Plenum Publishing Corp.

Janerich, D. T., R. G. Skalko, and I. H. Porter, ed. 1974. *Congenital defects. New directions in research*. New York: Academic Press Inc.

Karp, L. E. 1981. The elephant man. *American Journal of Medical Genetics* 10:1–3.

Kelly, S. 1977. *Biochemical methods in medical genetics*. Springfield, IL: Charles C. Thomas, Publisher.

Kelly, S., E. B. Hook, D. T. Janerich, and I. H. Porter, ed. 1976. *Birth defects. Risks and consequences*. New York: Academic Press Inc.

Klein, J. 1980. *Woody Guthrie: A life*. New York: Alfred A. Knopf Inc.

Kolata, G. B. 1980. Thalassemias: Models of genetic diseases. *Science* 210:300–302.

Lipkin, M., Jr., and P. T. Rowley, ed. 1974. *Genetic responsibility*. New York: Plenum Press.

McKusick, V. A. 1965. The royal hemophilia. *Scientific American* 213:88–95.

Milunsky, A. 1977. *Know your genes*. New York: Avon Books.

Montagu, A. 1979. *The elephant man. A study in human dignity*. New York: E. P. Dutton Inc.

Nora, J. J., and F. C. Fraser. 1981. *Medical genetics: Principles and practice*. 2d ed. Philadelphia: Lea and Febiger.

Novitski, E. 1977. *Human genetics*. New York: Macmillan Publishing Co. Inc.

Perutz, M. F. 1978. Hemoglobin structure and respiratory transport. *Scientific American* 239:92–125.

Porter, I. H., and E. B. Hook, ed. 1979. *Service and education in medical genetics*. New York: Academic Press Inc.

Porter, I. H., and R. G. Skalko, ed. 1973. *Heredity and society*. New York: Academic Press Inc.

Roberts, J. F. 1978. *An introduction to medical genetics*. 7th ed. New York: Oxford University Press.

Robison, W. L., and M. S. Pritchard, ed. 1979. *Medical responsibility*. Clifton, NJ: Humana Press.

Saunders, R. 1980. Liposomes as carriers in biological systems. *South African Journal of Science* 76:297–298.

Scriver, C. R., C. Laberge, C. L. Clow, and F. C. Fraser. 1978. Genetics and medicine: An evolving relationship. *Science* 200:946–952.

Shaw, M. W. 1977. Perspectives on today's genetics and tomorrow's progeny. *The Journal of Heredity* 68:274–279.

Thompson, J. S., and M. W. Thompson. 1980. *Genetics in medicine*. Philadelphia: W. B. Saunders Co.

Weissman, G., ed. 1979. *The biological revolution. Applications of cell biology to public welfare*. New York: Plenum Press.

Wills, C. 1970. Genetic load. *Scientific American* 222:98–107.

Chapter 8

Bloom, A. D., and L. S. James, ed. 1981. *The fetus and the newborn*. New York: Alan R. Liss Inc.

Carr, D. H. 1971. Chromosomes and abortion. *Advances in Human Genetics* 2:201–257.

Carr, D. H. 1971. Genetics of abortion. *Annual Review of Genetics* 5:65–80.

Cavalli-Sforza, L. L. 1977. *Elements of human genetics*. Menlo Park, CA: W. A. Benjamin Inc.

de la Chapelle, A. 1981. The etiology of maleness in XX men. *Human Genetics* 58:105–116.

de la Cruz, F. F., and P. S. Gerald. 1981. *Trisomy-21 (Down syndrome): Research perspectives*. Baltimore: University Park Press.

Ford, E. H. R. 1973. *Human chromosomes*. New York: Academic Press Inc.

German, J., ed. 1974. *Chromosomes and cancer*. New York: John Wiley & Sons Inc.

Gordon, J. W., and F. H. Ruddle. 1981. Mammalian gonadal determination and differentiation. *Science* 214:1265–1271.

Hamerton, J. L. 1971. *Human cytogenetics. Clinical cytogenetics*. New York: Academic Press Inc.

Hannah-Alava, A. 1960. Genetic mosaics. *Scientific American* 202:118–130.

Harsanyi, Z., and R. Hutton. 1981. *Genetic prophecy: Beyond the double helix*. New York: Rawson-Wade Publishers.

Haseltine, F. P., and S. Ohno. 1981. Mechanisms of gonadal differentiation. *Science* 214:1272–1278.

Holmes, H. B., B. B. Hoskins, and M. Gross, ed. 1981. *The custom-made child? Women-centered perspectives*. Clifton, NJ: Humana Press.

Hook, E. B., and P. K. Cross. 1983. Spontaneous abortion and subsequent Down syndrome livebirth. *Human Genetics* 64:267–270.

Hook, E. B., and I. H. Porter, ed. 1977. *Population cytogenetics. Studies in humans*. New York: Academic Press Inc.

Hook, E. B., D. T. Janerich, and I. H. Porter, ed. 1971. *Monitoring, birth defects and environment. The problem of surveillance*. New York: Academic Press Inc.

Hossald, T., N. Chen, J. Funkhouser, T. Jooss, B. Manuel, J. Matsuura, A. Matsuyama, C. Wilson, J. A. Yamane, and P. A. Jacobs. 1980. A cytogenetic study of 1000 spontaneous abortions. *Annals of Human Genetics* 44:151–178.

Hsu, T. C. 1979. *Human and mammalian cytogenetics. An historical perspective*. New York: Springer-Verlag.

Janerich, D. T., R. G. Skalko, and I. H. Porter, ed. 1974. *Congenital defects. New directions in research*. New York: Academic Press Inc.

Kelly, S., E. B. Hook, D. T. Janerich, and I. H. Porter, ed. 1976. *Birth defects. Risks and consequences*. New York: Academic Press Inc.

Kim, D.-S., and I. H. Porter. 1973. Indications for chromosome studies in clinical practice. *Pediatrics Annals* 2:26–51.

Lewandowski, R. C., and J. J. Yunis. 1975. New chromosomal syndromes. *American Journal of Disabled Children* 129:515–529.

Masters, W. H., and V. E. Johnson. 1979. *Homosexuality in perspective*. Boston: Little, Brown and Co.

484

McDermott, A. 1975. *Cytogenetics of man and other animals*. New York: John Wiley & Sons Inc.

Michels, V. V., C. Medrano, V. L. Venne, and V. M. Riccardi. 1982. Chromosome translocations in couples with multiple spontaneous abortions. *American Journal of Human Genetics* 34:507–513.

Milunsky, A. 1977. *Know your genes*. New York: Avon Books.

Milunsky, A., ed. 1979. *Genetic disorders and the fetus*. New York: Plenum Press.

Mittwoch, U. 1963. Sex differences in cells. *Scientific American* 209:54–62.

Morton, N. E., and C. S. Chung, ed. 1978. *Genetic epidemiology*. New York: Academic Press Inc.

Neri, G., A. Serra, M. Campana, and B. Tedeschi. 1983. Reproductive risks for translocation carriers: Cytogenetic study and analysis of pregnancy outcome in 58 families. *American Journal of Medical Genetics* 16:535–561.

Porter, I. H. 1976. The clinical side of cytogenetics. *Journal of Reproductive Medicine* 17:3–18.

Porter, I. H., and E. B. Hook, ed. 1980. *Human embryonic and fetal death*. New York: Academic Press Inc.

Porter, I. H., and R. G. Skalko, ed. 1973. *Heredity and society*. New York: Academic Press Inc.

Riccardi, V. M. 1977. *The genetic approach to human disease*. New York: Oxford University Press.

Roth, M. P., J. Feingold, A. Baumgarten, P. Bigel, and C. Stoll. 1983. Reexamination of paternal age effect in Down's syndrome. *Human Genetics* 63:149–152.

Russell, L. B., ed. 1978. *Genetic mosaics and chimeras in mammals*. New York: Plenum Press.

Sankaranarayanan, K. 1979. The role of non-disjunction in aneuploidy in man. An overview. *Mutation Research* 61:1–28.

Silvers, W. K., and S. S. Wachtel. 1977. H-Y antigen: Behavior and function. *Science* 195:956–959.

Simpson, J. L. 1976. *Disorders of sexual differentiation*. New York: Academic Press Inc.

Therman, E. 1980. *Human chromosomes. Structure, behavior, effects*. New York: Springer-Verlag.

Thompson, J., and M. W. Thompson. 1980. *Genetics in medicine*. Philadelphia: W. B. Saunders Co.

Tobach, E., and B. Rosoff, ed. 1978. *Genes and gender: I*. Staten Island, NY: Gordian Press Inc.

Vallet, H. L., and I. H. Porter, ed. 1979. *Genetic mechanisms of sexual development*. New York: Academic Press Inc.

van Niekerk, W., and A. E. Retief. 1981. The gonads of human true hermaphrodites. *Human Genetics* 58:117–122.

Wachtel, S. S. 1977. H-Y antigen and the genetics of sex determination. *Science* 198:797–799.

Wolf, U. 1981. Genetic aspects of H-Y antigen. *Human Genetics* 58:25–28.

485

Chapter 9

Bodmer, W. F., and L. L. Cavalli-Sforza. 1976. *Genetics, evolution, and man*. San Francisco: W. H. Freeman & Co. Publishers.

Childs, B. 1977. Persistent echoes of the nature-nurture argument. *American Journal of Human Genetics* 29:1–13.

Cravens, H. 1978. *The triumph of evolution*. Philadelphia: University of Pennsylvania Press.

Dobzhansky, T. 1973. *Genetic diversity and human equality*. New York: Basic Books Inc., Publishers.

Feldman, M. W., and R. C. Lewontin. 1975. The heritability hang-up. *Science* 190:1163–1168.

Halsey, A. H., ed. 1977. *Heredity and environment*. New York: The Free Press.

Porter, I. H., and R. G. Skalko, ed. 1973. *Heredity and society*. New York: Academic Press Inc.

Chapter 10

Bergsma, D., ed. 1976. *Cancer and genetics*. New York: Alan R. Liss Inc.

Bishop, J. M. 1982. Oncogenes. *Scientific American* 246:80–92.

Cairns, J. 1975. The cancer problem. *Scientific American* 233:64–78.

Cairns, J. 1978. *Cancer. Science and society*. San Francisco: W. H. Freeman & Co. Publishers.

Cairns, J. 1981. The origin of human cancer. *Nature* 289:353–357.

Croce, C. M., and H. Koprowski. 1978. The genetics of human cancer. *Scientific American* 238:117–125.

Emery, A. E. H. 1976. *Methodology in medical genetics*. New York: Churchill Livingstone Inc.

Farber, S. L. 1981. *Identical twins reared apart. A reanalysis*. New York: Basic Books Inc., Publishers.

Fraser, F. C. 1981. The genetics of familial disorders—major genes or multifactorial? *Canadian Journal of Genetics and Cytology* 23:1–8.

German, J., ed. 1974. *Chromosomes and cancer*. New York: John Wiley & Sons Inc.

Harsanyi, Z., and R. Hutton. 1981. *Genetic prophecy: Beyond the double helix*. New York: Rawson-Wade Publishers.

Heston, W. E. 1974. Genetics of cancer. *The Journal of Heredity* 65:262–272.

Knudson, A. G., Jr., L. C. Strong, and David E. Anderson. 1973. Heredity and cancer in man. *Progress in Medical Genetics* 9:113–158.

Land, H., L. F. Parada, and R. A. Weinberg. 1983. Cellular oncogenes and multistep carcinogenesis. *Science* 222:771–778.

Lehmann, A. 1979. Genetic clues to cancer. *New Scientist* 84:686–688.

Levan, A., G. Levan, and F. Mitelman. 1977. Chromosomes and cancer. *Hereditas* 86:15–30.

Miller, J. A. 1982. Spelling out a cancer gene. *Science News* 122:316–317.

Mulvihill, J. J., R. W. Miller, and J. F. Fraumeni, Jr., ed. 1977. *Genetics of human cancer.* New York: Raven Press.

Murphree, A. L., and W. F. Benedict. 1984. Retinoblastoma: Clues to human oncogenesis. *Science* 223:1028–1033.

Pitot, H. C. 1978. *Fundamentals of oncology.* New York: Marcel Dekker Inc.

Rowley, J. D. 1980. Chromosome abnormalities in human leukemia. *Annual Review of Genetics* 14:17–39.

Ryder, L. P., A. Svejgaard, and J. Dansett. 1981. Genetics of HLA disease association. *Annual Review of Genetics* 15:169–187.

Sandberg, A. A. 1980. *The chromosomes in human cancer and leukemia.* New York: Elsevier North-Holland.

Slamon, D. J., J. B. de Kernion, I. M. Verma, and M. J. Cline. 1984. Expression of cellular oncogenes in human malignancies. *Science* 224:256–262.

Temin, H. M. 1972. RNA-directed DNA synthesis. *Scientific American* 226:24–33.

Thompson, J., and M. W. Thompson. 1980. *Genetics in medicine.* Philadelphia: W. B. Saunders Co.

Yunis, J. J., C. D. Bloomfield, and K. Ensrud. 1981. All patients with acute nonlymphocytic leukemia may have a chromosomal defect. *The New England Journal of Medicine* 305:135–139.

Chapter 11

Aberle, D. F., U. Bronfenbrenner, E. H. Hess, D. R. Miller, D. M. Schneider, and J. N. Spuhler. 1963. The incest taboo and the mating patterns of animals. *American Anthropologist* 65:253–265.

Bodmer, W. F., and L. L. Cavalli-Sforza. 1976. *Genetics, evolution, and man.* San Francisco: W. H. Freeman & Co. Publishers.

Browning, D. H., and B. Boatman. 1977. Incest: Children at risk. *American Journal of Psychiatry* 134:69–72.

Cavalli-Sforza, L. L. 1969. Genetic drift in an Italian population. *Scientific American* 221:30–37.

Glass, H. B. 1953. The genetics of the Dunkers. *Scientific American* 189:76–81.

Hall, W. K., K. R. Hawkins, and G. P. Child. 1950. The inheritance of alkaptonuria in a large American family. *The Journal of Heredity* 41:23–25.

Herman, J. L. 1981. *Father-daughter incest.* Cambridge, MA: Harvard University Press.

Lukianowicz, N. 1972. Incest I: Paternal incest. *British Journal of Psychiatry* 120:301–313.

MacCluer, J. W. 1980. Inbreeding and human fetal death. In *Human embryonic and fetal death*, ed. I. H. Porter and E. B. Hook. New York: Academic Press Inc.

487

Meiselman, K. C. 1978. *Incest. A psychological study of causes and effects with treatment recommendations*. San Francisco: Vossey-Bass, Publishers.

Schull, W. J. 1958. Empirical risks in consanguineous marriages: Sex ratio, malformation, and viability. *American Journal of Human Genetics* 10:294–343.

Schull, W. J., and J. V. Neel. 1965. *The effects of inbreeding on Japanese children*. New York: Harper & Row, Publishers Inc.

Schull, W. J., and J. V. Neel. 1972. The effects of parental consanguinity and inbreeding in Hirado, Japan. V. Summary and interpretation. *American Journal of Human Genetics* 24:425–453.

Wallace, B. 1972. *Disease, sex, communication, behavior*. Englewood Cliffs, NJ: Prentice-Hall, Inc.

Wills, C. 1970. Genetic load. *Scientific American* 222:98–107.

Chapter 12

Angell, R. R., R. J. Aitken, P. F. A. van Look, M. A. Lumsden, and A. A. Templeton. 1983. Chromosome abnormalities in human embryos after *in vitro* fertilization. *Nature* 303:336–338.

Ann Arbor Science for the People Editorial Collective. 1977. *Biology as a social weapon*. Minneapolis: Burgess Publishing Co.

Arehart-Treichel, J. 1982. Fetal ultrasound: How safe? *Science News* 121:396–397.

Biggers, J. D. 1981. When does life begin? *The Sciences* 21:10–14.

Bloom, A. D., and L. S. James, ed. 1981. *The fetus and the newborn*. New York: Alan R. Liss Inc.

Chedd, G. 1981. Who shall be born? *Science 81* 2:32–40.

Committee for the Study of Inborn Errors of Metabolism. 1975. *Genetic screening: Programs, principles, and research*. Washington, DC: National Academy of Sciences.

Damme, C. J. 1982. Legal implications of monitoring workers for carcinogenic and mutagenic risk. *Teratogenesis, Carcinogenesis, and Mutagenesis* 2:211–219.

Davenport C. B. 1913. *Heredity in relation to eugenics*. New York: Henry Holt and Co.

Dunn, L. C. 1962. Cross currents in the history of human genetics. *American Journal of Human Genetics* 14:1–13.

Epstein, C. J., and M. S. Golbus. 1977. Prenatal diagnosis of genetic diseases. *American Scientist* 65:703–711.

Fraser, F. C. 1974. Genetic counseling. *American Journal of Human Genetics* 26:636–659.

Friedmann, T. 1971. Prenatal diagnosis of disease. *Scientific American* 225:34–42.

488

Fuchs, F. 1980. Genetic amniocentesis. *Scientific American* 242:47–53.

Fuhrman, W., and F. Vogel. 1983. *Genetic counseling*. New York: Springer-Verlag.

Gastel, B., J. E. Hadlow, J. C. Fletcher, and A. Neale, ed. 1980. *Maternal serum alpha-fetoprotein. Issues in the prenatal screening and diagnosis of neural tube defects*. Washington, DC: U.S. Government Printing Office.

Grobstein, C. 1982. Rights in the womb. *The Sciences* 22:6–8.

Guthrie, R. 1972. Mass screening for genetic defects. *Hospital Practice*, June, 93–100.

Harper, P. S. 1981. *Practical genetic counselling*. Baltimore: University Park Press.

Harris, A. 1981. On the track of cystic fibrosis. *New Scientist* 90:167–169.

Harris, H. 1974. *Prenatal diagnosis and selective abortion*. Cambridge, MA: Harvard University Press.

Holmes, H. B., B. B. Hoskins, and M. Gross, ed. 1981. *The custom-made child? Women-centered perspectives*. Clifton, NJ: Humana Press.

Hook, E. B. 1983. Chromosome abnormalities and spontaneous fetal death following amniocentesis: Further data and associations with maternal age. *American Journal of Human Genetics* 35:110–116.

Hsia, Y. E., K. Hirschhorn, R. Silverberg, and L. Godmilow, ed. 1979. *Counseling in genetics*. New York: Alan R. Liss Inc.

Humber, J. M., and R. F. Almeder, ed. 1979. *Biomedical ethics and the law*. New York: Plenum Press.

Janerich, D. T., R. G. Skalko, and I. H. Porter, ed. 1974. *Congenital defects. New directions in research*. New York: Academic Press Inc.

Kamin, L. 1974. *The science and politics of I.Q.* New York: Halsted Press.

Kelly, S., E. B. Hook, D. T. Janerich, and I. H. Porter, ed. 1976. *Birth defects. Risks and consequences*. New York: Academic Press Inc.

Kieffer, G. H. 1979. *Bioethics: A textbook of issues*. Reading, MA: Addison-Wesley Publishing Co. Inc.

Kolata, G. 1983. First trimester prenatal diagnosis. *Science* 221: 1031–1032.

Lappe, M. 1979. *Genetic politics*. New York: Simon & Schuster.

Lebel, R. R. 1978. Ethical issues arising in the genetic counseling relationship. In *Birth Defects: Original Article Series* 14(9). White Plains, NY: March of Dimes Birth Defects Foundation.

Lerner, I. M. 1968. *Heredity, evolution, and society*. San Francisco: W. H. Freeman & Co. Publishers.

Lipkin, M., Jr., and P. T. Rowley, ed. 1974. *Genetic responsibility. On choosing our children's genes*. New York: Plenum Press.

Lubs, H. A., and F. de la Cruz, ed. 1977. *Genetic counseling*. New York: Raven Press.

McCormack, M. K., S. Leiblum, and A. Lazzarini. 1983. Attitudes regarding utilization of artificial insemination by donor in Huntington disease. *American Journal of Medical Genetics* 14:5–13.

489

Milunsky, A., ed. 1979. *Genetic disorders and the fetus. Diagnosis, prevention, and treatment*. New York: Plenum Press.

Milunsky, A., and G. J. Annas, ed. 1980. *Genetics and the law II*. New York: Plenum Press.

Omenn, G. S. 1978. Prenatal diagnosis of genetic disorders. *Science* 200:151–157.

Oosthuizen, G. C., H. A. Shapiro, and S. A. Strauss, ed. 1980. *Genetics and society*. Cape Town: Oxford University Press.

Osborn, F. 1974. History of the American Eugenics Society. *Social Biology* 21:115–126.

Porter, I., and E. B. Hook, ed. 1979. *Service and education in medical genetics*. New York: Academic Press Inc.

Porter, I. H., and R. G. Skalko, ed. 1973. *Heredity and society*. New York: Academic Press Inc.

Raine, D. N., ed. 1976. *Medico-social management of inherited metabolic disease*. Baltimore: University Park Press.

Reed, S. 1964. *Parenthood and heredity*. Philadelphia: W. B. Saunders Co.

Reed, S. 1980. *Counseling in medical genetics*. New York: Alan R. Liss Inc.

Reilly, P. 1977. *Genetics, law and social policy*. Cambridge, MA: Harvard University Press.

Seidenfeld, M. J., and R. M. Antley. 1981. Genetic counseling: A comparison of counselee's genetic knowledge before and after. Part III. *American Journal of Medical Genetics* 10:107–112.

Siggers, D. C. 1978. *Prenatal diagnosis of genetic disease*. Oxford: Blackwell Scientific Publications.

Slater, E. 1936. German eugenics in practice. *Eugenics Review* 27:285–295.

Steele, M. W., and W. R. Berg. 1966. Chromosome analysis of human amniotic-fluid cells. *Lancet* 1:383–385.

U.S. Department of Health, Education, and Welfare. 1979. *Antenatal diagnosis*. National Institutes of Health Publication No. 80-1973.

Verlinsky, Y., L. Chu, and E. Pergament. 1984. Chromosomal preparations of chorionic villi samples (CVS). *Karyogram* 10:1–3.

Wegner, R.-D., G. Obe, and M. Meyenburg. 1980. Has diagnostic ultrasound mutagenic effects? *Human Genetics* 56:95–98.

Wright, E. E. 1980. Invited editorial essay. The legal implications of refusing to provide prenatal diagnosis in low-risk pregnancies or solely for sex selection. *American Journal of Medical Genetics* 5:391–397.

Chapter 13

American Medical Association, Committee on Transfusion and Transplantation and American Bar Association, Section on Family Law, Committee on Standards for the Judicial Use of Scientific Evidence in the Ascertainment of Paternity. 1976. Joint AMA-ABA guidelines: Present status of serological testing in problems of disputed parentage. *Family Law Quarterly* 10(3):247–285.

Blake, E. T., and G. F. Sensabaugh. 1976. Genetic markers in human semen: A review. *Journal of Forensic Science* 21:784–796.

Chakraborty, R., M. Shaw, and W. J. Schull. 1974. Exclusion of paternity: The current state of the art. *American Journal of Human Genetics* 26:477–488.

Given, B. W. 1976. Sex-chromatin bodies in penile washings as an indicator of recent coitus. *Journal of Forensic Sciences* 21:381–386.

Hodges, R. L. 1973. Judges, genes and man. *Stadler Symposium* 5:179–186.

Jasanoff, S., and D. Nelkin. 1981. Science, technology, and the limits of judicial competence. *Science* 214:1211–1215.

Milunsky, A., and G. J. Annas, ed. 1980. *Genetics and the law, II.* New York: Plenum Press.

Reilly, P. 1977. *Genetics, law, and social policy.* Cambridge, MA: Harvard University Press.

Reilly, P. 1981. HLA tests in the courts. *American Journal of Human Genetics* 33:1007–1008.

Soulier, J. P., O. Prou-Wartelle, and J. Y. Muller. 1974. Paternity research using the HL-A system. *Haematologia* 8:249–265.

Sussman, L. N. 1973. Blood grouping tests for non-paternity. *Journal of Forensic Sciences* 18:287–289.

Sussman, L., ed. 1976. *Paternity testing by blood grouping.* Springfield, IL: Charles C. Thomas, Publishers.

Terasaki, P. I. 1977. Resolution by HLA testing of 1000 paternity cases not excluded by ABO testing. *Journal of Family Law* 16:543–555.

Wiener, A. S., and W. W. Socha. 1976. Methods available for solving medicolegal problems of disputed parentage. *Journal of Forensic Science* 21:42–64.

William, J., J. D. Curran, and S. M. Hyg. 1976. Forensic medical science: The continued problem of judges and juries. *The New England Journal of Medicine* 294:1042–1043.

Chapter 14

Abel, E. L., ed. 1982. *Fetal alcohol syndrome.* Boca Raton, FL: CRC Press.

Ames, B. N. 1979. Identifying environmental chemicals causing mutations and cancers. In *The biological revolution. Applications of cell biology to public welfare*, ed. G. Weissman. New York: Plenum Press.

Ames, B. N. 1983. Dietary carcinogens and anticarcinogens. *Science* 221:1256–1264.

Baer, A. S., ed. 1977. *Heredity and society. Readings in social genetics.* New York: Macmillan Publishing Co. Inc.

Brookes, P. 1978. Environmental factors and cancer in man. *Biologist* 25:189–194.

Brusick, D. 1980. *Principles of genetic toxicology.* New York: Plenum Press.

Cairns, J. 1975. Mutation selection and the natural history of cancer. *Nature* 225:197–200.

491

Cairns, J. 1975. The cancer problem. *Scientific American* 233:64–78.

Cairns, J. 1978. *Cancer and society*. San Francisco: W. H. Freeman & Co. Publishers.

Cassens, R. G., T. Ito, M. Lee, and D. Buege. 1978. The use of nitrite in meat. *BioScience* 28:633–637.

Clarren, S. K., and D. W. Smith. 1978. The fetal alcohol syndrome. *The New England Journal of Medicine* 298:1063–1067.

Corbet, T. H. 1977. *Cancer and chemicals*. Chicago: Nelson-Hall.

Committee 17 of The Council of the Environmental Mutagen Society. 1975. Environmental mutagenic hazards. *Science* 187:503–514.

Devoret, R. 1979. Bacterial tests for potential carcinogens. *Scientific American* 241:40–49.

Dixon, B. 1981. Where there's smoke. *The Sciences* 21:29–30.

Drake, J. W. 1975. Environmental mutagenesis: Evolving strategies in the USA. *Mutation Research* 33:65–72.

Dunlap, T. R. 1981. *DDT. Scientists, citizens, and public policy*. Princeton, NJ: Princeton University Press.

Garmon, L. 1982. Puzzled over PCB's. A symposium beset by controversy looked at some industrial chemicals turned environmental contaminants. *Science News* 121:361–363.

Hartman, P. E. 1983. Mutagens: Some possible health impacts beyond carcinogenesis. *Environmental Mutagenesis* 5:139–152.

Heinonen, O. P., D. Stone, and S. Shapiro. 1977. *Birth defects and drugs in pregnancy*. Littleton, MA: Publishing Sciences Group.

Hollaender, A., ed. 1973. *Chemical mutagens. Principles and methods for their detection*, vol. 3. New York: Plenum Press.

Hook, E. B. 1980. Human germinal mutations: Monitoring for environmental effects. *Social Biology* 26:104–116.

Hook, E. B., D. T. Janerich, and I. H. Porter, ed. 1971. *Monitoring, birth defects, and environment. The problem of surveillance*. New York: Academic Press Inc.

Kalter, H. 1975. Some relations between teratogenesis and mutagenesis. *Mutation Research* 33:29–36.

Kelly, S., E. B. Hook, D. T. Janerich, and I. H. Porter, ed. 1976. *Birth defects. Risks and consequences*. New York: Academic Press Inc.

Kramers, P. G. N. 1975. The mutagenicity of saccharin. *Mutation Research* 32:81–92.

Mason, S. 1984. The left hand of nature. *New Scientist* 101:10–14.

Matter, B. E. 1976. Problems of testing drugs for potential mutagenicity. *Mutation Research* 38:243–258.

Nagao, M., and T. Sugimura. 1978. Environmental mutagens and carcinogens. *Annual Review of Genetics* 12:117–159.

Naismith, R. W. 1980. Genetic toxicology: An overview. *The American Biology Teacher* 42:457–461.

Omenn, G. S. 1984. A framework for reproductive risk assessment and surveillance. *Teratogenesis, Carcinogenesis, and Mutagenesis* 4:1–14.

492

Petersen, R. C. 1979. Health implications of marijuana use: A review. *The American Biology Teacher* 41:526–529.

Pitot, H. C. 1978. *Fundamentals of oncology.* New York: Marcel Dekker Inc.

Public Health Service. 1979. *Smoking and health. A report of the surgeon general.* Washington, DC: U.S. Department of Health, Education and Welfare.

Scott, R. 1977. Asbestos: Can we get away from it? *Environmental Action* 26(March):3–5.

Shaw, C. R., ed. 1981. *Prevention of occupational cancer.* Boca Raton, FL: CRC Press Inc.

Shaw, M. W. 1970. Human chromosome damage by chemical agents. *Annual Review of Medicine* 21:409–432.

Streissguth, A. P., S. Landesman-Dwyer, J. C. Martin, and D. W. Smith. 1980. Teratogenic effects of alcohol in humans and laboratory animals. *Science* 209:353–361.

Sutton, H. E., and M. I. Harris, ed. 1972. *Mutagenic effects of environmental contaminants.* New York: Academic Press Inc.

Tinklenberg, J. R., ed. 1975. *Marijuana and health hazards.* New York: Academic Press Inc.

Trimble, B. K., and M. E. Smith. 1977. The incidence of genetic disease and the impact on man of an altered mutation rate. *Canadian Journal of Genetics and Cytology* 21:375–385.

U.S. Army Medical Department. 1980. *LSD follow-up study report.* Washington, DC: U.S. Government Printing Office.

Waldbott, G. L. 1973. *Health effects of environmental pollutants.* St. Louis: C. V. Mosby Co.

Wallace, B. 1970. *Genetic load. Its biological and conceptual aspects.* Englewood Cliffs, NJ: Prentice-Hall, Inc.

Weisburger, J. H., and G. M. Williams. 1981. Carcinogen testing: Current problems and new approaches. *Science* 214:401–407.

Wills, C. 1970. Genetic load. *Scientific American* 222:98–107.

Zimmermann, F. K. 1982. Can we determine mutagenicity or only a mutagenic potential? *Mutation Research* 92:3–7.

Chapter 15

Adams, R., and S. Cullen, ed. 1981. *The final epidemic. Physicians and scientists on nuclear war.* Chicago: Educational Foundation for Nuclear Science, Inc.

Beck, H. L., and P. W. Krey. 1983. Radiation exposures in Utah from Nevada nuclear tests. *Science* 220:18–24.

Caldicott, H. M. 1982. The final epidemic. *The Sciences* 22:16–21.

Carlson, E. A. 1981. *Genes, radiation, and society. The life and work of H. J. Muller.* Ithaca, NY: Cornell University Press.

Cohen, B. L. 1979. What is the misunderstanding all about? *The Bulletin of the Atomic Scientists* 37:31–36.

493

Committee for the Compilation of Materials on Damage Caused by the Atomic Bombs in Hiroshima and Nagasaki. *Hiroshima and Nagasaki. The physical, medical, and social effects of the atomic bombs*. Translated from the Japanese by E. Ishikawa and D. L. Swain. New York: Basic Books Inc., Publishers.

Committee on the Biological Effects of Ionizing Radiation. National Academy of Sciences. 1980. *The effects on populations of exposure to low levels of ionizing radiation*. Washington, DC: National Academy Press.

Crow, J. 1959. Ionizing radiation and evolution. *Scientific American* 201:138–160.

Ehrlich, P. R., et al. 1983. Long-term biological consequences of nuclear war. *Science* 222:1293–1300.

Gofman, J. W. 1981. *Radiation and human health*. San Francisco: Sierra Club Books.

Hamilton, H. B. 1982. Genetic markers in the atomic bomb survivors and their children—Hiroshima and Nagasaki. *Japanese Journal of Human Genetics* 27:113–119.

Hollaender, A., and G. E. Stapleton. 1959. Ionizing radiation and the living cell. *Scientific American* 201:94–100.

Hussain, S., and L. Ehrenberg. 1979. Mutagenicity of radiation at low doses. *Hereditas* 91:111–116.

Land, C. E., F. W. McKay, and S. G. Machado. 1984. Childhood leukemia and fallout from the Nevada nuclear tests. *Science* 223:139–144.

Laws, P. W. 1981. *Medical and dental X-rays. A consumer's guide to avoiding unnecessary radiation exposure*. Washington, DC: Health Research Group.

Miller, R. W. 1968. Radiation effects on atomic bomb survivors. *Science* 166:569–574.

Morgan, K. Z. 1978. Cancer and low level ionizing radiation. *The Bulletin of the Atomic Scientists* 34:30–41.

Muller, H. J. 1955. Radiation and human mutation. *Scientific American* 193:58–68.

National Academy of Sciences. 1980. *Proceedings of the 15th Annual Meeting of the National Council on Radiation Protection and Measurements. Perceptions of risks*. Washington, DC: National Council on Radiation Protection and Measurements.

National Academy of Sciences. 1981. *Proceedings of the 16th Annual Meeting of the National Council on Radiation Protection and Measurements. Quantitative risks in standard setting*. Washington, DC: National Council on Radiation Protection and Measurements.

Pizzarello, D. J., and R. L. Witcofski. 1982. *Medical radiation biology*. Philadelphia: Lea and Febiger.

Puck, T. T. 1960. Radiation and the human cell. *Scientific American* 202:142–153.

Raloff, J. 1979. Radiation: Can a little hurt? *Science News* 115:44–45.

Raloff, J. 1983. Compensating radiation victims. *Science News* 124:330–331.

494

Raloff, J. 1984. Silkwood—the legal fallout. *Science News* 125:74–79.

Rotblat, J. 1981. Hazards of low-level radiation—less agreement, more confusion. *The Bulletin of the Atomic Scientists* 37:31–36.

Schull, W. J., M. Otake, and J. V. Neel. 1981. Genetic effects of the atomic bombs: A reappraisal. *Science* 213:1220–1227.

Silberner, J. 1981. Hiroshima and Nagasaki. Thirty-six years later, the struggle continues. *Science News* 120:284–287.

Sturtevant, A. H. 1954. Social implications of the genetics of man. *Science* 120:405–407.

Upton, A. C. 1982. The biological effects of low-level ionizing radiation. *Scientific American* 246:41–49.

Waldbot, G. L. 1973. *Health effects of environmental pollutants.* St. Louis: C. V. Mosby Co.

Chapter 16

Barash, D. P. 1983. *Aging. An exploration.* Seattle: University of Washington Press.

Behnke, J. A., C. E. Finch, and G. B. Moment, ed. 1978. *The biology of aging.* New York: Plenum Press.

Burnet, Sir M. 1974. *Intrinsic mutagenesis: A genetic approach to ageing.* New York: John Wiley & Sons Inc.

Comfort, A. 1979. *The biology of senescence.* New York: Elsevier North-Holland.

Fries, J. F., and L. M. Crapo. 1981. *Vitality and aging.* San Francisco: W. H. Freeman & Co. Publishers.

Hayflick, L. 1968. Human cells and aging. *Scientific American* 218:32–37.

Hayflick, L. 1975. Cell biology of aging. *BioScience* 25:629–636.

Hayflick, L. 1977. The biology of aging. *Natural History* 86:22, 26–27, 30.

Hayflick, L. 1980. The cell biology of human aging. *Scientific American* 242:58–65.

Kendig, F., and R. Hutton. 1979. *Life spans or how long things last.* New York: Holt, Rinehart & Winston General Book.

Kirkwood, T. B. L. 1977. Evolution of ageing. *Nature* 270:301–303.

Knook, D. L. 1982. Aging of cells: Accident or programme? *Endeavour* 6:162–167.

Macieira-Coelho, A. 1981. Tissue culture in aging research: Present status and prospects. *Experientia* 37:1050–1053.

MacLeod, W. 1966. Progeria. *British Journal of Radiology* 39:224–226.

Martin, G. M., C. A. Sprague, and C. J. Epstein. 1970. Replicative life-span of cultivated human cells. *Laboratory Investigations* 23:86–92.

Moment, G. B. 1975. The Ponce de Leon trail today. *BioScience* 25:623–628.

Oota, K., T. Makinodan, M. Iriki, and L. S. Baker, ed. 1980. *Aging phenomena. Relationships among different levels of organization.* New York: Plenum Press.

Panati, C. 1980. *Breakthroughs.* Boston: Houghton Mifflin Co.

Pitskhelauri, G. Z. 1982. *The longevity of Soviet Georgia*. Translated from the Russian and edited by G. Lesnoff-Caravoglia. New York: Human Sciences Press, Inc.

Rockstein, M., and M. Sussman. 1979. *Biology of aging*. Belmont, CA: Wadsworth Inc.

Schneider, E. L., ed. 1978. *The genetics of aging*. New York: Plenum Press.

Timiras, P. S. 1978. Biological perspectives on aging. *American Scientist* 66:605–613.

Chapter 17

Ann Arbor Science for The People Editorial Collective. 1977. *Biology as a social weapon*. Minneapolis: Burgess Publishing Co.

Applewhite, P. B. 1981. *Molecular gods. How molecules determine our behavior*. Englewood Cliffs, NJ: Prentice-Hall, Inc.

Baldwin, J. D., and J. I. Baldwin. 1981. *Beyond sociobiology*. New York: Elsevier.

Barash, D. P. 1977. *Sociobiology and behavior*. New York: Elsevier.

Barash, D. 1981. *The whisperings within: Evolution and the origin of human nature*. New York: Penguin Books.

Bateson, P. P. G., and P. H. Klopfer, ed. 1978. *Perspectives in ethology*. New York: Plenum Press.

Bennion, L. J., and T.-K. Li. 1976. Alcohol metabolism in American Indians and whites. *The New England Journal of Medicine* 294:9–13.

Benzer, S. 1973. The genetic dissection of behavior. *Scientific American* 229:24–37.

Block, N. J., and G. Dworkin, ed. 1976. *The IQ controversy*. New York: Pantheon Books Inc.

Bodmer, W. F., and L. L. Cavalli-Sforza. 1970. Intelligence and race. *Scientific American* 223:19–29.

Bouchard, T. J., Jr., and M. McGue. 1981. Familial studies of intelligence: A review. *Science* 212:1055–1059.

Boyd, W. C. 1950. *Genetics and the races of man*. Boston: Little, Brown and Co.

Broadhurst, P. L. 1978. *Drugs and the inheritance of behavior*. New York: Plenum Press.

Caplan, A. L., ed. 1978. *The sociobiology debate*. New York: Harper & Row, Publishers Inc.

Clutton-Brock, T. H., and P. H. Harvey, ed. 1978. *Readings in sociobiology*. San Francisco: W. H. Freeman & Co. Publishers.

Cooper, J. O. 1981. *Measuring behavior*. Columbus: Charles E. Merrill Publishing Co.

Davis, B. D. 1982. The importance of human individuality for sociobiology. *Perspectives in Biology and Medicine* 26:1–18.

Dawkins, R. 1976. *The selfish gene*. New York: Oxford University Press.

Dawkins, R. 1982. *The extended phenotype*. San Francisco: W. H. Freeman & Co. Publishers.

Dawkins, R. 1982. The myth of genetic determinism. *New Scientist*, Jan., 27–30.

Dobzhansky, T. 1973. *Genetic diversity and human equality*. New York: Basic Books Inc., Publishers.

Douglas, J. H. 1978. Intelligence: The hundred years' war. *Science News* 114:370–371.

Ehrlich, P. R., and S. S. Feldman. 1977. *The race bomb: Skin color, prejudice, and intelligence*. New York: Quadrangle/The New York Times Book Co.

Ehrman, L., and P. A. Parsons. 1976. *The genetics of behavior*. Sunderland, MA: Sinauer Associates Inc.

Eriksson, C. J. P., and M. A. Schuckit. 1980. Elevated blood acetaldehyde levels in alcoholics and their relatives: A reevaluation. *Science* 207:1383–1384.

Farber, S. L. 1981. *Identical twins reared apart. A reanalysis*. New York: Basic Books Inc., Publishers.

Feldman, M. W., and R. C. Lewontin. 1975. The heritability hang-up. *Science* 190:1163–1168.

Fieve, R. R., D. Rosenthal, and H. Brill, ed. 1975. *Genetic research in psychiatry*. Baltimore: The Johns Hopkins University Press.

Flynn, J. R. 1980. *Race, IQ, and Jensen*. London: Routledge & Kegan Paul.

Foster, H. L., J. D. Small, and J. G. Fox, ed. 1981. *The mouse in biomedical research*. New York: Academic Press Inc.

Frazer, A., and A. Winokur, ed. 1977. *Biological bases of psychiatric disorders*. New York: Spectrum Publications.

Freedman, D. G. 1979. *Human sociobiology. A holistic approach*. New York: The Free Press.

Fuller, J. L., and W. R. Thompson. 1978. *Foundations of behavior genetics*. St. Louis: C. V. Mosby Co.

Furlong, F. W. 1981. Determinism and freewill: Review of the literature. *American Journal of Psychiatry* 138:435–439.

Gambill, J. D. 1981. The relevance of sociobiology for mental illness. *Perspectives in Biology and Medicine*, Autumn, 155–165.

Goldsby, R. 1971. *Race and races*. New York: Macmillan Publishing Co. Inc.

Goodwin, D. W. 1979. Alcoholism and heredity. *Archives of General Psychiatry* 36:57–61.

Gottesman, I. I., and J. Shields. 1973. Genetic theorizing and schizophrenia. *British Journal of Psychiatry* 122:15–30.

Gottesman, I. I., and J. Shields. 1982. *Schizophrenia. The epigenetic puzzle*. Cambridge: Cambridge University Press.

Gould, S. J. 1981. *The mismeasure of man*. New York: W. W. Norton & Co., Inc.

Gould, S. J. 1982. A nation of morons. *New Scientist*, May, 349–352.

497

Gregory, M. S., A. Silvers, and D. Sutch, ed. 1978. *Sociobiology and human nature*. San Francisco: Jossey-Bass Inc., Publishers.

Halsey, A. H., ed. 1977. *Heredity and environment*. New York: The Free Press.

Harwood, J. 1983. The I.Q. in history. *Social Studies of Science* 13:465–477.

Hearnshaw, L. S. 1979. *Cyril Burt, psychologist*. Ithaca, NY: Cornell University Press.

Hinde, R. A. 1974. *Biological bases of human social behaviour*. New York: McGraw-Hill Book Co.

Hofer, M. A. 1981. *The roots of human behavior*. San Francisco: W. H. Freeman & Co. Publishers.

Holden, C. 1978. The criminal mind: A new look at an ancient puzzle. *Science* 199:511–514.

Hook, E. B., and K. M. Healy. 1977. Height and seriousness of crime in XYY men. *Journal of Medical Genetics* 14:10–12.

Hook, E. B., and D.-S. Kim. 1970. Prevalence of XYY and XXY karyotypes in 337 nonretarded young offenders. *The New England Journal of Medicine* 283:410–411.

Hubbard, R., and M. Lowe, ed. 1979. *Genes and gender. II. Pitfalls in research on sex and gender*. Staten Island, NY: Gordian Press Inc.

Hull, D. L. 1978. Sociobiology: Scientific bandwagon or traveling medicine show? *Society* 15:50–59.

Jakab, I., ed. 1982. *Mental retardation*. New York: S. Karger.

Jensen, A. R. 1972. *Genetics and education*. New York: Harper & Row, Publishers Inc.

Kaij, H. L. 1972. Definitions of alcoholism and genetic research. *Annals of The New York Academy of Science* 197:110–113.

Kaij, L. 1960. *Alcoholism in twins. Studies on the etiology and sequels of abuse of alcohol*. Stockholm: Almquist & Wiksell.

Kamin, L. 1974. *The science and politics of IQ*. New York: Halsted Press.

Kaplan, A. R., ed. 1972. *Genetic factors in schizophrenia*. Springfield, IL: Charles C. Thomas, Publisher.

Karlsson, J. L. 1978. *Inheritance of creative intelligence*. Chicago: Nelson-Hall Publishers.

King, J. C. 1971. *The biology of race*. New York: Harcourt Brace Jovanovich Inc.

Lewontin, R. C. 1970. Race and intelligence. *Bulletin of the Atomic Scientists* 26:2–8.

Lieber, C. S. 1976. The metabolism of alcohol. *Scientific American* 234:25–33.

Loehlin, J. C. 1980. Recent adoption studies of I.Q. *Human Genetics* 55:297–302.

Loehlin, J. C., G. Lindzey, and J. N. Spuhler. 1975. *Race differences in intelligence*. San Francisco: W. H. Freeman & Co. Publishers.

Lopreato, J. 1981. Toward a theory of genuine altruism in *Homo sapiens*. *Ethology and Sociobiology* 2:113–126.

498

Lumsden, C. J., and E. O. Wilson. 1981. *Genes, mind, and culture*. Cambridge, MA: Harvard University Press.

MacAlpine, I., and R. Hunter. 1969. Porphyria and King George III. *Scientific American* 221:38–46.

Mackintosh, N. J. 1981. A new measure of intelligence? *Nature* 289:529–530.

Mead, M., T. Dobzhansky, E. Tobach, and R. E. Light, ed. 1969. *Science and the concept of race*. New York: Columbia University Press.

Mednick, S. A., and K. O. Christiansen, ed. 1977. *Biosocial bases of criminal behavior*. New York: Gardner Press Inc.

Mednick, S. A., W. F. Gabrielli, Jr., and B. Hutchings. 1984. Genetic influences in criminal convictions: Evidence from an adoption cohort. *Science* 224:891–893.

Montagu, A. 1976. *The nature of human aggression*. New York: Oxford University Press.

Osborne, R. T. 1980. *Twins: Black and white*. Athens, GA: Foundation for Human Understanding.

Plomin, R., J. C. DeFries, and G. E. McClearn. 1980. *Behavioral genetics. A primer*. San Francisco: W. H. Freeman & Co. Publishers.

Porter, I. H., and R. G. Skalko, ed. 1973. *Heredity and society*. New York: Academic Press Inc.

Rainer, J. D. 1976. Genetics of intelligence: Current issues and unsolved questions. *Research Communications in Psychology, Psychiatry and Behavior* 1:607–618.

Reynolds, V. 1976. *The biology of human action*. San Francisco: W. H. Freeman & Co. Publishers.

Rieder, R. O. 1973. The offspring of schizophrenic parents: A review. *Journal of Nervous and Mental Disease* 157:179–190.

Rosenthal, D. 1970. *Genetic theory and abnormal behavior*. New York: McGraw-Hill Book Co.

Rosenthal, D. 1971. *Genetics of psychopathology*. New York: McGraw-Hill Book Co.

Ruse, M. 1979. *Sociobiology: Sense or nonsense?* Boston: D. Reidel Publishers.

Schuckit, M. A., D. W. Goodwin, and G. Winokur 1972. A study of alcoholism in half siblings. *American Journal of Psychiatry* 128:122–126.

Schuckit, M. A., and V. Rayses. 1979. Ethanol ingestion: Differences in blood acetaldehyde concentrations in relatives of alcoholics and controls. *Science* 203:54–55.

Scott, J. P. 1975. *Aggression*. Chicago: University of Chicago Press.

Shean, G. 1978. *Schizophrenia. An introduction to research and theory*. Cambridge, MA: Winthrop Publishers Inc.

Singer, P. 1981. *The expanding circle. Ethics and sociobiology*. New York: Farrar, Straus & Giroux.

Smith, J. M. 1978. The evolution of behavior. *Scientific American* 239:176–192.

499

Smith, R. D. 1979. Twins and science. *The Sciences* 19:10–14.

Sociobiology Study Group of Science for the People. 1976. Sociobiology—Another biological determinism. *BioScience* 26:182, 184.

Stent, G. S. 1980. *Morality as a biological phenomenon*. Berkeley: University of California Press.

Sulkowski, A. 1983. Psychobiology of schizophrenia: A neo-Jacksonia detour. *Perspectives in Biology and Medicine* 26:205–218.

Tariverdian, G., and B. Weck. 1982. Non-specific X-linked mental retardation—A review. *Human Genetics* 62:95–102.

Tsuang, M. T., and R. Vandermey. *Genes and the mind. Inheritance of mental illness*. New York: Oxford University Press.

Vale, J. R. 1980. *Genes, environment, and behavior. An interactionist approach*. New York: Harper & Row, Publishers Inc.

Van Abeelen, J. H. F. 1974. *The genetics of behavior*. Amsterdam: North-Holland Publishing Co.

Vernon, P. E. 1979. *Intelligence: Heredity and environment*. San Francisco: W. H. Freeman & Co. Publishers.

Watson, P. 1981. *Twins*. New York: The Viking Press.

Wilson, E. O. 1975. *Sociobiology: The new synthesis*. Cambridge, MA: Harvard University Press.

Wilson, E. O. 1976. Academic vigilantism and the political significance of sociobiology. *BioScience* 26:183–190.

Wilson, E. O. 1981. Epigenesis and the evolution of social systems. *The Journal of Heredity* 72:70–77.

Wilson, E. O. 1982. Sociobiology, individuality and ethics: A response. *Perspectives in Biology and Medicine* 26:19–29.

Wilson, R. S. 1983. Human behavioral development and genetics. *Acta Geneticae Medicae et Gemellologiae* 32:1–16.

Witkin, H. A., S. A. Mednick, F. Schulsinger, E. Bakkestrom, K. O. Christiansen, D. R. Goodenough, K. Hirschhorn, C. Lundsteen, D. R. Owen, J. Philip, D. B. Rubin, and M. Stocking. 1976. Criminality in XYY and XXY men. *Science* 193:547–555.

Wray, H. 1982. Melancholy genes. *Science News* 121:108–109.

Chapter 18

Alexander, R. D. 1978. Evolution, creation, and biology teaching. *The American Biology Teacher* 40:99–104.

Alexander, R. D. 1979. *Darwinism and human affairs*. Seattle: University of Washington Press.

Appleman, P. 1979. *Darwin*. New York: W. W. Norton & Co. Inc.

Ayala, F. J. 1978. The mechanisms of evolution. *Scientific American* 239:56–69.

Bajema, C. J., ed. 1971. *Natural selection in human populations. The measurement of ongoing genetic evolution in contemporary societies*. New York: John Wiley & Sons Inc.

500

Bannister, R. C. 1979. *Social Darwinism. Science and myth in Anglo-American social thought*. Philadelphia: Temple University Press.

Barghoorn, E. S. 1971. The oldest fossils. *Scientific American* 224:30–42.

Beck, S. D. 1982. Natural science and creationist theology. *BioScience* 32:738–742.

Berry, R. J. 1977. Evolution by natural selection—or not? *Biologist* 24:236–238.

Bishop, J. A., and L. M. Cook. 1975. Moths, melanism, and clean air. *Scientific American* 232:90–99.

Brush, S. G. 1981. Creationism/evolution: The case against "equal time." *The Science Teacher* 48:29–33.

Callaghan, C. A. 1980. Evolution and creationist arguments. *The American Biology Teacher* 42:422–425, 427.

Chancellor, J. 1973. *Charles Darwin*. New York: Toplinger Publishing Co.

Clarke, B. 1975. The causes of biological diversity. *Scientific American* 233:50–60.

Cloud, P. 1977. "Scientific creationism"—A new inquisition brewing? *The Humanist* 37:6–15.

Cravens, H. 1978. *The triumph of evolution*. Philadelphia: University of Pennsylvania Press.

Darwin, C. R. 1859. *On the origin of species by means of natural selection, or the preservation of favored races in the struggle for life*. London: John Murray.

Darwin, C. 1966. *The origin of species by means of natural selection*. Abridged and edited by C. Irvine and W. Irvine. New York: Frederick Ungar Publishing Co. Inc.

Darwin, F., ed. (1892) 1958. *The autobiography of Charles Darwin and selected letters*. D. Appleton and Co. Reprint. New York: Dover Publications Inc.

de Camp, L. S. 1968. *The great monkey trial*. Garden City, NY: Doubleday & Co. Inc.

Dickerson, R. E. 1978. Chemical evolution and the origin of life. *Scientific American* 239:70–86.

Dobzhansky, T. 1960. The present evolution of man. *Scientific American* 203:206–217.

Dobzhansky, T. 1973. *Genetic diversity and human equality*. New York: Basic Books Inc., Publishers.

Dobzhansky, T. 1973. Nothing in biology makes sense except in the light of evolution. *The American Biology Teacher* 35:125–130.

Eckhardt, R. B. 1972. Population genetics and human origins. *Scientific American* 226:94–103.

Eiseley, L. C. 1956. Charles Darwin. *Scientific American* 194:62–70.

Eiseley, L. 1979. *Darwin and the mysterious Mr. X*. New York: E. P. Dutton Inc.

Eldredge, N. 1982. *The monkey business*. New York: Washington Square Press.

Fowler, D. R. 1982. The creationist movement. *The American Biology Teacher* 44:528–530, 539–542.

Frye, R. M., ed. 1983. *Is God a creationist?* New York: The Scribner Book Companies Inc.

Futuyma, D. J. 1983. *Science on trial.* New York: Pantheon Books Inc.

Gillespie, N. C. 1979. *Charles Darwin and the problem of creation.* Chicago: University of Chicago Press.

Ginger, R. 1958. *Six days or forever. Tennessee vs. John Thomas Scopes.* New York: The New American Library Inc.

Gish, D. T. 1978. *Evolution? The fossils say no!* San Diego: Creation-Life Publishers.

Glick, T. F., ed. 1972. *The comparative reception of Darwinism.* Austin: University of Texas Press.

Godfrey, L. R. 1981. The flood of antievolutionism. *Natural History* 90:4–10.

Godfrey, L. R., ed. 1983. *Scientists confront creationism.* New York: W. W. Norton & Co., Inc.

Gould, S. J. 1977. *Ever since Darwin. Reflections in natural history.* New York: W. W. Norton & Co., Inc.

Gould, S. J. 1980. *The panda's thumb. More reflections in natural history.* New York: W. W. Norton & Co., Inc.

Gould, S. J. 1982. Darwinism and the expansion of evolutionary theory. *Science* 216:380–387.

Howells, W. W. 1966. *Homo erectus. Scientific American* 215:46–53.

Hughes, S. W. 1982. The fact and the theory of evolution. *The American Biology Teacher* 44:25–32.

Hull, D. L. 1973. *Darwin and his critics.* Cambridge, MA: Harvard University Press.

Hutchinson, C. 1980. The scientific status of the theory of evolution. *Biologist* 27:252–256.

Kaveski, S., and L. Margulis. 1983. The "sudden explosion" of animal fossils about 600 million years ago: Why? *The American Biology Teacher* 45:76–82.

Keosian, J. 1964. *The origin of life.* New York: Reinhold Publishing Corp.

Kettlewell, H. B. D. 1959. Darwin's missing evidence. *Scientific American* 200:48–53.

Kimura, M. 1979. The neutral theory of molecular evolution. *Scientific American* 241:98–126.

Kitcher, P. 1982. *Abusing science: The case against creationism.* Cambridge, MA: The MIT Press.

Lack, D. 1953. Darwin's finches. *Scientific American* 188:66–72.

Lerner, I. M. 1968. *Heredity, evolution, and society.* San Francisco: W. H. Freeman & Co. Publishers.

Levinton, J. S. 1982. Charles Darwin and Darwinism. *BioScience* 32:495–500.

Lewin, R. 1981. Do jumping genes make evolutionary leaps? *Science* 213:634–636.

Lewontin, R. C. 1978. Adaptation. *Scientific American* 239:212–230.

Lloyd, E. A. 1983. The nature of Darwin's support for the theory of natural selection. *Philosophy of Science* 50:112–129.

Margulis, L. 1982. *Early life*. New York: Van Nostrand Reinhold Co. Inc.

Mayr, E. 1978. Evolution. *Scientific American* 239:47–55.

Mayr, E. 1982. *The growth of biological thought. Diversity, evolution, and inheritance*. Cambridge, MA: Belknap Press of Harvard University Press.

Mayr, E., and Provine, W. B. 1980. *The evolutionary synthesis*. Cambridge, MA: Harvard University Press.

Miller, K. R. 1982. Special creation and the fossil record: The central fallacy. *The American Biology Teacher* 44:85–89.

Milne, D. H. 1981. How to debate with creationists—and "win." *The American Biology Teacher* 43:235–245, 266.

Moore, J. A. 1975. On giving equal time to the teaching of evolution and creation. *Perspectives in Biology and Medicine* 18:405–418.

Moore, J. A. 1979. Dealing with controversy: A challenge to the universities. *The American Biology Teacher* 41:544–547.

Moore, J. N. 1973. Evolution, creation, and the scientific method. *The American Biology Teacher* 35:23–26.

National Association of Biology Teachers. 1978. *A compendium of information on the theory of evolution and the evolution-creationism controversy*. Reston, VA: National Association of Biology Teachers.

Nelkin, D. 1976. The science-textbook controversies. *Scientific American* 234:33–39.

Newell, N. D. 1982. *Creation and evolution. Myth or reality?* New York: Columbia University Press.

Numbers, R. L. 1982. Creationism in 20th-century America. *Science* 218:538–544.

Olby, R. C. 1967. *Charles Darwin*. London: Oxford University Press.

Oldroyd, D. R. 1982. *Darwinian impacts. An introduction to the Darwinian revolution*. Atlantic Highlands, NJ: Humanities Press Inc.

Ospovat, D. 1981. *The development of Darwin's theory*. Cambridge: Cambridge University Press.

Raloff, J. 1982. Of God and Darwin. *Science News* 121:12–13.

Raloff, J. 1982. They call it creation science. *Science News* 121:44–46.

Richards, E. 1983. Will the real Charles Darwin please stand up. *New Scientist*, 22–29 Dec., 884–887.

Ridley, M. 1981. Who doubts evolution? *New Scientist* 90:830–832.

Root-Bernstein, R., and D. L. McEachron. 1982. Teaching theories: The evolution-creation controversy. *The American Biology Teacher* 44:413–420.

Ruse, M. 1979. *The Darwinian revolution*. Chicago: University of Chicago Press.

Ruse, M. 1982. A philosopher at the monkey trial. *New Scientist* 93:317–319.

Ruse, M. 1982. Creation science: The ultimate fraud. *New Scientist*, May, 586–591.

Ruse, M. 1982. *Darwinism defended. A guide to evolution controversies*. Reading, MA: Addison-Wesley Publishing Co. Inc.

Russett, C. E. 1976. *Darwin in America. The intellectual response. 1865–1912*. San Francisco: W. H. Freeman & Co. Publishers.

503

Scheer, B. T. 1979. Evolution: Help for the confused. *BioScience* 29:238–241.

Schopf, J. W. 1978. The evolution of the earliest cells. *Scientific American* 239:111–138.

Schwartz, A. W. 1983. Chemical evolution: The first stages. *Die Naturwissenschaften* 70:373–377.

Schweber, S. S. 1978. The genesis of natural selection—1838: Some further insights. *BioScience* 28:321–325.

Simons, E. L. 1964. The early relatives of man. *Scientific American* 211:50–62.

Simons, E. L. 1967. The earliest apes. *Scientific American* 217:28–35.

Simons, E. L. 1977. Ramapithecus. *Scientific American* 236:28–35.

Smith, J. M., ed. 1982. *Evolution now. A century after Darwin*. San Francisco: W. H. Freeman & Co. Publishers.

Stanley, S. M. 1981. Darwin done over. *The Sciences* 21:18–23.

Stanley, S. M. 1981. *The new evolutionary timetable. Fossils, genes, and the origin of species*. New York: Basic Books Inc., Publishers.

Stanley, S. M. 1982. Macroevolution and the fossil record. *Evolution* 36:460–473.

Stebbins, G. L. 1971. *Processes of organic evolution*. Englewood Cliffs, NJ: Prentice-Hall, Inc.

Stebbins, G. L. 1972. Evolution as the central theme of biology. *Biological Sciences Curriculum Study Newsletter* 49:4.

Stebbins, G. L. 1982. *Darwin to DNA, molecules to humanity*. San Francisco: W. H. Freeman & Co. Publishers.

Stone, I. 1941. *Clarence Darrow for the defense*. Garden City, NY: Doubleday & Co. Inc.

Stone, I. 1980. *The origin. A biographical novel of Charles Darwin*. Garden City, NY: Doubleday & Co. Inc.

Stone, I. 1981. Darwin's theory: An exercise in science. *New Scientist* 90:828–830.

Stone, I. 1981. The "unfinished naturalist" of the *Beagle. New Scientist* 92:902–905.

Stone, I. 1982. The death of Darwin. *New Scientist*, April, 91–93.

Taylor, G. R. 1983. *The great evolution mystery*. New York: Harper & Row, Publishers Inc.

Volpe, E. P. 1981. *Understanding evolution*. Dubuque, IA: Wm. C. Brown Group.

Wallace, B. 1966. *Chromosomes, giant molecules, and evolution*. New York: W. W. Norton & Co., Inc.

Wallace, B. 1972. Science, biology, and evolution. *Biological Sciences Curriculum Study Newsletter* 49:2–3.

Washburn, S. L. 1960. Tools and human evolution. *Scientific American* 203:62–75.

Washburn, S. L. 1978. The evolution of man. *Scientific American* 239:194–208.

Washburn, S. 1982. Fifty years of studies on human evolution. *The Bulletin of the Atomic Scientists* 38:37–43.

504

Weinberg, S. L. 1978. Two views on the textbook watchers. *The American Biology Teacher* 40:541–545, 560.

Woese, C. R. 1984. The origin of life. *Carolina Biology Reader 13*, ed. J. J. Head. Burlington, NC: Carolina Biological Supply Co.

Young, D. A. 1982. *Christianity and the age of the earth.* Grand Rapids, MI: Zondervan Publishing House.

Young, L. B., ed. 1970. *Evolution of man.* New York: Oxford University Press.

Chapter 19

Abelson, J. 1980. A revolution in biology. *Science* 209:1319–1321.

Abelson, P. H. 1983. Biotechnology: An overview. *Science* 219:611–613.

Anderson, W. F., and E. G. Diacumakos. 1981. Genetic engineering in mammalian cells. *Scientific American* 245:106–121.

Baer, A. S. 1977. *Heredity and society. Readings in social genetics.* New York: Macmillan Publishing Co. Inc.

Barton, K. A., and W. J. Brill. 1983. Prospects in plant genetic engineering. *Science* 219:671–676.

Beckwith, J. 1977. Recombinant DNA: Does the fault lie within our genes? *Science for the People* 9:14–17.

Berg, P., D. Baltimore, S. Brenner, R. O. Roblin III, and M. Singer. 1975. Asilomar conference on recombinant DNA molecules. *Science* 188:991–994.

Blank, R. H. 1981. *The political implications of human genetic technology.* Boulder, CO: Westview Press Inc.

Bodde, T. 1982. Genetic engineering in agriculture: Another green revolution. *BioScience* 32:572–575.

Bodde, T. 1982. Interferon: Will it live up to its promise? *BioScience* 32:13–15.

Brown, D. D. 1973. The isolation of genes. *Scientific American* 229:20–29.

Campbell, A. M. 1976. How viruses insert their DNA into the DNA of the host cell. *Scientific American* 235:102–113.

Chargaff, E. 1976. On the dangers of genetic meddling. *Science* 192:938–940.

Chilton, M.-D. 1983. A vector for introducing new genes into plants. *Scientific American* 248:51–59.

Cohen, S. N. 1975. The manipulation of genes. *Scientific American* 233:24–33.

Cohen, S. N. 1977. Recombinant DNA: Fact and fiction. *Science* 195:654–657.

Cohen, S. N., and J. A. Shapiro. 1980. Transposable genetic elements. *Scientific American* 242:40–49.

Culliton, B. J. 1981. Biomedical research enters the marketplace. *The New England Journal of Medicine* 304:1195–1201.

Facklam, M., and H. Facklam. 1979. *From cell to clone.* New York: Harcourt Brace Jovanovich Inc.

Fraser, T. H. 1980. The future of recombinant DNA technology in medicine. *Perspectives in Biology and Medicine* 23:499–512.

505

Green, H. P. 1978. The recombinant DNA controversy: A model of public influence. *Bulletin of the Atomic Scientists* 34:12–16.

Grobstein, C. 1977. The recombinant-DNA debate. *Scientific American* 237:22–33.

Grobstein, C. 1979. *A double image of the double helix. The recombinant-DNA debate.* San Francisco: W. H. Freeman & Co. Publishers.

Hamilton, M. P., ed. 1972. *The new genetics and the future of man.* Grand Rapids, MI: Wm. B. Eerdmans Publishing Co.

Hanson, E. D. 1982. Recombinant DNA and responsible intervention. *BioScience* 32:730–732.

Hanson, E. D., ed. 1983. *Recombinant DNA research and the human prospect.* Washington, DC: American Chemical Society.

Harsanyi, Z., and R. Hutton. 1981. *Genetic prophecy: Beyond the double helix.* New York: Rawson-Wade Publishers, Inc.

Helling, R. B., and S. L. Allen. 1976. Freedom of inquiry and scientific responsibility. *BioScience* 26:609–610.

Hickman, L., and A. Al-Hibri, ed. 1981. *Technology and human affairs.* St. Louis: C. V. Mosby Co.

Hopson, J. L. 1976. Genetic sabotage in the public interest. *Science News* 109:188–190.

Hoskins, B. B., J. T. O'Connor, T. A. Shannon, R. Widdus, and J. F. Danielli. 1977. Application of genetic and cellular manipulation to agricultural and industrial problems. *BioScience* 27:188–191.

Howard, T., and J. Rifkin. 1977. *Who should play God?* New York: Dell Publishing Co. Inc.

Hubbard, R. 1976. Gazing into the crystal ball. *BioScience* 26:608,611.

Humber, J. M., and R. F. Almedar, ed. 1979. *Biomedical ethics and the law.* New York: Plenum Press.

Johnson, I. S. 1983. Human insulin from recombinant DNA technology. *Science* 219:632–637.

Karp, L. E. 1976. *Genetic engineering: Threat or promise.* Chicago: Nelson-Hall Publishers.

Kieffer, G. H. 1979. *Bioethics: A textbook of issues.* Reading, MA: Addison-Wesley Publishing Co. Inc.

Krimsky, S. 1982. *Genetic alchemy. The social history of the recombinant DNA controversy.* Cambridge, MA: The MIT Press.

Lipkin, M., Jr., and P. T. Rowley, ed. 1974. *Genetic responsibility. On choosing our children's genes.* New York: Plenum Press.

Mertens, T. R. 1975. *Human genetics. Readings on the implications of genetic engineering.* New York: John Wiley & Sons Inc.

Miller, J. A. 1980. Spliced genes get down to business. *Science News* 117:202–205.

Miller, J. A. 1983. Toward gene therapy: Lesch-Nyhan syndrome. *Science News* 124:90–91.

Miller, J. A. 1984. The clergy ponder the new genetics. *Science News* 125:188–190.

506

Morris, G. 1984. Genetic engineering and the jumping gene. *Science News* 125:264–268.

Motulsky, A. G. 1974. Brave new world? *Science* 185:653–663.

Motulsky, A. G. 1983. Impact of genetic manipulation on society and medicine. *Science* 219:135–140.

Office of Technology Assessment. 1981. *Impacts of applied genetics. Microorganisms, plants, and animals*. Washington, DC: U.S. Government Printing Office.

Office of Technology Assessment. 1982. *Genetic technology. A new frontier*. Boulder, CO: Westview Press Inc.

Oosthuizen, G. C., H. A. Shapiro, and S. A. Strauss. 1980. *Genetics and society*. New York: Oxford University Press.

Plant, D. W., N. J. Reimers, and N. D. Zinder, ed. 1982. *Patenting of life forms*. Cold Spring Harbor, NY: Cold Spring Harbor Laboratory.

Pollack, R. 1982. Biologists in pinstripes. *The Sciences* 22:35–37.

Research and Education Association. 1982. *Genetic engineering*. New York: Research and Education Association.

Santos, M. A. 1981. *Genetics and man's future. Legal, social, and moral implications of genetic engineering*. Springfield, IL: Charles C. Thomas, Publisher.

Science Policy Research Division, Congressional Research Service, Library of Congress. 1980. *Genetic engineering, human genetics, and cell biology. Evolution of technological issues. Biotechnology*, suppl. report III. Washington, DC: U.S. Government Printing Office.

Shepard, J. F., D. Bidney, T. Barsby, and R. Kremble. 1983. Genetic transfer in plants through interspecific protoplast fusion. *Science* 219:683–688.

Sinsheimer, R. 1975. Troubled dawn for genetic engineering. *New Scientist* 16:148–151.

Sprague, G. F., D. E. Alexander, and J. W. Dudley. 1980. Plant breeding and genetic engineering: A perspective. *BioScience* 30:17–21.

Sylvester, E. J., and L. C. Klotz. 1983. *The gene age*. New York: The Scribner Book Companies Inc.

Thomas, D., and G. Gellf. 1981. Enzyme engineering: Accomplishments and prospects. *Endeavour* 5:96–98.

Wade, N. 1976. Recombinant DNA: A critic questions the right to free inquiry. *Science* 194:303–305.

Wade, N. 1977. *The ultimate experiment: Man-made evolution*. New York: Walker and Co.

Watson, J. D., and J. Tooze. 1981. *The DNA story. A documentary history of gene cloning*. San Francisco: W. H. Freeman & Co. Publishers.

Chapter 20

Baer, A. S. 1977. *Heredity and society. Readings in social genetics*. New York: Macmillan Publishing Co. Inc.

Baer, A. S. 1977. *The genetic perspective*. Philadelphia: W. B. Saunders Co.

507

Barker, J. S. F., K. Hammond, and A. E. McClintock, ed. 1982. *Future developments in the genetic improvement of animals*. New York: Academic Press Inc.

Barton, K. A., and W. J. Brill. 1983. Prospects in plant genetic engineering. *Science* 219:671–676.

Beers, R. F., and E. G. Bassett, ed. 1977. *Recombinant molecules: Impact on science and society*. New York: Raven Press.

Berlin, J. 1984. Plant cell cultures—a future source of natural products? *Endeavour*, new ser. 8:5–8.

Borlaug, N. E. 1983. Contributions of conventional plant breeding to food production. *Science* 219:689–693.

Caplan, A., L. Herrera-Estrella, D. Inzé, E. Van Haute, M. Van Montagu, J. Schell, and P. Zambryski. 1983. Introduction of genetic material into plant cells. *Science* 222:815–821.

Carlson, P. S., and J. C. Polacco. 1975. Plant cell cultures: Genetic aspects of crop improvement. *Science* 188:622–625.

Carson, R. 1962. *Silent spring*. Boston: Houghton Mifflin Co.

Chaleff, R. S. 1983. Isolation of agronomically useful mutants from plant cell cultures. *Science* 219:676–682.

Curtin, M. E. 1983. Harvesting profitable products from plant tissue culture. *Biotechnology*, Oct., 649–656.

Curtis, B. C., and D. R. Johnson. 1969. Hybrid wheat. *Scientific American* 220:21–29.

Davidson, G. 1974. *Genetic control of insect pests*. New York: Academic Press Inc.

Day, P. R. 1977. Plant genetics: Increasing crop yield. *Science* 197:1334–1339.

Dethier, V. G. 1976. *Man's plague? Insects and agriculture*. Princeton, NJ: The Darwin Press Inc.

Duvick, D. N. 1984. Genetic diversity in major farm crops on the farm and in reserve. *Economic Botany* 38:161–178.

Ennis, W. B., Jr., W. M. Dowler, and W. Klassen. 1975. Crop protection to measure food supplies. *Science* 188:593–598.

Feldman, M., and E. R. Sears. 1981. The wild gene resources of wheat. *Scientific American* 244:102–112.

Hardy, R. W. F., and U. D. Havelka. 1975. Nitrogen fixation research: A key to world food? *Science* 188:633–643.

Harlan, J. R. 1975. Our vanishing genetic resources. *Science* 188:618–621.

Heiser, C. B., Jr. 1973. *Seeds to civilization. The story of man's food*. San Francisco: W. H. Freeman & Co. Publishers.

Horsch, R. B., R. T. Fraley, S. G. Rogers, P. R. Sanders, A. Lloyd, and N. Hoffman. 1984. Inheritance of functional foreign genes in plants. *Science* 223:496–498.

Hoskins, B. B., J. T. O'Connor, T. A. Shannon, R. Widdus, and J. F. Danielli. 1977. Applications of genetic and cellular manipulations to agricultural and industrial problems. *BioScience* 27:188–190.

508

Jennings, P. R. 1976. The amplification of agricultural production. *Scientific American* 235:180–194.

Luck, R. F., R. van den Bosch, and R. Garcia. 1977. Chemical insect control—A troubled pest management strategy. *BioScience* 27:606–611.

Mangelsdorf, P. C. 1951. Hybrid corn: Its genetic basis and its significance in human affairs. In *Genetics in the 20th century*, ed. L. C. Dunn. New York: Macmillan Publishing Co. Inc.

Matthews, G. 1983. Can we control insect pests? *New Scientist*, 12 May, 368–372.

National Academy of Sciences. 1972. *Genetic vulnerability of major crops*. Washington, DC: National Academy of Sciences Printing and Publishing Office.

National Academy of Sciences. 1975. *Contemporary pest control practices and prospects*. Washington, DC: National Academy of Sciences Printing and Publishing Office.

National Academy of Sciences. 1976. *Pest control and public health*. Washington, DC: National Academy of Sciences Printing and Publishing Office.

National Plant Genetic Resources Board. U.S. Department of Agriculture. 1978. *Plant genetic resources. Conservation and use*. Washington, DC: U.S. Government Printing Office.

Office of Technology Assessment. 1982. *Genetic technology. A new frontier*. Boulder, CO: Westview Press Inc.

Perutz, M. 1981. Why we need science. *New Scientist* 92:530–536.

Plucknett, D. L., N. J. H. Smith, J. T. Williams, and N. M. Anishetty. 1983. Crop germplasm and conservation in developing countries. *Science* 220:163–169.

Reitz, L. P. 1970. New wheats and social progress. *Science* 169:952–955.

Richardson, R. H., J. R. Ellison, and W. W. Averhoff. 1982. Autocidal control of screwworms in North America. *Science* 215:361–370.

Riley, R. 1980. Agricultural research. *Biologist* (London) 27:229–232.

Ruttan, V. W. 1982. Agricultural scientists as reluctant revolutionaries. *Interdisciplinary Science Reviews* 7:170–177.

Shepard, J. F., D. Bidney, T. Barsby, and R. Kemble. 1983. Genetic transfer in plants through interspecific protoplast fusion. *Science* 219:683–688.

Sigurbjornsson, B. 1971. Induced mutations in plants. *Scientific American* 224:86–95.

Smith, H. H. 1974. Model systems for somatic cell plant genetics. *BioScience* 24:269–276.

Smith, R. H., and R. C. von Borstel. 1972. Genetic control of insect populations. *Science* 178:1164–1174.

Sprague, G. F., D. E. Alexander, and J. W. Dudley. 1980. Plant breeding and genetic engineering. *BioScience* 30:17–21.

Stakman, E. C., R. Bradfield, and P. C. Mangelsdorf. 1967. *Campaigns against hunger*. Cambridge, MA: Belknap Press of Harvard University Press.

Streeter, C. P. 1970. The wheat breeder who won the peace prize. *Farm Journal*, Dec., 16–17, 29.

509

Tudge, C. 1983. The future of crops. *New Scientist*, 26 May, 547–550.

U.S. Department of Agriculture. *Will there be enough food? The 1981 Yearbook of Agriculture*. Washington, DC: U.S. Government Printing Office.

van Emden, H. F. 1974. *Pest control and its ecology*. London: Edward Arnold Publishers.

Wade, N. 1975. International agricultural research. *Science* 188:585–589.

Walsh, J. 1981. Genetic vulnerability down on the farm. *Science* 214:161–164.

Chapter 21

Ann Arbor Science for the People Editorial Collective. 1977. *Biology as a social weapon*. Minneapolis: Burgess Publishing Co.

Blank, R. H. 1981. *The political implications of human genetic technology*. Boulder, CO: Westview Press Inc.

Brill, H. 1975. Presidential address: Nature and nurture as political issues. In *Genetic research in psychiatry*, ed. R. R. Fieve, D. Rosenthal, and H. Brill. Baltimore: The Johns Hopkins University Press.

Carlson, E. A. 1981. *Genes, radiation, and society: The life and work of H. J. Muller*. Ithaca, NY: Cornell University Press.

Caspari, E. W., and R. E. Marshak. 1965. The rise and fall of Lysenko. *Science* 149:275–278.

Djerassi, C. 1979. *The politics of contraception*. New York: W. W. Norton & Co., Inc.

Dobzhansky, T. 1968. On genetics and politics. *Social Education* 32:142–146.

Dobzhansky, T. 1973. *Genetic diversity and human equality*. New York: Basic Books Inc., Publishers.

Ferguson, J. R. 1983. Scientific freedom, national security, and the first amendment. *Science* 221:620–624.

Feyerabend, P. K. 1982. Science—political party or instrument of research? *Speculations in Science and Technology* 5:343–352.

Gilpin, R., and C. Wright, ed. 1964. *Scientists and national policy-making*. New York: Columbia University Press.

Gonzalez, H. B. 1984. Scientists and congress. *Science* 224:127–129.

Gould, S. J. 1981. *The mismeasure of man*. New York: W. W. Norton & Co., Inc.

Greenberg, D. S. 1967. *The politics of pure science*. New York: The New American Library Inc.

Haldane, J. B. S. 1938. *Heredity and politics*. New York: W. W. Norton & Co., Inc.

Holmes, H. B., B. B. Hoskins, and M. Gross, ed. 1981. *The custom-made child? Women-centered perspectives*. Clifton, NJ: Humana Press.

Hubbard, R., and M. Lowe. 1979. *Genes and gender II*. Staten Island, NY: Gordian Press Inc.

Jensen, A. R. 1973. *Genetics and education*. New York: Harper & Row, Publishers Inc.

Kamin, L. 1974. *The science and politics of I.Q.* New York: Halsted Press.

FURTHER READING

Lappe, M. 1979. *Genetic politics. The limits of biological control*. New York: Simon & Schuster.

Lubrano, L. L., and S. G. Solomon, ed. 1980. *The social context of Soviet science*. Boulder, CO: Westview Press Inc.

Massie, R. K. 1967. *Nicholas and Alexandra*. New York: Atheneum Publishers.

McKusick, V. A. 1965. The royal hemophilia. *Scientific American* 213:88–95.

Medvedev, Z. A. 1969. *The rise and fall of T. D. Lysenko*. Translated from the Russian by I. M. Lerner. New York: Columbia University Press.

Milunsky, A., and G. J. Annas, ed. 1980. *Genetics and the law II*. New York: Plenum Press.

Perutz, M. 1981. Why we need science. *New Scientist*, 19 Nov., 530–536.

Stover, E., and T. Eisner. 1982. Human rights abuses and the role of scientists. *BioScience* 32:871–875.

Wade, N. 1976. Recombinant DNA: A critic questions the right to free inquiry. *Science* 194:303–305.

Wade, N. 1977. Gene splicing: Senate bill draws charges of Lysenkoism. *Science* 197:348–349.

Watson, J. D., and J. Tooze. 1981. *The DNA story*. San Francisco: W. H. Freeman & Co. Publishers.

Wiesner, J. B. 1965. *Where science and politics meet*. New York: McGraw-Hill Book Co.

Glossary

ABH blood system. A major blood system whose identification is based on the occurrence of agglutination of the red blood cells when bloods from incompatible groups are mixed. The blood types (A, B, AB, and O) are named based on the particular antigens found on the red blood cells and the antibodies in the plasma.

Abiogenesis. A concept stating that life can be produced spontaneously from nonliving substances such as mud, ocean foam, and decaying wood.

ABO blood group. See ABH blood system.

Abortion. The expulsion of a fetus from the uterus before it is sufficiently developed to survive.

Abortus. A fetus that is expelled from the uterus before it is sufficiently developed to survive.

Achondroplasia. A dominant form of dwarfism in humans.

Acute radiation syndrome. The physical damage to human health brought about by a large amount of radiation received within a very brief period of time.

Adaptation. Any characteristic of an organism, or population of organisms, that aids in its adjustment to a particular environment.

Additive effects. The situation in which a group of genes affects the same phenotypic characteristic. Each gene adds to the effects of the others.

Adenine. One of the purine bases found in nucleic acids.

Adenosine triphosphate (ATP). One of the most important molecules utilized in the cell for energy transfers in metabolic processes.

513

Aflatoxin B$_1$. A carcinogen affecting the liver. Produced by certain fungal species.

Agglutination. The clumping effect brought upon antigenic structures, or cell components associated with them, by antibodies.

AID. The abbreviation for artificial insemination by a donor. Artificial insemination is the impregnation of the female without direct sexual contact.

AIH. The abbreviation for artificial insemination by the husband. Artificial insemination is the impregnation of the female without direct sexual contact.

Albinism. In animals, the absence of the pigment, melanin, in the skin, hair, and eyes. In plants, the absence of chlorophyll.

Albino. A person characterized by the absence of the pigment, melanin, in the skin, hair, and eyes.

Alcoholism. Alcoholism is difficult to define exactly. Generally, it is the condition resulting from the excessive and persistent use of alcoholic beverages.

Alkaptonuria. An inherited metabolic disorder that is due to an autosomal recessive gene. Alkaptonurics have excessive amounts of homogentisic acid in their urine because they lack the specific enzyme for breaking down the substance. The colorless homogentisic acid is oxidized to a black pigment; the urine darkens when exposed to the air.

Alleles. Two or more alternative forms of a gene that occupy the same relative sites, or loci, on homologous chromosomes.

Allelism. The relationship between alleles; that is, alternative forms of a gene that occupy the same locus or site on homologous chromosomes.

Alpha-1-antitrypsin. An enzyme normally found in the body; its deficiency can cause cirrhosis of the liver, pulmonary emphysema, or enlargement of the heart.

Alpha-fetoprotein. A protein that is formed by the fetal liver and identifiable in the amniotic fluid. Increased amounts of the protein are found in the amniotic fluid of fetuses with neural-tube closure defects. Therefore, amniocentesis techniques can be used to detect the defect before birth.

Alpha radiation. Radiation consisting of helium nuclei, which are composed of two protons and two neutrons.

Ames test. Also known as the *Salmonella*-mammalian microsome assay, an assay to determine the mutagenicity of various chemicals. It uses a special strain of *Salmonella* bacteria and a mammalian-liver extract. Reverse mutation rate is calculated in the *Salmonella*, and the liver-extract component is used as an activator of mutagens.

Amino acid. The basic chemical units of polypeptides that are organic compounds containing an amino group (NH_2), a carboxyl group (COOH), and a variable side chain.

514

Aminoacyl adenylate. One of the intermediate molecules in the formation of a chemical bond between an amino acid and its tRNA adaptor molecule.

Aminoacyl-tRNA synthetase. Any one of the specific enzymes that catalyze the reactions that activate the 20 amino acids used in protein synthesis.

Amniocentesis. A technique in which a sample of amniotic fluid is removed from the amniotic sac of a pregnant woman by needle puncture. This fluid and the suspended cells are then used for prenatal detection of fetal disorders. The term applies to both the procedure for obtaining the fluid and its analysis.

Amniotic fluid. The liquid contents of the amniotic sac within which the fetus develops.

Amniotic sac. The sac that contains the amniotic fluid and encloses the fetus.

Amphiploid. An organism that is diploid for two genomes, each from a different species.

Anaphase. A major stage of nuclear division, during which the daughter chromosomes in mitosis and meiosis II or the homologous chromosomes in meiosis I leave the equatorial plate and move to the opposite ends of the dividing cell.

Anaphase lag. The irregular chromosome movement at anaphase in which one or more chromosomes fail to move normally with the others. Such chromosomes are not included in the resulting nuclei and are usually lost.

Anencephaly. A congenital defect in which the brain and head are severely reduced, or even absent.

Aneuploid. A chromosome number in a cell or individual that is not an exact multiple of the haploid or basic number. The term is also used to describe an individual with an aneuploid chromosome number.

Angstrom (Å). A unit of length that is equal to 1/10,000th of a micrometer (micron), or 10^{-10} meter.

Ankylosing spondylitis. A condition in which vertebrae are abnormally stiff causing a forward-like spinal deformity.

Antibody. A protein substance, called an immunoglobulin, in a tissue or fluid of the body that is produced in response to a foreign substance known as an antigen. Different antibody molecules can recognize and specifically react with different antigens.

Anticodon. A sequence of three nucleotides in a tRNA molecule that pairs in a complementary way with a specific three-nucleotide codon in the mRNA molecule during polypeptide synthesis. This leads to the correct positioning of an amino acid into a growing polypeptide chain.

Antigen. Any foreign substance or large molecule that can stimulate the production of specific neutralizing antibodies when introduced into the tissues of a vertebrate organism.

515

Antiparallel.   Describes two molecules, or two parts of a molecule, that lie parallel to each other, but point in opposite directions. The two polynucleotide strands of a DNA molecule serve as a good example of an antiparallel structure.

Antisense strand.   The particular polynucleotide strand of a DNA molecule that is complementary to the sense strand. The sense strand defines the amino acid sequence of a polypeptide, while the sole function of the antisense strand is to generate another sense strand in the next generation of the cell lineage.

Artificial insemination.   The impregnation of the female without direct sexual contact.

Artificial selection.   The type of selection in which humans choose the genotypes of a given species that will contribute to the gene pool of succeeding generations.

Ascertainment bias.   Bias that is inherent in some methods of identifying and selecting families and individuals for inclusion in a study of human genetics.

Ascorbic acid.   Also known as vitamin C. This substance is essential to normal metabolism and prevents scurvy.

Asexual reproduction.   Any mode of reproduction that does not involve the union of genetic material from two sexes or mating types. Resulting progeny have the same genotype as the parent organism.

A-site.   A concept of protein synthesis that holds that the ribosome contains a site (called A) that accepts the incoming aminoacyl tRNA molecule.

Assortative mating.   Nonrandom mating in which like phenotypes or unlike phenotypes tend to mate with each other more often than expected with random mating.

Ataxia telangiectasia.   An autosomal recessive disease marked by the development of cerebellar ataxia (uncoordination of muscles) and telangiectasia (dilation of groups of capillaries). Severe immunodeficiency often accompanies this disease.

Atom.   The smallest particle of an element that is capable of taking part in chemical reactions.

ATP.   See adenosine triphosphate.

Autosomal dominant.   Describes a gene that lies within one of the nonsex chromosomes and expresses itself to the exclusion of the other allele when in the heterozygous state.

Autosomal recessive.   Describes a gene that lies within one of the nonsex chromosomes and does not express itself or affect the phenotype when in the heterozygous state.

Autosome.   All of the chromosomes, other than those designated as sex chromosomes.

Avirulent.   Lacking the ability to cause disease.

Background radiation.   The radiation that is emitted into our environment from naturally radioactive substances.

Bacterial-liver extract test. See Ames test.

Balanced translocation. A chromosome rearrangement in which two non-homologous chromosomes have exchanged segments, but all the genetic material is still present in the correct dose.

Barr body. The inactive condensed X chromosome generally observed appressed to the nuclear membrane in the nuclei of mammalian somatic cells. The number of Barr bodies is one less than the total number of X chromosomes present.

Base pair deletion. The absence of a single pair of nucleotides in the genetic material.

Base pair insertion. The addition of a single pair of nucleotides in the genetic material.

Base pair substitution. A mutational event in which one base pair in the nucleic acid is exchanged for another base pair.

Benign tumor. An abnormal, localized population of proliferating cells (neoplasm) in an animal. The tumor does not spread from the original site to another part of the body.

Beta particle. An electron emitted from the nucleus of an atom undergoing radioactive decay.

Bilateral retinoblastoma. A malignant condition of the cells in both retinas that eventually progresses along the optic nerve.

Biogeography. The branch of biology that deals with the geographic distribution of plants and animals.

Biological determinism. The controversial concept that humans are very much the products of their genes.

Biological pest control. A reduction in the population of an unwanted organism by the introduction of a parasite, predator, or disease-causing organism.

Biological science. The science that includes the study of living things.

Biome. A large community of living organisms having a certain dominant vegetation and characteristic animals.

Biopsy. The diagnostic examination of small pieces of tissue that have been removed from a living subject.

Biosphere. The part of the entire earth and its atmosphere that consists of living organisms.

Biosynthesis. The synthesis of the substances and components of the cell from simple precursors as the result of genetic direction.

Biotin. A member of the vitamin B complex essential for the growth of some organisms.

Bivalent. The configuration produced by two synapsed homologous chromosomes during meiosis I. Each of the chromosomes consists of two chromatids at this stage.

Black urine disease. See alkaptonuria.

Blastocyst. An early stage of mammalian development in which the cell mass is a hollow ball-shaped structure.

517

Bombay phenotype.   A blood phenotype in which a substance called H is absent. This substance is believed to be a precursor in the synthesis of the A and B blood antigens.

Brachydactyly.   An abnormality in humans caused by a dominant gene. The phenotype consists of markedly shortened fingers.

Broad heritability.   A measure of the extent to which the total phenotypic variance of a trait is the result of genetic factors, regardless of their mode of expression.

Buccal cell.   A cell from the mouth cavity or the inner cheek.

Budding.   A type of asexual reproduction in which a protuberance on an organism grows and develops into another organism.

Capsule.   The outermost structure of many kinds of microorganisms, including blue-green algae and bacteria. The capsule is external to the cell wall, but in close contact with it.

Carcinogen.   A substance capable of inducing cancer in an organism.

Carcinogenesis.   The origin and development of cancer.

Carrier.   An individual who is heterozygous for a normal gene and one of its alleles. The latter is not expressed because of complete dominance of the normal gene.

Cataract.   The partial or complete opacity of the crystalline lens of the eye or of its capsule.

Cat-cry syndrome.   A condition of multiple congenital malformations in humans. The syndrome is due to a deficiency in the short arm of chromosome 5. Infants with this condition produce a cry that sounds very much like a cat mewing.

Caucasoid.   Designates one of the major groups of people. Often referred to as the white race.

Cell.   The basic structural and functional unit of life, and the smallest membrane-bound unit of protoplasm produced by independent reproduction.

Cell cycle.   The life cycle of an individual cell; more specifically, the timed sequence of events occurring in a cell in the period between mitotic divisions.

Cell fusion.   The experimental formation of a single hybrid cell by the physical union of two somatic cells.

Cell membrane.   The thin surface component that surrounds a cell. Also called the plasma membrane.

Cellulose.   A carbohydrate constituent of cell walls of green plants and some fungi.

Cell wall.   The hard rigid material lying outside the cell membrane and surrounding plant cells.

Central dogma. The concept describing the basic relationships that exist among DNA, RNA, and protein. DNA serves as a template for its own replication and the synthesis of RNA, while RNA, in turn, is translated into protein. Today, it is known that RNA can also serve as the template for DNA synthesis in some organisms.

Centriole. The paired cellular organelles found near the nucleus and involved in the formation of the spindle fibers during nuclear division.

Centromere. The region of the eukaryotic chromosome involved in spindle fiber attachment during cell division; important in the movement of chromosomes to the poles of the cell. Also known as a kinetochore or primary constriction.

Cerebral palsy. Paralysis occurring as a result of a lesion in the cerebrum of the brain.

Character. A contraction of the word characteristic. Any one of the many details of structure, function, and substance of an organism that contributes to its phenotype.

Chargaff's rules. Analysis of the nucleotide base composition of DNA that showed the molar proportions of the bases to be such that adenine always equaled thymine and cytosine always equaled guanine.

Chimera. An organism having a mixture of tissues of genetically different constitutions. The individual is composed of cells derived from different zygotes.

Chorionic villus biopsy. A method of prenatal diagnosis in which chorionic villi (small epidermal projections) are obtained from the uterus and analyzed for genetic disorders.

Chromatid. One of the two daughter strands of a chromosome that has replicated. A chromosome, on the other hand, is composed of two chromatids held together at the centromeric region during prophase and metaphase of cell division.

Chromatin. The DNA-protein complex of eukaryotic chromosomes making up the substance of the nucleus during interphase.

Chromosomal aberration. Any change of chromosome number or basic chromosome structure.

Chromosome. In eukaryotes, a DNA-protein complex that is the structure for an array of genes. In prokaryotes, the chromosome is essentially a molecule of DNA, or RNA in a few viruses; generally, each species has a characteristic number of chromosomes.

Chromosome theory. The theory that describes chromosomes as the vehicles for genes and genetic information.

Cilium. A fine cytoplasmic thread projecting from the surface of certain cells.

Class. One of the groupings used in the classification of organisms; consists of subdivisions called orders.

Clastogen. Any agent that is capable of breaking chromosomes.

Cleavage. The process in which a fertilized egg cell divides mitotically to give rise to all of the initial cells of the developing embryo.

Cleft lip. Congenital condition in which the developing two halves of the face fail to unite at the midline, resulting in a fissure of the palate.

Cleft palate. See cleft lip.

Clone. A population in which all of the descendants, cells, or individual organisms are derived by asexual means from a single original cell. The members of the population, therefore, are genetically identical.

Cloning. The technique of developing large numbers of cells or individuals from a single ancestral cell by asexual means.

Clubfoot. A congenital condition in which the forefoot is deformed.

Cockayne syndrome. A disease characterized by dwarfism, mental retardation, and senility, among other symptoms; generally, the life span of such persons is greatly shortened.

Codominance. The situation in which alleles produce an independent effect. In a heterozygote there will be full expression of both alleles. The products from both alleles can be detected, and neither a masking effect nor a dilution effect exists.

Codon. A sequence of three adjacent ribonucleotides in an mRNA molecule, or three nucleotides in a DNA molecule, that specify a single amino acid placement in the synthesis of a polypeptide chain.

Coefficient of inbreeding. The probability that the two alleles at a particular locus are identical by descent.

Coefficient of relationship. The probability that two particular persons have inherited a certain gene from a common ancestor. Also, the proportion of all their genes that have been inherited from common ancestors.

Colchicine. A poisonous alkaloid found in some plants of the genus *Colchicum*. It arrests spindle formation and thereby interrupts mitosis.

Collagen. An abundant fibrous protein in animals, synthesized mostly by fibroblasts, and found in tissues such as skin, tendons, cartilage, and bone.

Complete dominance. The situation in which a gene expresses itself such that it yields the same phenotype when it is heterozygous as it does when it is homozygous.

Complete medium. A sterile nutritive substance containing most supplements that an organism requires for growth and development. Ingredients vary depending upon the organism.

Concordance. A term used in twin studies to describe the situation in which both members of a twin pair exhibit a certain trait.

Concordance rate. See concordance.

Congenital. Existing prior to or at the time of birth. The particular trait may have a genetic or an environmental cause.

Connective tissue. Those tissues that connect and bind other tissues together, such as bone, cartilage, ligaments, and tendons.

Consanguinity. Genetic relationship by descent resulting from a common ancestry, at least through the preceding few generations. Usually refers to matings between persons having these genetic relationships.

520

Continuous trait. See quantitative inheritance.

Contraception. Any means that prevents the fertilization (conception) of the egg by sperm.

Copulation. In the broad sense, a sexual act. More technically, the union of sexual units.

Corona radiata. A zone of follicular cells surrounding the ovum, persisting until fertilization.

Corpus luteum. A small yellow body in the space left in an ovary after an ovum ruptures. It functions as a gland.

Correlation coefficient. The measure of the intensity of an association, given as r; it is an unbiased estimate of the corresponding degree of association between two variables in a population.

Cosmic radiation. Variously charged particles that reach the earth from outer space.

Cowper's gland. An accessory sex gland in male mammals involved in secretions that are mixed with spermatozoa.

Creationism. The belief that the world was created by a supernatural power with all of its life forms intact, just as they are now.

Cri du chat. See cat-cry syndrome.

Cross-fertilization. The fusion of male and female gametes from different organisms.

Crossover. The reciprocal exchange of corresponding segments of genetic material between two homologous non-sister chromatids. This results in the recombination of alleles.

Cryobanking. See cryopreservation.

Cryopreservation. The treatment and storage of cells by freezing at very cold temperatures to preserve them for later use.

Cystic fibrosis. An autosomal recessive hereditary disease characterized by the production and accumulation of large amounts of viscous mucous in the lungs.

Cystinuria. The presence of an abnormal amount of cystine in the urine.

Cytogenetics. The science that deals with the problems that interrelate genetics and the cell (cytology).

Cytokinesis. The division of the cytoplasm resulting in two daughter cells from a parent cell. It is distinct from nuclear division, which is called karyokinesis.

Cytology. The branch of biology that deals with the study of the structure, development, function, and reproduction of cells.

Cytoplasm. The protoplasm of the cell (outside of the nucleus) containing the organelles of the cell. The protoplasm of the nucleus is called nucleoplasm.

Cytoplasmic male sterility. In plants, pollen abortion that occurs as a result of factors in the cytoplasm that are maternally transmitted. These genetic factors will only act in the absence of pollen-restoring genes.

Cytoplast. An enucleated cell; that is, one in which the nucleus has been removed and only the cytoplasm remains.

521

Cytosine. One of the pyrimidine nitrogenous bases found in DNA and RNA molecules.

Daughter cells. The two cells resulting from division of a single cell.

DDT. An insecticide, now banned from general use, chemically known as dichloro diphenyl trichloroethane.

Deficiency. Same as a deletion. The absence of a segment of the genetic material in a chromosome that may range from a single nucleotide to a significant part of a chromosome.

Degrees of freedom. The number of items of data that are free to vary independently; in any set of data only $n - 1$ items are free to vary when a mean is specified. The value of one item is rigidly determined by the other values and the mean.

Deletion. See deficiency.

Demography. Mostly, the field of human population analysis, but can also apply to the study of other animal populations.

Denaturation. The loss of the three-dimensional native configuration of a macromolecule as a result of various physical or chemical treatments. It is usually accompanied by the loss of biological activity. With regard to DNA, denaturation is the separation of the two strands of the double helix as a result of disruption of the hydrogen bonds.

Deoxyribonucleic acid. A molecule composed of nucleotide units, usually double stranded and helically coiled. Abbreviated DNA, the molecule is the principal material for the storage of genetic information in most organisms.

Deoxyribonucleotide. Any of the four different compounds that make up the backbone of the DNA molecule. These compounds consist of a purine or a pyrimidine bonded to the sugar, 2-deoxyribose, which, in turn, is bonded to a phosphate group.

Deoxyribose. A sugar molecule that is one of the basic constituents of DNA.

DES. See diethylstilbesterol.

Diabetes mellitus. A condition characterized by the failure of the pancreas to produce the insulin needed by body tissues to oxidize carbohydrates at the normal rate.

Diethylstilbesterol. A chemical, also known as DES, used in the past to prevent miscarriages.

Differentiation. A sequence of changes that are involved in the progressive specialization and diversification of cell types.

Dihybrid. Any individual that is heterozygous for two pairs of alleles at two different loci under consideration.

Diploid. The state of having two complete sets of chromosomes (two genomes) that are homologous to each other. It is the typical number of chromosomes for the species or group. Generally symbolized as $2n$.

Diploidization. The process of a haploid becoming a diploid by the doubling of the chromosomes.

522

Discordance. The situation, usually in twin studies, in which one member of a twin pair expresses a particular trait and the other does not.

Dizygotic twins. Twins that are the products of two ova fertilized by two separate spermatozoa at approximately the same time. Also called fraternal twins.

DNA. See deoxyribonucleic acid.

DNAase. Deoxyribonuclease. Any enzyme that degrades DNA.

Dominance. The situation in which one member of an allelic pair expresses itself to the exclusion of the other allele when in the heterozygous state.

Dopamine. An amine that serves as one of several chemical transmitters, essential to the normal nerve activity in the central nervous system.

Dosage compensation. A genetic mechanism of gene regulation that compensates for genes that are present in two doses in the homogametic sex, and in just one dose in the heterogametic sex.

Double cross. The technique sometimes used for producing hybrid seed for field corn. Inbred lines are first crossed; then the single-cross hybrids are crossed to gain the double cross for commercial agriculture.

Double helix. Two strands that coil around each other like two spirals. Double helix describes the overall structure of the DNA molecule.

Double trisomy. The aneuploid condition in which two chromosomes are additional to the normal complement, and each of the two is homologous to a different chromosome of the genome. This amounts to two separate trisomic situations in the same organism.

Down syndrome. A disorder in humans caused by the presence of an extra chromosome 21. Characterized by mental retardation, short stature, a fold on the eyelids, and a number of other physical malformations. Sometimes called mongolism.

Drumstick. A small protrusion (drumstick-like appendage) from the nucleus of the human polymorphonuclear leukocyte. This is sex chromatin, and it is observed in a certain percentage of the female cells, but not in male cells.

Duchenne muscular dystrophy. A disease characterized by muscular wasting.

Duplex. A description of a DNA molecule; it has two polynucleotide chains running in opposite directions and chemically bonded to each other.

Duplication. A chromosomal aberration in which a chromosome segment occurs twice in the same chromosome or genome.

Ecology. The study of how organisms interrelate to each other and their physical environment.

ECO-RI. A restriction enzyme that cleaves DNA at sites having a particular nucleotide base sequence, leaving cohesive ends on the DNA molecule.

Ecosystem. An ecological system composed of a group of interacting organisms and the abiotic components of the environment. It is the basic ecological unit, more or less self sustaining and self regulating.

523

Edwards syndrome. A highly lethal disorder resulting from the presence of three doses of chromosome 18. It is characterized by mental retardation, heart disease, and a variety of other symptoms.

Ejaculation. The forcible and sudden ejection of semen from the male genital organ.

Electromagnetic radiation. Long-wavelength radiation consisting of bundles of energy called photons.

Electrophoresis. A technique that allows for the migration of suspended molecules or particles in an electric field and the detection of these molecular migrations.

Emasculation. In a botanical context, the removal of the anthers from a flower.

Embryo. A young organism in the early stages of development arising from a fertilized egg. In animals, the embryo is usually considered as the stage incapable of supporting a separate existence.

Embryology. The science that is concerned with the origin, structure, and development of the embryo.

Embryo transfer. The removal of an embryo from one organism and the subsequent placement of the embryo into another organism.

Endometrium. The mucous membrane lining the uterus of mammals.

Endonuclease. An enzyme that can internally cut the polynucleotide strands of nucleic acids.

Endoplasmic reticulum. Network of membranes within the cytoplasm of eukaryotic cells that provides sites for ribosomes and protein synthesis. The network may also provide channels for transport within the cell.

Entomology. The study of insects.

Enucleation. The removal of the nucleus from a cell.

Environmentalist. An extreme proponent of the concept that experience is the most important ingredient in the expression of certain traits, especially behavioral traits.

Environmental variance. Denotes that portion of phenotypic variation that is due to the environment.

Enzyme. A protein that acts as a catalyst. It regulates the rate of a specific chemical reaction in the cells of an organism, without being consumed in the reaction.

Epidemiology. The study of disease, particularly with the intent to determine the cause of the disease.

Epididymis. The portion of the sperm duct of the male vertebrate that receives spermatozoa from the testis tubules, stores, and delivers them to the vas deferens.

Epilepsy. A disorder of the central nervous system in which the person may suffer transient episodes of convulsions, psychic dysfunction, and unconsciousness.

524

Epithelium. The tissue that acts as a covering or lining for any organ or organism. The cells comprising an epithelium form a continuous compact sheet.

Equatorial plane. The plane on which the chromosomes align at metaphase of mitosis and meiosis.

Erg. A unit that describes the amount of work accomplished by a system.

Error catastrophe. A hypothesis about aging that suggests that errors in the synthesis proteins have a cascading effect upon cell products. Such errors would result in a widening effect and eventually cause cell malformations.

Erythrocyte. Hemoglobin-rich red blood cell involved in oxygen transport.

Estrogen. Any hormone that stimulates uterine growth and regulates secondary sex characteristics.

Eugenics. Activity that entails the application of genetics toward the improvement of hereditary qualities of future generations of humans.

Euphenics. Improvement of the phenotype of a person by such means as chemical administration, surgery, or other medical methods.

Euploidy. Refers to chromosome numbers in which the cell or individual has the monoploid (haploid) number or an exact multiple of the monoploid number, that is, the number of chromosomes in the basic complement.

Euthenics. The improvement of humans by the control of the physical, biological, and social environments.

Exon. Those base sequences in DNA that result in mRNA for translation into a polypeptide, after the initial transcribed RNA is processed by the cell.

Expressivity. The variation in the phenotype associated with a particular genotype. The severity with which a trait is expressed.

Extrachromosomal inheritance. Describes the hereditary processes that are not due to the standard chromosomes, which are nuclear in eukaryotes; rather, this inheritance is usually cytoplasmic.

Extracorporeal gestation. The period in which an environment is provided in some artificial way for developing offspring after conception, such as within a machine.

F_1. The first filial generation; that is, the generation of individuals produced by the first parental generation being considered in any pedigree or mating. Filial pertains to any generation following the parental generation.

F_2. The second filial generation; that is, the generation of individuals resulting from self-fertilization or intercrossing of the F_1 individuals. Filial pertains to any generation following the parental generation.

Fabry's disease. One of the lysosomal storage diseases in which the victim cannot carry out normal cellular digestion because of a deficiency for the enzyme ceramide trihexosidase. It is an X-linked recessive disorder.

525

Factor. A term that was initially used in genetics to designate a hereditary determinant. Today, factors are called genes.

Fallopian tube. The oviduct in the human female that conveys the egg from the ovary to the uterus.

Family. One of the types of groupings used in the classification of organisms; consists of subdivisions called genera.

Fanconi's anemia. A type of anemia that is autosomally inherited and characterized by chromosome fragility due to a DNA repair deficiency.

Fertilization. The union of a male gamete with a female gamete to produce a zygote.

Fetal alcohol syndrome. Teratogenic effects upon newborns resulting from alcohol consumption by the mother during pregnancy.

Fetoscopy. A prenatal diagnosis technique in which direct visualization of the fetus is possible by means of a slender tube containing an optical arrangement. A fetoscope is maneuvered into the amniotic sac in order to view the fetus.

Fetus. The prenatal stage, usually in the human, which refers to the time between the embryonic stage and birth. The period begins when the body shape has been formed.

Fibroblasts. Spindle-shaped differentiated cells involved in fiber formation in connective tissues. They also grow well in tissue culture.

Filial. Describes any generation following a particular parental generation.

Fission. Biologically, the word means to undergo division by cell cleavage into two daughter cells.

Forensic genetics. Genetics in its relation to law.

Fossil. The remains of an organism (or part of an organism) or any direct evidence of its presence.

Founder principle. The establishment of a new population by a very small number of individuals who isolate themselves from the parent population in some manner. These individuals may not be representative of the gene frequencies expressed within the larger parent population.

Fragile X chromosome. A particular X chromosome that is relatively brittle at a specific site, resulting in frequent breaks; associated with mental retardation.

Free radical. Any one of a group of unstable molecules produced by the interaction of ionizing radiation and water.

Galactosemia. A recessive genetic disorder in humans resulting from an inability to metabolize galactose.

Gamete. A mature reproductive cell capable of fusing with a cell of similar origin, but of opposite sex (ova and spermatozoa).

Gametic pool. The sum total of the reproductive cells contributed by the males and the females of a population. The gametic pool will form the zygotes of the next generation.

Gamma ray. A high-energy electromagnetic ray of short wavelength emitted from nuclei undergoing radioactive decay.

526

Hardy-Weinberg equilibrium. The description of the mathematical relationship between gene frequencies and genotype frequencies within populations. This relationship, under ideal conditions, is: For any given frequency of two alleles at equilibrium (for example A and a), the genotype frequencies will be p^2 (AA), $2pq$ (Aa), and q^2 (aa).

Hardy-Weinberg principle. The frequency of alleles and the resulting genotypes will remain constant within a large randomly mating population from one generation to another, unless certain disrupting forces are acting upon the population.

Hartnup's disease. A genetic disorder characterized by a scaly rash, neuromuscular problems, and possibly mild mental retardation.

Hashish. The concentrated form of marijuana.

Hayflick limit. The concept that a population of embryonic cells will undergo a specific number of divisions, and therefore doublings, before becoming senescent and dying. In humans, the limit is estimated to be approximately 50 cell doublings.

HeLa cell. An established cell line that was originally obtained from a human carcinoma of the cervix, and has been maintained as a standard human cell line for tissue culture work and experimentation.

Hematopoietic. Refers to the blood-forming system of the bone marrow.

Heme. An iron-containing component of the hemoglobin molecule that is involved in the binding of oxygen.

Hemizygous. The condition in which only one allele of a pair is present in the diploid organism, as in the case of X-linked genes in the mammalian male. The term can also apply to other chromosome conditions that result in a single gene dose, rather than two. Heterozygous and homozygous do not apply in these cases.

Hemoglobin. Protein molecule found in the red blood cells and a transporter of oxygen. The molecule is composed of two different pairs of identical polypeptide strands and an iron-containing heme group.

Hemophilia. A hereditary disorder that is due to an X-linked recessive gene. The disorder manifests itself as a tendency to bleed profusely, even from small wounds, because of a lack of normal blood-clotting mechanisms.

Hemophilia-A. A form of the genetic disorder in humans in which blood clotting is defective.

Hemophilia-B. Another form of the genetic disorder in which blood clotting is defective. Also known as Christmas disease.

Hereditarian. A term sometimes used to describe those persons who are extreme proponents of heredity as the underlying basis of traits, especially behavioral traits.

Heredity. The overall process of transmitting traits from parents to offspring, resulting in a resemblance among individuals related by descent.

Heritability. A measure of the extent to which the total phenotypic variance of a trait is the result of genetic factors.

529

Hermaphrodite. An individual with both male and female reproductive organs. More technically, both ovarian and testicular tissue are present in the one individual.

Heterochromatin. Those parts of the interphase chromosome composed of coiled and compact chromatin that is not transcribed into RNA. Usually these regions stain very deeply at all stages of the cell cycle.

Heterogametic. Describes an organism, or one of the sexes of a species, that produces unlike gametes. The term is usually used to denote the differences in gametes that are due to the differences in sex chromosomes.

Heterosis. The state in which being heterozygous for one or more genes is superior to being either of the homozygotes. Heterosis usually results in hybrids that have more vigor and are more fit than the homozygotes.

Heterozygote. An individual that carries both members of a pair of alleles (alternative forms of a gene) at a given locus.

Heterozygous. The state in which an individual carries both members of a pair of alleles (alternative forms of a gene) at a given locus.

Hexosaminidase-A. The enzyme that is deficient in persons having the Tay-Sachs metabolic disorder.

Histidinemia. A genetic disease characterized by mental retardation and speech defects; caused by a deficiency of the enzyme histidase.

Histocompatibility. The immunological property that allows one tissue to tolerate a transplanted tissue from another source.

Histocompatibility leukocyte antigens. See human-leukocyte-associated antigens.

Hit theory. See target theory.

HLA. See human-leukocyte-associated antigens.

Holandric. Refers to Y-linked genes in an XY system, such as for hypertrichosis in the human. Such genes would be transmitted directly from fathers to sons, and they would appear only in males.

Hominoid. A member of the primate family or subgroup that includes humans.

Homocystinuria. A hereditary disease in humans that is caused by a deficiency of the enzyme cystathionine synthase.

Homogametic. Describes an individual or the sex of a species in which gametes of a single type are produced, especially relative to the sex chromosomes that they contain.

Homogentisic acid. A compound derived from the metabolic breakdown of the amino acid tyrosine. A buildup of this substance occurs in the urine in those persons with the hereditary disorder alkaptonuria.

Homogentisic acid oxidase. The enzyme deficiency in persons having the genetic disorder alkaptonuria.

Homologous. Describes chromosomes that are similar in structure, pair during meiosis, and correspond identically relative to their gene loci.

Homologue.   Chromosomes that are similar in structure, pair during meiosis, and correspond identically relative to their gene loci.

Homo sapiens.   Human beings. The scientific name for the species of modern man.

Homosexuality.   The situation in which sexual desire is directed toward a member of the same sex.

Homozygote.   An individual who carries a pair of identical alleles at a particular locus.

Homozygous.   The state in which an individual carries a pair of identical alleles at a particular locus.

Homunculus.   The minute human being believed by early biologists to be present in its entirety in the sperm.

Hormone.   A chemical substance synthesized in one organ of the body that influences functional activity in cells of other tissues and organs.

H substance.   A substance that appears to be an intermediate substrate in the synthesis of the blood antigens A and B. The gene involved in the metabolism of the H substance is not allelic to the ABO locus.

Human-leukocyte-associated antigens.   Factors located on the surfaces of human leukocytes and concerned with the rejection or acceptance of foreign tissue; may also be important in the function of the immune system.

Huntington's disease.   A chronic progressive hereditary chorea characterized by irregular movements, disturbance of speech, and gradually increasing dementia. The age of onset is highly variable, with most victims showing some of these symptoms between 30 and 40 years of age.

Hurler syndrome.   A genetic disorder of connective tissue occurring in late infancy and causing mental retardation, stiff joints, and other physical problems.

Hutchinson-Gilford syndrome.   Same as progeria. A rare disorder characterized by a strikingly precocious aging, frequently causing death before 10 to 12 years of age. A small amount of data suggests that the disorder is recessive.

H-Y antigen.   Y-linked histocompatibility locus believed to be involved in the differentiation of sex in mammals.

Hybrid.   Several meanings have been adopted. An offspring resulting from a cross between two distinct species. Or an offspring resulting from a cross between parents of the same species that are genetically unlike. Also, the state of heterozygosity at one or more gene loci.

Hybridization.   The crossing of two different species, distinct races, or genetically different individuals within a species. The term can also refer to the molecular process of two DNA strands from different sources being fused.

Hybrid vigor.   The superiority of heterozygotes relative to the inbreds that formed them with regard to characteristics such as growth, development, and fertility.

531

Hypertrichosis. A genetic trait in humans that consists of long hairs growing from the pinna (external ear); believed to be Y-linked.

Hypocenter. The point on the ground lying directly below the center of a nuclear bomb explosion.

Hypophosphatasia. A defect in which an abnormally low concentration of phosphates is found in the blood, causing severe demineralization of the skeleton.

Hypothalamus. In vertebrates, a small part of the lower brain just behind the cerebral hemispheres. The structure is believed to influence many regulatory functions of the body.

Hypothyroidism. A condition of lowered metabolism that is due to the diminished activity of the thyroid gland and a resulting deficiency of the thyroid hormone.

Hypotonic. Refers to a solution in which the osmotic pressure is lower than some other solution; hence, the latter solution would draw water from the hypotonic solution if the two were separated by a permeable membrane.

Ichthyosis. An X-linked congenital abnormality characterized by a dry, hard, scaliness of the skin. The disease seems to be the most severe in cold weather.

Identical twins. See monozygotic twins.

Immune system. The system that responds to the presence of antigens by producing antibodies and/or by white blood cell activity.

Implantation. In a reproductive sense, the attachment of the mammalian embryo to the lining of the uterus. In primates, the embryo embeds itself into the endometrium of the uterine wall.

Inborn errors of metabolism. Genetically caused biochemical disorders in which a specific enzyme defect results in a metabolic block. Such blocks can, in turn, have pathological consequences.

Inbred. A progeny that results from matings between relatives, and therefore, is likely to have one or more situations in which two copies of a gene are identical by descent.

Inbreeding. The mating of closely related individuals, which results in increased homozygosity and uniformity of phenotypes among the individuals affected.

Inbreeding depression. The decrease in vigor that sometimes accompanies intensive inbreeding in organisms.

Incestuous. Describes some consanguineous matings; that is, matings between certain categories of closely related persons. Such matings are usually restricted by society.

Incomplete dominance. In an individual heterozygous for two alleles, the phenotypic expression is intermediate between the homozygous dominant expression and the homozygous recessive expression.

532

Incomplete penetrance. The situation in which not all of the individuals having a certain genotype express the specific trait that usually results from that genotype.

Independent assortment. The assortment or distribution of one pair of alleles at meiosis independently of other pairs of alleles.

Index case. Same as proband in human genetics. The person with a particular character through whom a family is located and incorporated into a genetic study of some kind.

Inheritance. The transmission of genetic information from parents to their progeny.

Initiation site. The point on a DNA strand where transcription begins to produce RNA.

Insemination. The process of injecting semen into the vagina of a female.

Insulin. A hormone manufactured by a small region of the pancreas and involved in the metabolism of carbohydrates. Its deficiency in the body results in diabetes mellitus.

Intelligence quotient. A numerical assignment given to a person or a group of persons that is meant to indicate the mental age. The designation is based on the performance shown on standardized intelligence tests.

Intercross. The mating of heterozygotes, or, in some cases, the mating of organisms that differ from each other in some other way, such as having different chromosomal structural aberrations.

Interferon. A class of proteins produced by cells in response to viral infection; interferon tends to inhibit viral replication.

Intermediary metabolism. The chemical reactions (metabolic steps) required to change precursor substances into products that are useful to the cell and the organism.

Interphase. The stage in the cell cycle during which the cell is not dividing.

Intersex. An individual with sexual characteristics intermediate between those of males and females, or individuals may show secondary sex characteristics of both sexes.

Interstitial cells. Cells that lie between ovarian follicles or between the tubules of the testes; in the latter case, the cells secrete the hormone testosterone.

Intrauterine device (IUD). Any one of a number of devices inserted into the uterus as a contraceptive measure.

Intron. Those base sequences in DNA that do not result in mRNA after the initial transcribed RNA is processed by the cell.

Inversion. A chromosomal rearrangement such that an internal chromosome segment has been completely turned around end-to-end. As a result, the linear sequence of genes for that segment is reversed.

In vitro. Pertaining to experiments performed outside the organism, that is, inside glass vessels.

In vitro **fertilization.** The union of an egg and a spermatozoan under artificial circumstances, that is, within a glass vessel.

533

In vivo.   Pertaining to experiments that are carried out such that the living organism is left intact.

Ion.   An electrically charged atom or group of atoms.

Ionizing radiation.   Electromagnetic or corpuscular radiation that generates charged molecules in the material that it strikes.

IQ.   See intelligence quotient.

Irradiation.   The process of exposing something to a form of radiation.

Isoagglutinin locus.   That site within the genetic material that is responsible for the production of the antigens in the ABO blood system.

Isochromosome.   An abnormal chromosome with two identical arms; this is the result of a transverse division at the centromere rather than a longitudinal division.

Isolating mechanism.   Any barrier, regardless of whether it is a feature of the organism, geographic, or geologic, that disallows successful interbreeding between two or more related groups of organisms. Isolating mechanisms prevent gene flow between different populations.

IUD.   See intrauterine device.

Karyokinesis.   Division of the nucleus of the cell, as distinguished from the cytoplasmic division (cytokinesis).

Karyotype.   The composite view of the somatic chromosomes of an individual (usually defined by the morphology, size, and number as they appear in mitotic metaphase) arranged in a particular sequence considered to be standard. The morphological criteria include centromere location and other physical landmarks. More recently, special staining techniques and fluorescent microscopy have been used to gain additional banding patterns to aid in the process of identifying the chromosomes.

Kinetochore.   Same as the centromere, also known as the primary constriction. The region of the eukaryotic chromosome involved in spindle fiber attachment during cell division, and important in the movement of chromosomes to the poles of the cell.

Klinefelter syndrome.   A set of abnormal conditions in the phenotype of a human male that occasionally occurs due to the presence of more than one X chromosome (XXY).

Lamarckism.   Inheritance of acquired characteristics proposed by Lamarck. The concept that organic evolution is due to the environmental induction of adaptive changes in living organisms that are transmitted to the offspring. Now believed to be erroneous.

Laparoscope.   A medical instrument for viewing inside the abdominal cavity. It consists of a long tube equipped with lenses.

Law of independent assortment.   See independent assortment.

Law of segregation.   See segregation.

LD_{50}.   See lethal dose-50.

Legume.   Members of a plant family (Leguminosae) distinguished by their dry fruit; that is, a pod containing seeds that splits along longitudinal sutures.

LET.   See linear energy transfer.

Lesch-Nyhan disease.   A hereditary disorder that causes mental retardation and spastic cerebral palsy, among other things. It is inherited as an X-linked recessive. The individual with Lesch-Nyhan syndrome lacks the enzyme hypoxanthine guanine phosphoribosyl transferase.

Lethal dose-50.   The radiation dose necessary to kill half of a population of organisms within a specified period of time.

Lethal equivalent.   A gene carried in the heterozygous state, which would be lethal if homozygous. Also, the combination of two genes in the heterozygous state, each of which, if homozygous, would cause the death of 50 percent of these homozygotes. Any equivalent situation can be calculated in this manner.

Lethal gene.   A gene that causes the death of the individual who carries it sometime before the age of reproduction.

Leukemia.   A group of diseases considered to be a form of cancer; characterized by a marked increase in the number of leukoctyes in the blood.

Leukocyte.   White blood cells that are important agents in protecting the organism against infectious diseases.

L-gulonolactone oxidase.   An enzyme involved in the synthesis of vitamin C and found in many animals, but not in humans.

Life expectancy.   The probable length of life of an individual, based upon accumulated statistics.

Life span.   The maximum length of life regarded as biologically possible in a group of organisms.

Linear energy transfer (LET).   The energy dissipated in a tissue traversed by an ionizing particle such that a stream of electrons and ions are left to take part in subsequent chemical reactions.

Linkage.   An association between the inheritance of two or more characters such that the parental combinations of them appear more often in the offspring than the nonparental combinations. This occurs because the genes for these characters do not assort independently; they are located on the same chromosome.

Lipids.   A group of compounds that can be extracted by fat-solvent substances such as ether and alcohol and are very weakly soluble in water. Consist partly of fatty acids.

Lipopolysaccharide.   A substance that forms a covering in some bacteria.

Liposome.   Small vesicles that form when water-insoluble fatty substances are mixed with water. These artificial bodies are being investigated as a possible means to transfer other substances into the body's cells.

L locus.   The locus of the alleles for the MN blood group. Named after Landsteiner.

Locus.   The specific position occupied by a given gene, or any of its alleles, on a chromosome.

535

Lymph. A diluted blood plasma (mostly water with a little protein and a few lymphocyte cells) found in lymph spaces and vessels.

Lymphocyte. One of the major types of white blood cells found in the lymph nodes, spleen, and blood.

Lyon hypothesis. The hypothesis that in any given somatic cell of mammals, all of the X chromosomes except one will become inactive during very early development and take the form of a Barr body. The process is one of dosage compensation, and it can lead to female mosaicism.

Lysenkoism. The belief in the inheritance of acquired characteristics rather than the present gene concept. Such doctrines were advocated by Lysenko and his collaborators in the Soviet Union during the period of 1932 to 1965.

Lysis. The rupturing and subsequent destruction of a cell by the dissolution of its cell membrane.

Lysosomal storage disease. A group of diseases in which a particular enzyme that is necessary for normal cellular digestion is lacking.

Lysosome. A membrane-bound vesicle of the cell containing various hydrolytic (digestive) enzymes.

Macroevolution. Evolution that relates to the gross morphological changes in organisms and the origin of new genera and even higher taxonomic categories.

Macromolecule. A very large molecule such as DNA, RNA, and some proteins.

Major gene. A gene associated with a readily identifiable phenotype, that is, a gene with pronounced effects.

Malignancy. The condition in which tumor cells have the ability to grow progressively and often can kill their host.

Manic depression. A mental disorder in persons that is characterized by periods of depression alternating with periods of excitement.

Manifesting heterozygote. A female heterozygous for an X-linked recessive disorder who expresses the trait to approximately the same degree as the affected males who are hemizygous. This is believed to be due to a case of Lyonization whereby most of the cells by chance inactivate the normal X chromosome rather than the X chromosome carrying the gene for the disorder.

Maple syrup urine disease. A genetic disorder in humans caused by a deficiency of the enzyme keto acid decarboxylase.

Marfan syndrome. An autosomal dominant genetic disorder in which affected persons are usually tall, long-limbed, loose jointed, and gaunt-faced.

Marijuana. Dried leaves and stems of the plant *Cannabis sativa*; contains a drug that is intoxicating and habit forming.

Marker. See genetic marker.

Maximum life span. See life span.

536

Mean. The arithmetic average that is the sum of all the values of a group of measurements divided by the total number of individual measurements in the group.

Medicolegal science. Science as it relates to medicine and law.

Meiosis. The process of two successive nuclear divisions while the chromosomes replicate only once; by this means, diploid precursor cells (germ cells) form haploid sex cells.

Meiosis I. The first nuclear division of the meiotic process, through which the homologous pairs of chromosomes are separated.

Meiosis II. The second nuclear division of the meiotic process, through which the sister chromatids are separated.

Melanin. A group of pigments ranging from black to brown to yellow and found in the cells of skin, hair, iris of the eye, other animal tissues, and some plants.

Mendelian trait. Refers to Mendelian inheritance; that is, the theory of inheritance stating that hereditary traits are carried in cells and transmitted from one generation to another in the form of particles (genes).

Menstrual cycle. The periodic discharge of blood and fluid from the uterus of the nonpregnant female.

Menstruation. See menstrual cycle.

Messenger RNA (mRNA). The ribonucleic acid product of transcription; serves as a chemical message in that it is translated into a specific polypeptide sequence.

Metabolic block. A mutation that has the effect of preventing the synthesis of an enzyme that is necessary for some critical step in a metabolic pathway.

Metabolism. The sum total of the various chemical reactions that are involved in the life functions occurring in the living cell.

Metaphase. The stage of nuclear division (mitosis and meiosis) in which the chromosomes are aligned at the equatorial plate of the cell.

Metastasis. The situation in which a disease, such as cancer, is transferred to another part of the body.

Methylmalonic acidemia. A hereditary disorder in humans which is caused by a deficiency of the enzyme methylmalonyl coenzyme A carboxymutase.

Microbial genetics. The branch of genetics that studies the heredity of microbial organisms and uses them to elucidate genetic and molecular principles.

Microcephaly. A type of congenital defect in which the head and brain are abnormally small.

Microevolution. The term applied to genetic changes in the population that result in subtle changes, such as gene frequency changes or the formation of new subspecies and species.

Microsomal fraction. A subcellular cytoplasmic fraction of small particles that consists of ribosomes and broken fragments of the endoplasmic reticulum.

537

Microvillus.   A very minute protuberance of a cell surface.

Mid-parent value.   Refers to the mean of the two parents for some measurable characteristic.

Migration.   Movement of one or more individuals from one geographic population to another, which may result in a change in gene frequencies for either or both of the two populations involved.

Millirem.   1/1000th of a rem; a measure of radiation that takes into account the damage done to tissue by different types of radiation, even though the dose may be the same.

Minimal medium.   A nutritive substance used to grow organisms under laboratory conditions. It provides only those nutrients essential for the growth and reproduction of the wild-type organism.

Mitochondrion.   Small DNA-containing organelles in the cytoplasm of eukaryotic cells; involved in many of the critical steps of cellular respiration.

Mitosis.   The process of nuclear division in which a replication of the chromosomes is followed by separation of the replication products and their incorporation into two daughter nuclei that are genetically identical to the original nucleus.

Mixoploid.   An organism whose cells are mosaic for different chromosome numbers as a result of nondisjunction or other mitotic irregularities.

MN blood system.   One of the blood systems in humans; involves two different antigens, called M and N. The mode of inheritance of the blood types (MM, MN, and NN) is one of codominance.

Molecular genetics.   The branch of genetics concerned with the chemical and molecular bases of genetics.

Mongolism.   See Down syndrome.

Monogenic.   Character differences that are controlled by the alleles of one particular gene.

Monohybrid.   A cross between parents differing in one pair of allelic genes; also, an individual heterozygous with respect to the alleles for one gene locus.

Monosomic.   Cytogenetically describes a diploid individual who lacks one chromosome of the normal set. This is a type of aneuploidy, designated as $2n - 1$.

Monozygotic twins.   Same as identical twins. The members of the twin pair are genetically identical because they develop from one fertilized egg (zygote). At some time in early development, cleavage occurs that results in two embryos.

Mortality rate.   Death rate.

Morula.   A solid mass of cells formed by cleavages of an ovum in the early stages of embryonic development.

Mosaicism.   The situation in which an organism is composed of cells of two or more different types with respect to genes and chromosomes.

mRNA.   See messenger RNA.

Multifactorial.   See multigenic.

538

Multigenic. The mode of inheritance by which a trait is determined by two or more nonallelic pairs of genes.

Multihit hypothesis. The concept that a cellular target must be hit more than once by radiation in order to inactivate the cell.

Multiple allelism. The situation in which three or more different alleles (alternative forms of a gene) exist for a particular locus among the organisms of a population.

Multiple genes. See multigenic.

Muscular dystrophy. See Duchenne muscular dystrophy.

Mutagen. A chemical or physical agent that significantly increases the rate of occurrence of mutations.

Mutagenesis. The production of mutations.

Mutagenicity. Refers to the potential of chemical and physical agents causing mutations.

Mutant. A cell or an individual organism that shows an observable change brought about by one or more mutations.

Mutation. The alteration of genetic material. In the broad sense, mutation refers to any heritable change not due to segregation or genetic recombination, therefore, including small genetic changes and larger chromosomal changes.

Myelogenous leukemia. Leukemia that is a result of disease of the bone marrow.

Natural selection. A differential reproduction of genotypes that comes about through the forces of natural processes favoring the individuals that are better adapted. This process has a general tendency to eliminate those unfitted to their environment.

Negative correlation. A relationship between paired values in which the high values of one of the variables are generally associated with the low values of the other variable and vice versa.

Negative eugenics. Refers to the removal or reduction of undesirable traits by activities that entail the application of genetics.

Negroid. The black race.

Neoplasm. An abnormal proliferation of cells in an organism.

Neurofibromatosis. A condition in which multiple tumors (neurofibromas) are present in the skin and along nerves.

Neutralist. One who is a proponent of the neutrality theory relative to certain genes; that is, genes that are neutral with regard to the effects of natural selection.

Neutrality hypothesis. The idea that some genetic mutations do not confer any selective advantage or disadvantage on an organism; rather, their presence in the population is simply due to chance events.

Neuron. A nerve cell.

Nitrogen fixation. The changing of nitrogen gas into compounds of nitrogen that can be used by higher plants.

539

Nitrogenous base.   An aromatic nitrogen-containing molecule with the properties of a base. The important nitrogenous bases in cells are the purines and the pyrimidines found in the nucleic acids.

Nondisjunction.   The failure of either chromosomes or sister chromatids to separate normally during cell division, producing various types of aneuploid daughter cells.

Nonionizing radiation.   Electromagnetic radiation of long waves consisting of bundles of energy (photons). It cannot transfer sufficient energy to ionize atoms, that is, cause the atom to lose an electron.

Nonrandom mating.   Assortative mating in which the mating pairs have genotypes that are either more closely related or less closely related than the average relationship.

Nonreciprocal translocation.   A structural rearrangement of a chromosome section or sections in which there is not a mutual exchange between the nonhomologues involved.

Nonsense codon.   A codon that does not specify an amino acid. These are deemed to be polypeptide chain terminators in the translation process.

Normal distribution.   The commonly used probability distribution. The formula of the normal distribution produces a bell-shaped symmetrical curve when the data are plotted.

Nuclear envelope.   The envelope that bounds the nucleus of the cell; usually composed of two membranes, each of which resembles the structure of the plasma membrane.

Nucleic acid.   An organic compound that is a polymer composed of phosphoric acid, pentose sugar, and organic bases forming a chain-like molecule. RNA and DNA are nucleic acids.

Nuclein.   The crude nucleoprotein complex initially isolated from nuclei by early biochemical investigators.

Nucleolus.   Structure within the nucleus of some cells that is associated with specific chromosome regions, and serves as an area for the processing and storage of rRNA.

Nucleotide.   An organic compound composed of a purine or pyrimidine base, a pentose sugar, and a phosphate group. The polymeric molecule, a nucleic acid, consists of many of these monomeric units covalently held together in a chain by sugar-phosphate bonds.

Nucleus.   A membrane-bound organelle of the eukaryotic cell. Contains the major part of the DNA of the cell, that is, the genetic information.

Nullisomic.   Pertains to a diploid cell or organism in which both members of a pair of chromosomes are absent. Designated as $2n - 2$.

Ocular albinism.   A genetic disease in which there is an absence of the melanin pigment in the iris of the eye.

Offspring.   The same as progeny, that is, the descendants of parents.

Oncogene theory.   The concept that genes, which are part of the normal animal genome, may be determinants for certain cancers.

One gene-one enzyme hypothesis. The hypothesis that one gene is responsible for the production of one enzyme.

One gene-one polypeptide hypothesis. The hypothesis that one gene produces one polypeptide. Specifically, one unit of transcription produces one unit of mRNA.

Oogenesis. The formation of female gametes, from the germ cell to the mature ovum, by the meiotic process.

Oogonium. In females, active germ cell that gives rise to primary oocytes by mitosis.

Order. One of the types of groupings used in the classification of organisms; consists of subdivisions called families.

Organelle. Any discrete substructure of a eukaryotic cell that performs a specialized function.

Outbreeding. The mating of genetically unrelated individuals that are less closely related than average pairs that might be selected from the population at random.

Ovary. A female reproductive structure. In animals, the female reproductive gonad. In plants, the part of the pistil of the flower that contains the ovules.

Oviduct. The duct that carries eggs from the ovary to the uterus or to the exterior of the body.

Ovulation. The discharge of one or more mature eggs from the ovarian follicles of mammals.

Ovum. The mature female gamete, that is, an unfertilized egg.

Palate. The roof of the mouth.

Paleontology. The branch of science that deals with the study of plant and animal fossils.

Paracyclic ovulation. A discharge of an egg that is secondary to the regular ovulation of the cycle.

Parthenogenesis. The development of an embryo or an adult individual from an ovum without fertilization by a sperm.

Partial sex linkage. Refers to genes located in those regions of the X and Y chromosomes that are homologous; that is, they show pairing during meiosis. Such linkage has not yet been established in humans.

Patau syndrome. A chromosome 13 trisomy condition in humans that causes a well-defined group of congenital defects.

Pathogen. A disease-causing or toxin-producing organism.

Pedigree. A diagram of two or more generations of a kindred. The diagram uses symbols to depict the ancestral history of one or more traits for a given family.

Penetrance. The proportion of individuals in which a particular genotype is expressed phenotypically among all of the individuals that have this same genotype.

Penis. The muscular male organ in most higher animals associated with the duct from the testis; used to introduce sperm into the female.

541

Pentosuria. An inborn error in which the affected persons excrete large amounts of a pentose (5-carbon sugar) in their urine.

Peptic ulcer. A sore occurring in the stomach or first part of the small intestine, partly due to the digestive action of gastric juice.

Peptide bond. A covalent bond between amino acid subunits which links them together in a polypeptide chain. The NH_2 group of one amino acid is bonded to the COOH group of the second amino acid, with H_2O being eliminated.

Pernicious anemia. A serious, sometimes fatal, type of anemia in which there is a great reduction of red blood cells.

Phagocytosis. The invagination and engulfment of solid particles by cells.

Pharmacogenetics. The science that deals with the genetics of drug use and the variability of responses to them.

Phenotype. The actual appearance or other discernible characteristics of an organism produced by its genotype interacting with the environment.

Phenotypic variance. Denotes the total variation of a trait due to the combined effects of the environment, heredity, and genetic-environment interaction.

Phenylalanine. One of the 20 essential amino acids.

Phenylalanine hydroxylase. That enzyme deficient in persons having phenylketonuria.

Phenylketonuria (PKU). A recessive metabolic disorder characterized by an inability to oxidize the amino acid phenylalanine to tyrosine. If not treated early, the affected person will have severe mental retardation.

Phenylthiocarbamide (PTC). A chemical that usually tastes bitter to individuals carrying the dominant allele, but is tasteless to those who are homozygous recessive.

Pheromone. A substance released by members of the same species, such as a sex attractant, which affects their behavior.

Philadelphia chromosome. The designation given to the aberrant chromosome 22 in persons who have chronic myelogenous leukemia.

Photon. A bundle (quantum) of electromagnetic energy.

Phylum. One of the types of groupings used in the classification of organisms; consists of subdivisions called classes.

Pinna. In animals, sometimes used to refer to the outer ear of mammals.

Pinocytosis. The engulfment of liquids by a cell.

Pituitary gland. A complex endocrine gland that lies beneath the floor of the brain and within the skull of vertebrates.

PKU. See phenylketonuria.

Placenta. In animals that produce living young, the wall of tissue formed in the uterus by a combination of tissue of the uterine wall with tissue from the embryonic membranes. The structure makes a close contact between the blood of the embryo and the blood of the mother.

Plantlet. Often used to refer to young plants that are produced by differentiation of tissue cultured *in vitro*.

Plasma membrane. See cell membrane.

542

Plasmid. Autonomously replicating elements found in the cytoplasm of bacteria. Plasmids are nonessential DNA entities.

Pleiotropy. The situation in which a single gene or gene pair produces multiple phenotypic effects, apparently unrelated to each other.

Point mutation. A mutation that involves only one or a few nucleotide pairs. These changes in the DNA cannot be resolved by conventional microscopy.

Poisson distribution. The distribution expected for a particular event to occur, if the event occurs randomly (not clumped or uniform) and if the distribution follows certain defined conditions.

Polar body. One of the products of meiosis in the division of primary or secondary oocytes. Polar bodies are extruded as small bodies almost devoid of cytoplasm, and they are nonfunctional in the reproductive process.

Pollination. The transfer of pollen from the anther to the stigma within the same flower or from one flower to another.

Polydactyly. The occurrence of more than the normal number of fingers and/or toes.

Polygene. One of a group of genes that together control the inheritance of a quantitative trait.

Polymerase. The general term for any of the enzymes that can catalyze a polymer from monomers. Good examples in genetics are those that synthesize nucleic acids, either DNA or RNA, from nucleotide monomers.

Polymorphonuclear. Describes a cell, such as some of the white blood cells that have nuclei of varied shapes.

Polynucleotide ligase. An enzyme that catalyzes the chemical bonding of two segments of an interrupted strand on one side of the DNA duplex molecule.

Polynucleotide strand. A linear sequence of nucleotides chemically bonded together.

Polypeptide. A chain of two or more amino acid residues linked covalently (peptide bonds) which will usually assume further structural configurations.

Polyploid. A cell or an organism that has more than two complete sets of the basic haploid complement (genome) of chromosomes.

Polyposis. The condition whereby small stalked projections grow out from a mucous surface.

Polyribosome. A linear array of ribosomes held together by an mRNA molecule.

Population. A group of interbreeding individuals of one kind living at any particular time.

Population genetics. That branch of genetics that is involved with inheritance on the level of populations. The discipline strives to describe Mendelian inheritance and its many interactions and consequences in mathematical terms.

543

Position effect. A phenotypic change that can occur in an organism when the responsible gene or genes change their positions relative to other genes in the genetic material.

Positive correlation. A relationship between paired values in which the high values of one of the variables are generally associated with high values of the other variable, and low values are generally associated with low values.

Positive eugenics. Refers to the improvement of traits in humans by activities that entail the application of genetics.

Postnatal. The period immediately following birth.

Postzygotic. Describes a type of isolating mechanism that exerts its effect after fertilization has occurred. Such mechanisms keep populations reproductively isolated from each other because the hybrids will abort, be sterile, nonviable, or too weak to survive.

Precursor. An initial substance in metabolism that is used to synthesize other products.

Preformation. The discredited idea that one of the gametes or the zygote contained a miniature copy of all of the adult structures, which only needed to unfold during development.

Prenatal. Occurring before birth.

Presymptomatic genetic screening. Genetic screening to determine individuals who have a genetic disease, but have not yet displayed the physical symptoms.

Prezygotic. Any one of a number of isolating mechanisms that are active before the egg has become fertilized; that is, those that interfere with the fertilization of the egg. These mechanisms keep populations reproductively isolated from each other.

Primary oocyte. A cell in the female that is derived from an oogonium, undergoing the meiosis I division.

Primary sex. The sex of an organism as established directly by the type of gonads that are present in the organism.

Primary spermatocyte. A cell derived from a spermatogonium that undergoes the first meiotic division.

Probability. The likelihood of occurrence of an event; that is, the ratio of a specific event to the total number of possible events.

Proband. Also known as the index case or the propositus in human genetics. The person with a particular character through whom a family is located and incorporated into a genetic study.

Progeny. Same as offspring. The individuals that result from a particular cross.

Progeria. A rare disease, possibly with a genetic basis, in which the affected person undergoes a very precocious aging, often dying before age 20 as a result of old age causes.

Progesterone. A hormone secreted by the corpus luteum of the mammalian ovary; it is responsible for preparing the reproductive structures for implantation and pregnancy.

544

Prophase. The initial stage of a nuclear division, that is, mitosis, meiosis I, or meiosis II. It includes all of the events from the first appearance of chromosomes as discrete units following interphase to metaphase (migration of the chromosomes to the equator of the cell).

Propositus. See proband.

Prospective genetic screening. The genetic screening that attempts to identify heterozygotes (carriers) for deleterious recessive alleles before they have children.

Prostate gland. A male accessory sex gland of mammals that contributes substances to the semen.

Protein. Nitrogenous polymers composed of amino acids and usually of high molecular weight. Proteins form important structural and enzymatic constituents of the body.

Protoplast. All of the living contents of the cell, that is, consisting of the cytoplasm and the nucleus.

Pseudohermaphrodite. An individual with gonadal tissue of only one sex, but with some features of both sexes.

P-site. Also called the peptidyl site. This is one of two adjacent sites on the ribosome for complexing tRNA molecules during the process of genetic translation.

PTC. See phenylthiocarbamide.

Punctualistic model. A hypothetical model that describes evolutionary change as taking place with extensive branching of lineages, that is, speciation.

Punnett square diagram. A checkerboard grid designed to aid in determining all of the possible genotypes produced by a given cross. Genotypic and phenotypic ratios can be calculated from its use.

Purine. A class of nitrogenous bases with a double-ring configuration that occurs in DNA and RNA; commonly, adenine and guanine.

Pyloric stenosis. A congenital disorder in which a constriction of the valve exists between the stomach and the small intestine. The disease may have a hereditary basis.

Pylorus. The junction between the stomach and duodenum (first part of the small intestine); it encloses the tissue that serves as a valve during digestion.

Pyrimidine. A class of nitrogenous bases with a single-ring configuration that occurs in DNA and RNA; commonly, thymine, cytosine, and uracil.

Quantitative inheritance. The study of traits that require measurement in order to be described. These traits vary over a continuous distribution as a result of polygenes, each polygene producing a small measurable effect that is additive.

Quasicontinuous. Describes the type of variation shown in a trait that is governed by multiple genes and also has a threshold effect; therefore, the trait appears to have a discontinuous distribution.

Race.   A population having a gene pool that is distinct from other populations of the same species with regard to frequencies of certain genes.

Rad.   A unit of absorbed radiation, equal to an ionizing radiation dose that contains 100 ergs per gram of the exposed matter or organism.

Rad equivalent man.   The rem, which is a measure of radiation that takes into account the damage done to tissue by different types of radiation.

Radiation.   Process of emission or diffusion of radiant energy from a given source that broadly includes electromagnetic radiations and ionizing particles.

Radiation sickness.   The response to large doses of radiation characterized by diarrhea, nausea, vomiting, depression, and possibly death.

Radioactivity.   The spontaneous disintegration of atomic nuclei, thereby giving off radiant energy.

Radiology.   The science dealing with various forms of radiant energy, especially as used in medicine.

Random mating.   Mate selection in a population relative to a certain genotype that is determined by chance; that is, any individual of one sex has an equal probability of mating with any individual of the opposite sex.

RBE.   See relative biological effectiveness.

Receptor.   Any organ, cell, or structure of a cell designed to perceive external stimuli.

Recessive.   Any gene form (allele) that does not express or affect the phenotype when in the heterozygous state, that is, in the presence of another allele that masks it. Also describes any phenotype that will not manifest under these conditions.

Reciprocal translocation.   The mutual exchange of segments between nonhomologous chromosomes.

Recombinant DNA.   A DNA molecule that is made up of DNA from at least two different individuals, either of the same species or of different species. Such molecules are produced by enzymatically cutting the DNA into pieces from two different sources, and then bonding the pieces so that both sources reside within the same molecule.

Recombination.   The formation of combinations of alleles in the offspring that are not found in the parents. This can result either from a rearrangement following crossing over or from independent assortment.

Regeneration.   The regrowth of lost tissue or a whole part of an organism.

Relative biological effectiveness (RBE).   A correction factor for tissue differences as they apply to radiation measurements.

Rem.   See rad equivalent man.

Repair mechanism.   A repertoire of enzymes and possibly other factors that give the cell the ability to repair some forms of genetic damage.

Replication.   In the most commonly used sense, the synthesis of additional DNA from preexisting DNA by a process that involves the copying of templates in a precise manner.

Residue.   In one sense, each amino acid unit of a polypeptide chain is referred to in this way.

546

Restriction enzyme. One of a number of enzymes that internally cut unmodified DNA molecules at specific points in the molecule. These points are determined by a highly specific base pair sequence.

Retrospective genetic screening. See presymptomatic genetic screening.

Reverse transcriptase. The enzyme that is directly involved in the synthesis of DNA using an RNA template.

Ribonucleic acid (RNA). A polymer molecule that is usually single stranded, consisting of a chain of ribonucleotides in which the pentose sugar is D-ribose. The commonly found nitrogenous bases of the molecule are adenine, cytosine, guanine, and uracil (rather than thymine as in DNA).

Ribonucleotide. The subunit of the polymer RNA that is a purine or a pyrimidine base bonded to D-ribose. The ribose, in turn, is bonded with a phosphate group.

Ribose. The pentose (5-carbon) sugar component of ribonucleic acid (RNA).

Ribosomal RNA (rRNA). A type of RNA which, in association with a number of protein molecules, forms the structure of ribosomes.

Ribosome. Small cellular particles composed of two unequal subunits consisting of rRNA and proteins. Ribosomes are the sites of translation in the process of polypeptide synthesis.

Ring chromosome. A circular chromosome. In eukaryotes, this is an aberrant chromosome.

RNA. See ribonucleic acid.

RNA polymerase. Any enzyme that catalyzes the synthesis of RNA using single-stranded DNA as a template.

rRNA. See ribosomal RNA.

Saccharin. A substance used as a sugar substitute.

Salmonella-**mammalian liver test.** See Ames test.

Schizophrenia. A form of mental disorder in which a personality split seems to take place; often a withdrawal from the usual human relationships accompanies the disorder.

Scurvy. A deficiency disease resulting from a lack of ascorbic acid (vitamin C) in the diet.

Secondary oocyte. The larger of the two cells produced by the first meiotic division of a primary oocyte.

Secondary sex characteristics. Those physical traits, other than the genitalia themselves, by which members of one sex differ from members of the other sex.

Secondary spermatocyte. During spermatogenesis, these are the two cells that result from the first meiotic division of the primary spermatocyte.

Segregation. The separation of the pairs of chromosomes, usually at meiosis, and the consequent separation of the alleles on these chromosomes.

547

Selectionist.   One who advocates that all gene forms have either an advantage or a disadvantage for survival, regardless of how subtle the effects might be.

Selective medium.   A medium containing ingredients such that only certain cells or microbes will survive on it. All other types would fail to show viability and growth.

Self-fertilization.   The formation of a zygote by fertilization involving gametes that are produced by a single individual. Many plant species can self-fertilize in this manner.

Self-pollination.   The transfer of pollen to the stigmas of the same flower or to a flower on the same plant.

Semen.   The impregnating fluid from male animals that contains the spermatozoa.

Semiconservative replication.   The method of DNA replication in which the two polynucleotide sides become separated longitudinally, and each serves as a template for the synthesis of a new molecule. In this manner, the two resultant DNA molecules will be composed of a conserved polynucleotide strand and a newly synthesized polynucleotide strand.

Seminal vesicle.   A small sperm storage organ in the male.

Seminiferous tubules.   The mass of tubules that produce spermatozoa in the testes of an animal.

Senescence.   The process of aging.

Sense strand.   The one polynucleotide strand of the two that make up the DNA that is used as the template for transcription.

Serum.   The fluid of the blood remaining after coagulation.

Sex chromatin.   Refers to the Barr body. A chromatin mass in the nucleus of interphase cells of many mammalian species. The mass represents the X chromosome that is inactivated. All X chromosomes become Barr bodies except one.

Sex chromosome.   Those chromosomes that are at least partly concerned with sex differentiation, and show a difference in number and/or morphology between the sexes. For example, the X and Y chromosomes of mammals.

Sex linkage.   The association or linkage of genes that are located on the sex chromosomes in eukaryotic cells.

Sexual reproduction.   A mode of reproduction that usually involves the fusion of meiotically produced gametes and genetic recombination.

Sibs.   The brothers and sisters from the same family.

Sickle cell anemia.   A severe hereditary disorder caused by a homozygous recessive mutant gene that controls hemoglobin structure. The red blood cells become sickle shaped under low oxygen tensions because of the defective hemoglobin.

Sickle cell trait.   A mild hereditary condition caused by a recessive mutant gene in the heterozygous state. The red blood cells tend to form different shapes under low oxygen tensions, but the person does not normally experience an anemic condition.

Sister chromatid exchange.   An exchange of segments between sister chromatids, that is, the two chromatids of the same chromosome after replication has occurred in interphase.

Sister chromatids.   The two chromatids of the same chromosome after replication has occurred in interphase.

Sociobiology.   The discipline that seeks to establish that social behavior among animals has a biological basis.

Somatic.   Pertaining to body cells as opposed to reproductive cells in the germ line. Normally, these cells have two sets of chromosomes, one from each parent.

Somatic hybrid.   A hybrid organism resulting from the fusion of somatic (body) cells.

Somatic mutation.   Genetic changes that occur in body cells; that is, those cells not destined to become gametes.

Somatic mutation hypothesis.   The hypothesis that explains the process of aging as the result of the accumulation of mutations in the somatic cells.

Somatotropin.   A growth hormone.

Speciation.   The process of forming new species in which a population diverges into two or more separate groups that can no longer interbreed.

Species.   A population or populations of organisms that are phenotypically similar, make up of a common gene pool, and are reproductively isolated from other species.

Spermatid.   One of the haploid products of meiosis in the spermatogenesis of males. This cell product will subsequently mature into a spermatozoon.

Spermatogenesis.   The series of events by which maturation of the gametes of the male takes place.

Spermatogonium.   A primordial germ cell in a male that can mitotically produce more spermatogonia or grow to a stage that gives rise to a primary spermatocyte.

Spermatozoon.   A mature male gamete.

S phase.   That part of the cell cycle in which the DNA of the nucleus is replicated.

Spina bifida.   A congenital neural tube defect in which the spinal column fails to develop normally over the spinal cord.

Spindle apparatus.   A cellular structure made up of an aggregation of microtubules that has an important role in the movement of chromosomes during nuclear divisions.

Spore.   A reproductive cell in plants that is capable of developing asexually into a new individual. In higher plants, a spore is the haploid product of meiosis that gives rise to the male or female gametes.

Standard deviation.   A statistic that is a measure of the amount of variability in a group of observations. It is calculated for a sample by taking the square root of the sum of the squared deviations from the mean divided by one less than the number of observations.

Stem cell.   A cell from which other cells arise and undergo differentiation.

Sterile-male technique.   An insect control program whereby large numbers of sterilized males are released into the native population. The concept is based upon a high probability that most wild-type females will mate with the sterilized males yielding no progeny, rather than mating with the fertile males in the population.

Strain.   Describes a population of related individuals that is distinguished from other groups by certain characteristics. Often designated by geographic locations.

Subspecies.   A population that occupies part of the geographic range of a species. Also known as a geographic race.

Surrogate mother.   A substitute mother made possible by embryo transfer from one organism to another.

Symbiosis.   Two biologically dissimilar organisms living together in an intimate association with resulting benefit to one or the other, or both.

Synapsis.   The pairing of homologous chromosomes during prophase I of meiosis.

Syndrome.   A group of specific traits occurring together that are characteristic of a particular genetic condition, especially those considered abnormal.

Target theory.   Originally, a theory that was used to estimate the size of a gene based on the effects of ionizing radiation upon a small volume within the cell. One or more such ionizing events within this small sensitive volume are deemed to be necessary to cause a mutation.

Taxonomy.   The science of the classification of living things.

Tay-Sachs disease.   A type of genetic disorder that is inherited as an autosomal recessive and usually fatal at an early age. It is due to an accumulation of gangliosides in the brain.

Telophase.   The concluding stage in nuclear division in which the new complements of daughter chromosomes become uncoiled, thinner, and more elongate. New nuclear membranes are formed, and the daughter cells return to interphase.

Template.   With regard to the nucleic acids, the strand that dictates the complementary sequence of bases in the newly synthesized strand.

Teratogen.   Any agent that produces, or at least increases, the incidence of gross congenital malformations.

Teratogenesis.   The process by which gross congenital malformations are produced.

550

Testicular feminization. An abnormality in which a person with an XY genotype tends to have a female appearance with external female genitalia, but a blind vagina. The inguinal testes secretes estrogen instead of androgen.

Testis. Male gonad that produces spermatozoa and sex hormones.

Testosterone. A hormone secreted by interstitial cells of the testes that affects the development and maintenance of the male secondary sex characteristics.

Tetrad. The four haploid cells that result from the second meiotic division in plants. The term is also used to identify the four chromatids formed by the pairing of replicated homologous chromosomes.

Tetraploid. A polyploid cell or organism that contains four complete haploid sets of chromosomes, each of which is considered a genome.

Tetrasomic. Describes a condition in which a cell or an organism has one of its chromosomes of the basic complement represented by four homologues, rather than the usual two homologues found in diploids.

tfm. The gene responsible for testicular feminization.

Thalassemia. A group of genetic conditions in which an inadequate amount of hemoglobin is produced, resulting in an anemic condition.

Thalidomide. A drug, previously used as a tranquilizer, now determined to be highly teratogenic when taken by pregnant women.

Threshold. Describes certain phenotypic characters whose segregating distributions are discontinuous even though the inheritance is multigenic, which usually results in continuous characters.

Thymine. A pyrimidine base found regularly in DNA, but not normally found in RNA.

Thyroxin. A hormone produced by the thyroid gland and involved in the control of the metabolic rate of the body.

Tissue culture. The growth and maintenance of cells and tissues from higher organisms *in vitro*.

Totipotency. A property of a cell, or group of cells, such that they have the capacity to develop into a completely differentiated organism.

Toxin. A poisonous substance of organic origin.

Transcription. The synthesis of RNA from a DNA template by base pair complementarity.

Transsexualism. Refers to voluntary sex changes, usually effected by surgery and hormonal injections.

Transfer RNA (tRNA). A particular type of RNA molecule with a relatively low molecular weight. The tRNA transports activated amino acids to the ribosome where they are assembled into a polypeptide chain.

Transformation. The genetic modification of a cell or organism induced by the integration of purified DNA from the cells of one organism into the cells of another organism.

Transforming principle. The substance causing a transformation event, ultimately identified to be DNA.

551

Translation. The process whereby the genetic information of a particular mRNA nucleotide sequence is converted to a specific sequence of amino acids in a polypeptide chain.

Translocation. A structural chromosomal aberration in which a portion of one chromosome breaks away and becomes incorporated at a different site, either within the same chromosome or within a nonhomologous chromosome.

Transplantation. Transfer of an organ or part of an organism to another organism, or to another location within the same organism.

Trihybrid. An individual heterozygous for three pairs of genes.

Triple-X syndrome. See trisomy-X.

Triploid. Describes a polyploid cell or organism having three sets of the basic haploid chromosome complement, that is, genomes.

Trisomic. Describes a condition in which a cell or an individual has one extra chromosome; hence, three homologues exist for one of the chromosomes of the basic set rather than two. Designated as $2n + 1$.

Trisomy-13. Also known as Patau syndrome. A trisomic condition in humans in which chromosome 13 is represented three times. The infant is born with severe internal and external anomalies and mental retardation. Death usually occurs at a very early age.

Trisomy-18. A trisomic condition in humans in which chromosome 18 is represented three times. The defect results in a child with multiple congenital malformations. The infant usually dies within the first six months after birth.

Trisomy-21. A genetic defect in humans in which chromosome 21 is represented three times, rather than as a pair. The disorder is also called Down syndrome.

Trisomy-X. A trisomic condition in humans in which the X chromosome is represented three times. Also known as the triple-X female.

tRNA. See transfer RNA.

True hermaphrodite. See hermaphrodite.

True to type. Refers to true breeding. Describes the situation whereby parents have progeny with the same phenotype as themselves, that is, without segregation of the alternative forms of a gene among the progeny. It tends to designate homozygotes.

Trypsin. An enzyme that catalyzes the breakdown of proteins.

Tubal ligation. To bind or tie off a tube in the body.

Tumor. A local mass on or in any part of the body that arises by an abnormal growth of the tissues. It may or may not be malignant.

Turner syndrome. Also called XO. In humans, an abnormal condition in which the person is monosomic for the X chromosome and lacks an accompanying Y chromosome. The affected persons are sterile females.

Tyrosinemia. A rare hereditary disease of the liver due to an enzyme deficiency.

552

Ultrasonography. The use of high-frequency sound waves to gain an image of an internal body part or of a developing fetus. Often used as part of prenatal diagnoses.

Unbalanced translocation. Translocated chromosomes (structural changes that are due to a change in the position of chromosome segments) in which all of the chromosome segments are not present in the normal dosage.

Unit factor. The particles responsible for heredity as postulated by Mendel. Today, these are known as genes.

Universality. The concept that only one genetic code is to be found among all species of organisms.

Uracil. One of the pyrimidine bases found in RNA, but not in DNA.

Urethra. In mammals, the duct leading from the urinary bladder to the exterior of the body.

Uric aciduria. An excessive amount of uric acid in the urine.

Uterus. That part of the female mammalian reproductive system in which the embryo develops.

Vagina. That part of the female reproductive system that receives the male penis during copulation.

Variance. A statistic that provides an indication of the variability in a group of measurements. It is the square of the standard deviation.

Vas deferens. The duct that conveys spermatozoa from the epididymis of the testis to the urethra and the exterior of the body.

Vasectomy. The surgical excision of all or part of the vas deferens.

Vector. An agent consisting of a DNA molecule that can replicate autonomously in a cell, such that another DNA segment can be attached to it and be replicated at the same time.

Vertebrate. In a general sense, animals with backbones.

Vestigial. A part of an organism existing in a diminished condition compared with an earlier time in its ancestry, and no longer useful.

Virulent. Generally, the ability to cause disease. In viruses, it describes those causing lysis and thereby destroying the host cell.

Virus. A submicroscopic organism that is a noncellular obligate parasite composed of a nucleic acid and a protein shell.

Vitamin D-resistant rickets. A sex-linked hereditary disease that occurs regardless of adequate vitamin D in the diet.

Werner's syndrome. A syndrome in humans in which the symptoms of aging are manifested prematurely, usually in early adulthood.

Wilson's disease. A disease in humans, inherited as an autosomal recessive, that results in an abnormal copper metabolism.

Xeroderma pigmentosum. A skin disease characterized by the development of numerous pigmented spots. Frequently the condition develops into skin cancer.

553

X irradiation. Subjection to X rays, a high-energy form of radiation.

X-linkage. Genes located on the X chromosome that do not have alleles on the Y chromosome.

XO syndrome. See Turner syndrome.

X ray. A high-energy radiation that can penetrate body tissues, allowing for photographs of the internal regions for analysis.

X-ray diffraction. A technique in which diffraction patterns of X rays scattered from crystals are analyzed to determine the three-dimensional structure of molecules.

XXY karyotype. See Klinefelter syndrome.

XYY karyotype. The presence of an extra Y chromosome resulting in males who are usually taller than average, with an abnormal phenotype similar to that of Klinefelter syndrome.

Yellow body. See corpus luteum.

Y-linkage. Genes located on the Y chromosome, but without alleles on the X chromosome.

Zona pellucida. A thick membrane-like structure surrounding the mammalian egg; it disappears before the egg implants in the wall of the uterus.

Zygote. Usually a diploid cell formed by the fusion of two haploid gametes during fertilization. In higher forms of life, the fertilized egg before cleavage.

Answers
to Problems

CHAPTER 2

1. G G G C C A T T A C G

2. a. 50%
   b. 50%

3. a. Sample 1 = 20% guanine, 30% adenine, 30% thymine
      Sample 2 = 14% cytosine, 36% adenine, 36% thymine
      Sample 3 = 30% adenine, 20% guanine, 20% cytosine
      Sample 4 = 28% thymine, 22% guanine, 22% cytosine
   b. Yes; samples 1 and 3.

4. Based upon probability, $\dfrac{40 \times 2}{4} = 20$; however, the base pair proportions could be highly variable.

5. $20 \times 2 = 40$

6. a. $\dfrac{1.0}{0.8} = 1.25$
   b. $\dfrac{0.8 + 1.0}{1.0 + 0.8} = \dfrac{1.8}{1.8} = 1.0$

7. Single stranded are samples B, C, and D; double stranded is A. You cannot be absolutely sure of these answers because adenine being equal to

555

thymine and guanine being equal to cytosine could be a coincidence in a particular single-stranded DNA.

8. $4^6 = 4096$

9. d, b, c, a

10. a. 50%
    b. 25%
    c. 12.5%

CHAPTER 3

1. a. 2
   b. 4
   c. 1

2. a. *A/A', B/B', C/C', D/D'*
   b. *A, B, C, D*   *A, B', C, D*   *A', B, C, D*   *A', B', C, D*
      A, B, C, D'   *A, B', C, D'*   *A', B, C, D'*   *A', B', C, D'*
      A, B, C', D   *A, B', C', D*   *A', B, C', D*   *A', B', C', D*
      A, B, C', D'   *A, B', C', D'*   *A', B, C', D'*   *A', B', C', D'*

3. a. 1/4
   b. 1/2
   c. 1/16
   d. 1/2

4. a. 3/4
   b. None
   c. 1/4
   d. *AaBbDD* = 4 and *AABbdd* = 2

5. Due to both sides of the family. (The disorder is recessive.)

6. a. 50%
   b. 50%
   c. Yes, but astronomically improbable
   d. Yes, but astronomically improbable
   e. 50%

7. a. Man is non-wooly and normal pigmentation; woman is wooly and albino
   b. Man = 1/2 *wA* and 1/2 *wa*; woman = 1/2 *Wa* and 1/2 wa
   c. 1/4 *WwAa*, 1/4 *Wwaa*, 1/4 *wwAa*, 1/4 *wwaa*

8. $TT \times tt = F_1$: all *Tt*
   $Tt \times Tt = F_2$: $TT \xrightarrow{\text{selfed}} F_3$
   $Tt \xrightarrow{\text{selfed}} F_3$

556

$$Tt \xrightarrow{\text{selfed}} F_3$$

tt omitted from F_3 study

F_3:

TT × TT = all TT (no segregation)
Tt × Tt = 3/4 T−, 1/4 tt (segregation) } 2/3 segregated
Tt × Tt = 3/4 T−, 1/4 tt (segregation) } and 1/3 did
not segregate

9. 1/4

10. *Pp*; unknown, but probably *PP*. Both of her parents, however, would have to show the polydactyl trait in the latter case. If she is *Pp*, the probability of having nine children with polydactyly would be $(1/\frac{1}{2})^9 = 1/512$.

11. a. .40 × .40 = 16%
    b. .60 × .60 = 36%
    c. (.40 × .60) + (.60 × .40) = 48%

12. 1/128; about 8; about 30

13. a. 4C
    b. 2C
    c. Each telophase cell with 2C
    d. Each telophase cell with 1C
    e. 4C
    f. 4C
    g. Depends upon whether or not the DNA has replicated

CHAPTER 4

1. Possible. Bias is involved. One needs to note a child with the recessive disorder before incorporating the family into the study; hence, the percent of the children affected with the disease will tend to be higher than 25%. The sibships of 2 result in 78.1%; sibships of 3 = 55.1%; sibships of 4 = 46.7%, etc.

2. a. $\dfrac{1000}{20} = 50$
   b. 0

3. $4^2 = 16$; $4^4 = 256$

4. a. CCCATGGGCTACGATGGGAAG
   b. CCCAUGGGCUACGAUGGGAAG
   c. GGGUACCCGAUGCUACCCUUC
   d. Proline-mehionine-glycine-tyrosine-aspartic acid-glycine-lysine

5. 1/4 AA, 1/4 BB, and 1/2 AB

557

6. a. All phenylalanine
   b. Glutamic acid-tryptophan-serine-alanine-glycine
   c. None

7. a. 4 (CCA, ACA, GCA, and UUA)
   b. 3 (UCC, UCG, and UCU)
   c. 2 (UGA and UAA)

8. a. 4
   b. Yes
   c. Some enzymes require more than one type of polypeptide.

9. $d_1d_1D_2D_2$(dwarfism) $\times$ $D_1D_1d_2d_2$(dwarfism)

   All normal progeny: $D_1d_1D_2d_2$

10. Three. Individuals A and B have defects that are non-allelic; individuals C and D have defects that are allelic but different from A and B, as shown by the normal children from the subsequent marriages.

CHAPTER 5

1. 9.1%

2. a. $DD = 41.0\%$; $Dd = 46.1\%$; $dd = 12.9\%$
   b. $II = 3.6\%$; $Ii = 30.8\%$; $ii = 65.6\%$
   c. $GG = 65.6\%$; $Gg = 30.8\%$; $gg = 3.6\%$
   d. $KK = 98.01\%$; $Kk = 1.98\%$; $kk = .01\%$

3. 4%; 64%; 32%

4. No. The 2999:1001 ratio, which is extremely close to a 3:1 ratio, represents only phenotype frequencies of the entire population; it has nothing to do with the 3:1 ratio expected in the F_2 progeny from a particular pair of heterozygous parents.

5. a. .00000001 (1×10^{-8})
   b. .9998
   c. .0039

6. The proportion of individuals in the population expressing the dominant trait that are heterozygotes.

7. a. .0198   c. .000098
   b. .000392

8.
observed	expected
220*MM*	360*MM*
20*NN*	160*NN*
760*MN*	480*MN*

The population does not appear to be in Hardy-Weinberg equilibrium.

558

9. Example:
   $A = .50$; $a = .50$; $AA = .20$; $Aa = .60$; $aa = .20$

			Progeny		
Matings	Probabilities		AA	Aa	aa
$AA \times AA$	$.20 \times .20$	$= .04$	.04	—	—
$AA \times Aa$	$.20 \times .60 \times 2$	$= .24$	.12	.12	—
$AA \times aa$	$.20 \times .20 \times 2$	$= .08$	—	.08	—
$Aa \times Aa$	$.60 \times .60$	$= .36$	.09	.18	.09
$Aa \times aa$	$.60 \times .20 \times 2$	$= .24$	—	.12	.12
$aa \times aa$	$.20 \times .20$	$= .04$	—	—	.04
		1.00	.25	.50	.25

And $p^2 = .50^2 = .25$; $2pq = 2 \times .50 \times .50 = .50$; $q^2 = .50^2 = .25$

CHAPTER 6

1. Approximately 12 to 14 ovulations per year for 30 to 36 years; hence, 360 to 500 in a normal lifetime.

2. **a.** 2 (*ABC* and *abc*)
   b. 8 (*ABC, ABc, AbC, Abc, aBC, aBc, abC, abc*)

3. $2 \times 2 \times 2 \times 4 \times 2 \times 3 \times 2 = 384$ phenotypes. Chances that any two students will have the same phenotype would probably not be very high among only 25 students, but it depends upon the phenotype frequencies.

4. Unequal probabilities for the contrasting traits would result in relatively high probabilities for some phenotype combinations and low probabilities for other combinations; for example, the probability for being a taster, Rh+, and a non-albino is 59.5%, while the probability for being a non-taster, Rh−, and an albino is .000225%.

5. The offspring would have to be a female. The 12 tests that show identical results do not prove the situation, since this is only 12 of many traits.

6. $1 - (1/2)^8 = 255/256$; $8 \times 1 = 8$ years

7. **a.** No; only 1/2 of the genetic material was given to the egg by the original female.
   b. No; the eggs being used have different gene combinations due to meiosis.
   c. No; the resulting progeny do not have exactly the same genotype of the female parent.
   d. *Aa* female parent

   A a eggs

   AA aa made homozygous by the doubling procedure causing lethality.

8. 20 pregnancies/100 woman years; approximately twice as good as no attempt to control births; slightly more effective than the rhythm method of birth control.

CHAPTER 7

1. Unable to determine; inheritance could be either recessive or dominant.

2. 1/8 female, normal; 1/4 female, sickle cell trait; 1/8 female, sickle cell anemia; 1/8 male, colorblind; 1/4 male, colorblind, sickle cell trait; 1/8 male, colorblind, sickle cell anemia.

3. Recessive. II-4, III-2, III-3, and III-4 originate from two normal parents. All offspring of IV have the trait.

4. $X^D X^d \times X^d Y \longrightarrow$ 1/4 $X^D X^d$ (affected female)
   1/4 $X^D Y$ (affected male)
   1/4 $X^d X^d$ (normal female)
   1/4 $X^d Y$ (normal male)

5. a. Recessive
   b.

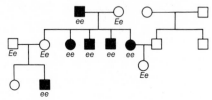

6. a. 1/4
   b. 1/2
   c. 1
   d. About 1/60

7. a. $3/4 \times 3/4 = 9/16$
   b. $2 \times 3/4 \times 1/4 = 6/16$
   c. $1/4 \times 1/4 = 1/16$

8. The probability of being a carrier is 1/100; the probability that both parents are carriers is $1/100 \times 1/100 = 1/10,000$; thus, the probability that a child is homozygous recessive for galactosemia is $1/10,000 \times 1/4 = 1/40,000$.

9. a. 1/2
   b. The probability of the fetus having Huntington's disease is $1/2 \times 1/2 = 1/4$; therefore, $1 - 1/4 = 3/4$ would be normal.

10. The probability of each normal parent being a carrier is 2/3; hence, $2/3 \times 2/3 = 4/9 \times 1/4 = 1/9$.

CHAPTER 8

1. Inversion; heterozygous; before replication

2. a. XXXY or XXXYY
   b. XO
   c. XXXX

3. XXY, because, despite two Xs like females, the presence of the Y causes maleness; XO, because, despite one X like males, the absence of the Y causes femaleness.

4. a. Either;

   b. Either;

   c. Spermatogenesis;

5. Nondisjunction in the male = X_gaX_g phenotype
   Nondisjunction in the female = X_gX_g phenotype

CHAPTER 9

1. a. $R_1R_1\ R_2R_2\ r_3r_3\ R_4r_4$
   b. 6
   c. 3
   d. $1 \times 1 \times 1/2 \times 1/4 = 1/8$

2. a. 5
   b. Maximum of 9

3.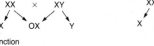

4.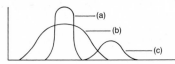

561

5. 2 gene pairs

6. 85 to 115

7. 20%

8. Cannot determine; the genotype-environmental interaction is not known.

9. Cannot determine; the genotype-environmental interaction is not known.

10. a. 20%
    b. 12%

CHAPTER 10

1. 25%

2. Parents: $A_1A_1\ A_2A_2\ A_3A_3 \times A_1A_1\ A_2A_2\ A_3A_3$

   most severe: $A_1A_1\ A_2A_2\ A_3A_3$
   (all children with the malformation)

   Parents: $a_1a_1\ A_2a_2\ a_3a_3 \times a_1a_1\ A_2a_2\ a_3a_3$

   most severe: $a_1a_1\ A_2A_2\ a_3a_3$
   (3/4 chance of receiving one or more of the alleles for the malformation)

3. a. If one pair of alleles, 0 cancer; if dominant gene, all cancer; if polygenic, depends upon the presence of a threshold
   b. Some of both types; numbers dependent upon a threshold

4. a. B
   b. C

5. The control group shows 54.2% of the persons with A or O blood type have O blood; however, 61.5% of the persons with A or O blood *and* peptic ulcer have O blood. A slight association is indicated.

CHAPTER 11

1. Inbreeding coefficient is 1/8.

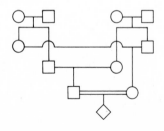

2. a. 1/32 **b.** 1/32 **c.** 1/32

3. a. 2/3
   b. 1/8
   c. 1/48

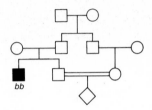

bb

4.

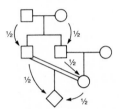

$(1/2)^5 = 1/32 \times 4$ alleles $= 1/8$ $(1/2)^4 = 1/16 \times 2$ alleles $= 1/8$

5. a.

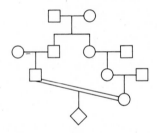

   b. 1/32

6. a. 1/16
   b. 1/64
   c. Depends upon the frequency of the allele in the population.

7. 1/12

8. Yes

   albino   *normal*

9. $F = (1/2)^4 \times 2$ alleles $= 1/8$

10. 1/1600

563

CHAPTER 12

1. a. 1/4   b. 1/4

2. a. Both parents homozygous for an autosomal recessive (or dominant) disease (*aa* × *aa*)
   b. Both parents heterozygous for a dominantly caused abnormality (*Aa* × *Aa*)
   c. One parent heterozygous for a dominant allele and the other parent homozygous for the recessive allele (*Aa* × *aa*)
   d. Both parents heterozygous for a recessive abnormality (*Aa* × *Aa*)

3. a. Not possible to determine; could be recessive or dominant
   b. Recessive
   c. Dominant; if recessive, the normal offspring would not be possible
   d. Recessive, either autosomal or X-linked
   e. Good probability of being autosomal recessive, but could be autosomal dominant
   f. Recessive, autosomal, or X-linked; X-linked is favored

4. A mutation has occurred; therefore, recurrence depends upon the mutation rate for that gene.

5. Genetically equivalent to first-cousin unions. For any deleterious gene the man carries, there is a 1/8 chance that his mother's half-sister also carries the allele, and a 1/32 chance that each of their children would be homozygous for such an allele. Risk for having genetically defective children would be about 6%.

6.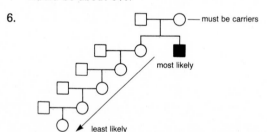

7. a. 1   b. 1/2, but 3 normal children diminishes the probability
   c. Theoretically 1/2, but 10 normal children greatly diminishes the probability

8. If the spouse (II-1) is not heterozygous for cystic fibrosis, their children could not have cystic fibrosis; this will most often be the case, and only one parent needs to be tested.

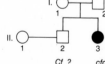

564

9. 6000/150000 = .04 = 2*pq*
   expected: 2*pq* × 2*pq* = .04 × .04 = .0016
   observed: 150000/2 = 75000 and 125/75000 = .00167
   Good fit exists for random mating.

10.

practice	disease frequency	allele frequency
a. improved treatment of the disease	remains the same	remains the same
b. selective abortion without replacing with another child	decreases	remains the same
c. limitation of family size in carrier marriages	decreases	decreases

CHAPTER 13

1. a. The child could be his offspring
   b. The child could not be his offspring

2. a. $I^O I^O$
   b. $I^B I^O$
   c. $I^A I^O$, $I^O I^O$, or $I^B I^O$
   d. $I^A I^B$, $I^A I^A$, or $I^B I^B$

3. 2; 3.

4.

	A	B	AB	O
A		X	X	X
B	X		X	X
AB	X	X		X
O	X	X	X	

5. a. 9; *MMSS, MMSs, MMss, NNSS, NNSs, NNss, MNSS, MNSs, MNss*
   b. 6
   c. 45

6. a. 12
   b. *A− M, A− MN, A− N, B− M, B− MN, B− N, AB M, AB MN, AB N, O M, O MN, O N*
   c. 2 × 12 = 24

565

7.

babies	parents	assignments
a. O	a. O and O	(a.) + (a.)
b. A	b. AB and O	(b.) + (b.)
c. B	c. A and B	(c.) + (d.)
d. AB	d. B and B	(d.) + (c.)

8. O M × B– M $\longrightarrow$ B M
   O M × A– MN $\longrightarrow$ A– MN

9. Case 1, no; case 2, no; case 3, no; case 4, no; case 5, no; case 6, excludes.

10. a. OO × A– $\longrightarrow$ B; Mother is O and he cannot be responsible for B
    b. Mutations are extremely rare
    c. hh (B)O or BB × HH AO $\longrightarrow$ Hh BO

CHAPTER 15

1. (a); (c);

safe
level

2. Genetic effects occur even at 0 radiation dose; due to other causes.

3. 3%; 18%

4. a. All are equal to each other.
   b. 300 rems as one dose; cells cannot keep up with repair because of "swamping" effect.

5. More than one "hit" necessary to cause the genetic effect.

6. About 319 rems.

CHAPTER 16

1. 36 doublings come close to 75 trillion; 37 doublings result in a number well over 75 trillion.

2. a. 58 for 1900; 65 for 1925; 72 for 1950; 76 for 1975
   b. They do not show survival to the maximum life span (about 100 to 110 years).

3. Some cells are dividing at a higher rate than others.

4. Species a; they have the same maximum life span.

566

5.

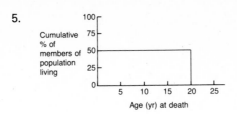

CHAPTER 17

1. a. Case 3
   b. Case 2
   c. Case 1

2. Cannot determine from the information given.

3. One can separate the genetic and environmental components in the study of various traits.

etc.

4. $574 \times 2 = 1148$ (dizygotic twins); $2000 - 1148 = 852$ (monozygotic twins)

5. a. 50.0%
   b. 8.0%
   c. There is a possible genetic component involved in this trait.

6. 100% for monozygotic twins and about 50% for dizygotic twins; 100% for monozygotic twins and about 25% for dizygotic twins.

7. 100%; very low; definitely a genetic causation.

8. parents: Aa Bb cc . . . *etc.* × Aa Bb CC . . . *etc.*

   identical   A B c . . . *etc.* ⟶ A b C . . . *etc.*
   spermatozoa   A B c . . . *etc.* ⟶ A b C . . . *etc.*   identical ova

   identical
   dizygotic   AA Bb Cc . . . *etc.*
   twin offspring   AA Bb Cc . . . *etc.*

9. parents: $A_1A_2\ B_1B_2$ . . . × $a_1a_2\ b_1b_2$ . . .
   assume the first genotype to be $A_1a_1\ B_1b_1$
   possibilities for the second genotype:

567

genes shared

$A_1a_1 B_1b_1 = 4/4$		$A_2a_1 B_1b_1 = 3/4$	
$A_1a_1 B_1b_2 = 3/4$		$A_2a_1 B_1b_2 = 2/4$	
$A_1a_1 B_2b_1 = 3/4$		$A_2a_1 B_2b_1 = 2/4$	
$A_1a_1 B_2b_2 = 2/4$		$A_2a_1 B_2b_2 = 1/4$	
$A_1a_2 B_1b_1 = 3/4$		$A_2a_2 B_1b_1 = 2/4$	
$A_1a_2 B_1b_2 = 2/4$		$A_2a_2 B_1b_2 = 1/4$	
$A_1a_2 B_2b_1 = 2/4$		$A_2a_2 B_2b_1 = 1/4$	
$A_1a_2 B_2b_2 = 1/4$		$A_2a_2 B_2b_2 = 0/4$	

average genes shared: $32/64 = 50\%$

10. A, D, B, C

11. a. 0
    b. No; pairing at random should not show any correlation in either case.

12. a. 1.0
    b. 0
    c. 0
    d. 0

13. From left to right: (c); (b); (a).

14. In the absence of extenuating circumstances, it appears that there could be a genetic component in tendencies toward criminality.

CHAPTER 18

1. a. Most, if not all of them.
   b. Neither. The experiment does not test for preadaptation versus postadaptation. The results could be due to either process; however, other experiments have indicated preadaptation to be the mode of adaptation.

2. Allow the two populations of flies the opportunity to mate with each other under conditions as natural as possible.

3.

	typica	carbonaria	ratio
unpolluted woodland	13.7%	4.7%	2.9:1
polluted woodland	25%	53.2%	0.47:1

Lesser recapture indicates greater degree of natural selection by predators.

568

CHAPTER 20

1. 8

2.
$$\underline{\text{Inbred \#1} \times \text{Inbred \#2}}$$
Hybrid #1/2

$$\underline{\text{Hybrid \#1/2} \times \text{Inbred \#3}}$$
Hybrid #1/2/3

3. single cross:
AA BB <u>CC DD × aa bb</u> cc dd

    all Aa Bb Cc Dd

double cross:
AA BB cc <u>dd × aa</u> bb cc dd   <u>aa bb CC DD × aa bb</u> cc dd

    Aa Bb cc dd   ×   aa bb Cc Dd

           Aa Bb Cc Dd
           Aa bb Cc Dd
           aa Bb Cc dd
             etc.

4. somatic fusion: sexual cross:
    AA <u>BB × CC</u> DD   If homozygous:
                      AA <u>BB × CC</u> DD
    AA BB CC DD

                       A B C D
           If not homozygous:
                      Aa <u>Bb × Cc</u> Dd

                        a b C D
                        A B c d
                        a b c d
                         etc.

5. somatic fusion: sexually produced plant:
    AA <u>BB × CC</u> DD   AA <u>BB × CC</u> DD

    AA BB CC DD

A B C D   meiosis   A B C D
        A B C D

                    A D   B C   meiosis
                        or
                    A C   B D
                       etc.
             (missing chromosomes)

569

To avoid sterility, chromosomes can be doubled:

AA B̲B̲ × C̲C̲ DD

A B C D

doubled

AA BB CC DD

meiosis

A B C D   A B C D

6. 66.7%; 90%

Index